AF786093

Smart Sensors, Measurement and Instrumentation 4

Series Editor

Subhas Chandra Mukhopadhyay
School of Engineering and Advanced Technology (SEAT)
Massey University (Turitea)
Palmerston North
New Zealand
E-mail: S.C.Mukhopadhyay@massey.ac.nz

For further volumes:
http://www.springer.com/series/10617

Subhas Chandra Mukhopadhyay
and Alex Mason (Eds.)

Smart Sensors for Real-Time Water Quality Monitoring

Springer

Editors
Prof. Subhas Chandra Mukhopadhyay
Massey University (Manawatu)
Palmerston North
New Zealand

Dr. Alex Mason
The School of Built Environment
Liverpool John Moores University
Liverpool
UK

ISSN 2194-8402 ISSN 2194-8410 (electronic)
ISBN 978-3-642-37005-2 ISBN 978-3-642-37006-9 (eBook)
DOI 10.1007/978-3-642-37006-9
Springer Heidelberg New York Dordrecht London

Library of Congress Control Number: 2013932787

© Springer-Verlag Berlin Heidelberg 2013
This work is subject to copyright. All rights are reserved by the Publisher, whether the whole or part of the material is concerned, specifically the rights of translation, reprinting, reuse of illustrations, recitation, broadcasting, reproduction on microfilms or in any other physical way, and transmission or information storage and retrieval, electronic adaptation, computer software, or by similar or dissimilar methodology now known or hereafter developed. Exempted from this legal reservation are brief excerpts in connection with reviews or scholarly analysis or material supplied specifically for the purpose of being entered and executed on a computer system, for exclusive use by the purchaser of the work. Duplication of this publication or parts thereof is permitted only under the provisions of the Copyright Law of the Publisher's location, in its current version, and permission for use must always be obtained from Springer. Permissions for use may be obtained through RightsLink at the Copyright Clearance Center. Violations are liable to prosecution under the respective Copyright Law.
The use of general descriptive names, registered names, trademarks, service marks, etc. in this publication does not imply, even in the absence of a specific statement, that such names are exempt from the relevant protective laws and regulations and therefore free for general use.
While the advice and information in this book are believed to be true and accurate at the date of publication, neither the authors nor the editors nor the publisher can accept any legal responsibility for any errors or omissions that may be made. The publisher makes no warranty, express or implied, with respect to the material contained herein.

Printed on acid-free paper

Springer is part of Springer Science+Business Media (www.springer.com)

Editorial

Potable water is a finite resource which is key to continued sustainable development. However, maintaining sufficient clean water to meet our growing needs is becoming increasingly challenging; factors such as climate change, natural disasters and our own industrial development are a significant threat to the cleanliness of water supplies. Some problems have been present for a long time but have only recently reached a critical level, while others are newly emerging. Whatever the cause, there is the growing need for the water industry to tighten its control and develop its understanding of what is happening to water resources in both fine detail and in real-time. In doing this, key information will be available, allowing industries and authorities to characterize and identify changes or trends in water properties over time, and identify new threats to water quality. Obviously having such information is only part of a solution to this struggle however, but it is envisaged that this information will allow:

(1) assisting in the setting of specific contaminant prevention or remediation programs;
(2) determination of whether current goals (e.g. legislation compliance) are being met;
(3) responding to issues as they happen, thus providing for increased real-time preventative measures rather than post-event remediation.

Although having such aspirations is admirable, it will not be possible without significant collaboration between industry, legislative bodies and researchers. From a legislative perspective, pressure must be maintained to ensure that the water industry and its user strive toward better management and distribution of water sources. In addition, factors impacting on water sources such as landfill, agriculture and industrial sites also require strict control to ensure that they continue to improve and observe best practice policies. Where contamination is a natural occurrence and remediation strategies are executed, careful monitoring of the process will be necessary to ensure that the strategy is appropriate and provide timely information for modification where necessary.

Ensuring that all concerned are playing their part as intended will require increased monitoring, and for this we need researchers to engage in developing appropriate *smart sensors and sensor systems*. Such systems will need to be low-cost so that they can be applied *in situ*, robust, maintenance free and easy to deploy. The sensors or sensor systems will also require some level of intelligence in order to determine whether the sensor data retrieved from the field is of importance, and perhaps also in what context it is important so that an appropriate automated or human response can be instigated. Furthermore, and perhaps the greatest challenge of all, sensors should provide information on a wide range of water contaminants in order to reduce wide array of instruments required currently for the purposes of quality monitoring.

In order to support the current requirements of the water industry to preserve our limited fresh water supplies, this book aims to promote the discussion on current trends in water quality monitoring, including topics such as sensors, the supporting technologies and also some of the issues which are emerging currently from a research perspective. After a peer-review processes we have selected 12 work presentations that cover various aspects of water quality monitoring.

The book focuses on three main themes; literature review, examples of technology and practice and supporting technologies/future trends. Thus the book begins in Chapter 1 by providing a very general overview of the current contaminants of concern for the wastewater industry and the available sensor technologies with which these contaminants have been detected to varying extents. Popular systems in use by industry (such as ultraviolet and mass spectroscopy) are discussed in addition to some technologies which are still within the research domain. This is continued in Chapter 2, which discusses the detection of phosphates in sea water, comparing the available spectrophotometric technologies for this purpose. The chapter then continues by discussing the future potential of electrochemical techniques, and this is supplemented by Chapter 3 which describes fluorescence and phosphorescence chemical sensors applied to water samples. Chapter 4, written by group in Spain well known for the work in the area of voltammetric electronic tongues, discusses the principle and application of this technology in the case of water contamination analysis. Finally for the literature review section, Chapter 5 discusses the automation of water analysis systems with a focus particularly on Sequential Injection Analysis.

While it is interesting to have a good background of the available technologies and how devices operate or perform, many also find it interesting to see the implementation or results of practical use of such systems. From such examples we gain experience of the challenges faced when deploying sensor systems, and truly get a feeling for the current state of technology. Thus, the next section of the book focuses on this area, beginning with Chapter 6, which discusses an optical system from initial design to industrial prototype and shares real-world results and associated experiences, including aspects such as biofouling which is a pressing concern for many long-term sensor systems in aquatic environments. Chapter 7 describes and critically analyses the results of a survey in Guyana, where a number of sites were tested using an ultraviolet spectroscopy method to determine if their nitrate contamination levels were within acceptable limits. Chapters 8 and 9 present the use of electromagnetic and microwave sensor devices respectively for water monitoring and analysis, offering an alternative to the almost standard optical methods which have seen reasonably widespread commercial adoption in recent years. This section is then rounded off with work from a group in China, who have been looking at water monitoring in an alternative sense to the other works presented in this book; i.e. considering the quality of water in fish farming.

The final section of this book is intended to offer a glimpse into the future of water monitoring and quality systems. Chapter 11 looks at how Information and Communication Technology (ICT) is or can assist in the water industry to become more responsive to contamination occurrence. While this chapter is not focused on sensors particularly, we must remember that sensors themselves are not standalone devices. Instead, they require a wide range of support from ICT including data transport and storage, data evaluation and also appropriate presentation. Thus, it must be recognized that ICT has an important role to play in the monitoring of water quality in order to ensure key objectives are met, including stakeholder participation, pollution and water quality control, monitoring for emergency prevention and detection, economic and financial management, information management and smart water distribution. Chapter 12 takes an alternative course, looking not at sensors or sensor systems, but rather at emerging contaminants. While this book is dominantly focused on sensors and is intended to help promote discussion regarding the technology, it is important that we remain focused on the challenges faced in the real world; the current ones, and the ones we are likely to face more and more in the future. Hence, Chapter 12 provides a timely and welcome conclusion to this book by helping us establish also the types of contaminants we need to be detecting in the future.

This book is written for academic and industry professionals working in the field of water quality monitoring, and is positioned to give an overview to the state of this discipline in relation to the sensors available and envisaged for the future, in addition to the issues faced in their implementation and the diverse range of contaminants that they must be able to detect.

We would like to express our appreciation to our distinguished authors of the chapters whose expertise and professionalism has certainly contributed significantly to this book.

Alex Mason, Guest Editor
Built Environment and Sustainable Technologies Research Institute
School of Built Environment
Liverpool John Moores University
Liverpool
United Kingdom
A.Mason1@ljmu.ac.uk

Subhas Chandra Mukhopadhyay, Guest Editor
School of Engineering and Advanced Technology (SEAT)
Massey University (Manawatu Campus)
Palmerston North
New Zealand
S.C.Mukhopadhyay@massey.ac.nz

Dr. Alex Mason graduated from the University of Liverpool, UK, with a first class honors degree in Computer and Multimedia Systems, after which he went on to complete a PhD in Wireless Sensor Networks and their Industrial Applications at Liverpool John Moores University (LJMU), UK. Upon completing his PhD in 2008, he concentrated for 2 years solely on research, working on aspects of non-invasive and non-destructive sensing for the healthcare, automotive and defense sectors.

Since 2010, Dr. Mason has held a position as a Senior Lecturer within the School of Built Environment at LJMU and has continued research in healthcare and defense, in addition to new areas such as water quality monitoring. Since becoming more involved in Built Environment issues, he has also developed an interest in Structural Health Monitoring and is currently working closely with the UK Defense Science and Technology Laboratories in this field.

Dr Mason is responsible for supervising a number of PhD students in the areas of sensing and renewable energy technologies, has coauthored over 90 publications (including 4 patents), has helped to organize national and international conferences, and give a number of invited talks on his work. He is also an active member of the IET in the UK.

Dr. Subhas Chandra Mukhopadhyay graduated from the Department of Electrical Engineering, Jadavpur University, Calcutta, India in 1987 with a Gold medal and received the Master of Electrical Engineering degree from Indian Institute of Science, Bangalore, India in 1989. He obtained the PhD (Eng.) degree from Jadavpur University, India in 1994 and Doctor of Engineering degree from Kanazawa University, Japan in 2000.

Currently, he is working as a Professor of Sensing Technology with the School of Engineering and Advanced Technology, Massey University, Palmerston North, New Zealand. His fields of interest include Smart Sensors and Sensing Technology, Wireless Sensors Network, Electromagnetics, control, electrical machines and numerical field calculation etc.

He has authored/co-authored over 270 papers in different international journals and conferences, edited nine conference proceedings. He has also edited ten special issues of international journals and twelve books with Springer-Verlag as guest editor.

He is a Fellow of IEEE, a Fellow of IET (UK), an associate editor of IEEE Sensors journal and IEEE Transactions on Instrumentation and Measurements. He is also a Technical Editor of IEEE Transactions on Mechatronics. He is a Distinguished Lecturer of IEEE Sensors council. He is in the editorial board of many international journals. He has organized many international conferences either as a General Chair or Technical programme chair.

Contents

Monitoring Pollutants in Wastewater: Traditional Lab Based versus Modern Real-Time Approaches

O. Korostynska, A. Mason*, and A.I. Al-Shamma'a

Built Environment and Sustainable Technologies (BEST) Research Institute,
Liverpool John Moores University,
Byrom Street, Liverpool, L3 3AF, UK
A.Mason1@ljmu.ac.uk

Abstract. Clean water – the most natural and yet the most precious natural resource that mankind needs. Current quality assessment methods of water parameters are mainly laboratory based, require fresh supplies of chemicals, trained staff and are time consuming. The ever-growing demand for simple, real-time and reliable techniques for the detection of pollutants and contaminants in the environment has sparked the development of remote detection and monitoring systems which include application specific sensors, instrumentation and signal processing. Real-time water quality monitoring is essential for National and International Health and Safety, as it can significantly reduce the level of damage and also the cost to remedy the problem. This book chapter critically compares the capabilities of traditional lab based and modern in situ methods for real-time wastewater quality monitoring and suggests further developments in this area.

Keywords: water quality monitoring, *in situ* analysis, optical methods, mass spectrometry, lab on chip sensors, electromagnetic waves, microwave sensors, solid-state sensors, amperometric detection.

1 Introduction

The increasing worldwide contamination of water sources with thousands of industrial and natural chemical compounds is one of the key environmental problems facing humanity. Approximately 3 billion people do not have access to safe drinking water, which is linked to over 35% of all deaths in developing countries. The World Health Organization (WHO) cites water borne illnesses as a major factor in 1.8 million deaths each year of which 88% are children in developing countries. Prediction models estimate that global consumption of water will double in the next 20 years, yet water quality assurance is singularly undervalued and poses a significant threat to global health. Water distribution systems are vulnerable to intentional and inadvertent contamination. Despite many technological advances, significant work remains to be accomplished for the purification of water.

---

* Corresponding author.

S.C. Mukhopadhyay & A. Mason (Eds.): *Real-Time Water Quality Monitoring*, SSMI 4, pp. 1–24.
DOI: 10.1007/978-3-642-37006-9_1 © Springer-Verlag Berlin Heidelberg 2013

Consumers expect water supply companies to deliver safe drinking water that meets both health quality standards and aesthetic requirements such as color, turbidity, taste and odor. The European Water Framework Directive (WFD) requires the monitoring of water systems in the European Community with the goal of maintaining and improving these systems [1]. Continuous assessment of source waters and the aquatic systems that receive inputs from industrial waste and sewage treatment plants, storm water systems, and runoff from urban and agricultural lands is vital. The implementation of the WFD has triggered the need for new methods and systems which enable the monitoring of chemical and biological pollutants in real time.

Water quality assessments are based on the analysis of the physical, chemical and bacteriological parameters and require customized apparatus and trained staff. Environmental water monitoring includes measurements of physical characteristics (e.g. pH, temperature, conductivity), chemical parameters (e.g. oxygen, alkalinity, nitrogen and phosphorus compounds), and abundance of certain biological taxa. Monitoring could also include assays of biological activity such as alkaline phosphatase, tests for toxins and direct measurements of pollutants such as heavy metals or hydrocarbons. In daily use there are up to 70,000 known [2] and emerging [3, 4] chemicals that might be present in various water resources, including for drinking water production. Notably, approximately 860 active compounds are currently formulated in pesticide products, and these compounds belong to more than 100 substance classes, among them benzoylureas, carbamatea, organophosphorous compounds, pyrethroids, sulfonylureas and triazines are the most important groups [5]. The chemical and physical properties of the pesticides can differ significantly; for example, they might include heteroatoms such as halogens, phosphorous, sulfur or nitrogen. These properties are relevant for the detection of pesticides by various analytical methods. In May 2007, the European Parliament proposed increasing from 33 to 61 the toxic products covered by European legislation on water quality. Forty five of these were classified as priority substances and should no longer be used by 2015.

There is an acute need in on-line monitors that are able to detect the excess of pollutants established by the official water quality regulations. Ultra-sensitive sensors are required, as standards are often close to the modern limits of detection, e.g. the EU pesticide standard of 0.1 μg/L [1]. Multi-parameter water quality monitors, or sensor panels, are mainly used in finished water, i.e. in water which has been treated and is ready for consumption. Typical parameters and techniques used in these monitors are listed in Table 1. Single probes or combinations of sensors are commercially available, enabling water utilities to monitor the quality of processed water. There are difficulties with independent validation of these systems as the methods and algorithms employed are commercially sensitive.

The quality of treated wastewater is primarily defined by the measurement of global parameters such as Biological Oxygen Demand (BOD), Chemical Oxygen Demand (COD), Total Organic Carbon (TOC) and Total Suspended Solids (TSS) [6]. However, wastewater is one of the most difficult sources of contaminants to deal with, as it carries micro-pollutants such as pharmaceuticals and hormonally active substances [7]. Table 2 lists current techniques in monitoring wastewater quality and their limitations [8].

Table 1. Most commonly measured water parameters and associated sensing technologies

Parameter being measured	Sensing technology
Aluminium	Colorimetry; Atomic Absorption Spectrometry
Antimony	Atomic Absorption Spectrometry
Ammonia	Colorimetric (Manual; Nessler's Reagent; Automated; Berthelot Reaction); Ion selective electrode
Chlorine	Colorimetric; Membrane electrode; Polarographic membrane; 3-electrode voltametric method
Conductance	Conductivity cell; annular ring electrode; nickel electrode; titanium or noble metal electrode
Dissolved oxygen	Membrane electrode; 3-electrode voltametric method; optical sensor
Ions (Cl-, NO3-, NH4+)	Ion-selective electrodes
Oxidation-Reduction Potential (ORP)	Potentiometric; platinum or noble metal electrode
pH	Titration with Sodium Hydroxide; proton selective glass bulb electrode, proton selective metal oxide; Ion Sensitive Field Effect Transistor (ISFET)
Phosphates	Manual or Automated Colorimetry
Temperature	Thermistor
Total Organic Carbon (TOC)	UV-persulfate digestion with near infrared detection or membrane conductometric detection of CO2
Turbidity	Optical sensor; Nephelometric (light scattering) method

Table 2. Status of current techniques in monitoring wastewater quality and their limitations

Parameter	Technique	Limitations
BOD	Non-specific sensor array (electronic nose)	Relationship is source/site specific and time dependent, further development needed
BOD, COD	Oxidation by hydrogen peroxide with UV light	Limited range and long (~55 min) measurement times, range and correlation are source dependent
OD, COD, TOC, TSS, Nitrates and anionic surfactants	UV spectral measurements and multivariate calibration	Sample handling is problematic, acquisition of reference spectra and calibration necessary for samples of different origin
BOD, nitrates, (TOC and COD)	Optical scattering (fluorescence)	Still in infancy, research needed, fluorescence affected by pH and temperature, correlation with BOD is plant sensitive.
BOD	UV adsorption (280 nm)	Poor sensitivity, uses only one wavelength, interferences from particles and toxic metals
COD, TOC	UV adsorption	Immerged sensor (fouling), influence of suspended particulate material
RQ value	Off-gas analysis (CO_2 and O_2)	Does not distinguish C-oxidation from N-removal, only big changes in nitrification activity can be monitored
COD, NH_4, NO_3	Artificial neural network + multi sensor (pH, temp, conductivity, redox potential DO, turbidity)	Approximate estimation, training needed, problems in case of sudden changes in wastewater composition, reliable for a short period only

This chapter provides a critical review of the currently available lab based and state-of-the-art real-time methods of water pollutants monitoring, in particular for monitoring of nitrates and phosphates. The presence of nutrients and heavy metals in water is a serious threat to human health [9]. Phosphorus is widely used as an agricultural fertilizer and within domestic detergents. Phosphates can exist in wastewater in several forms depending on the source/nature of the discharge but are generally grouped within three broad classes: orthophosphates, condensed phosphates (pyro-, meta- and poly-) and organic phosphorus [10].

From an environmental perspective, the concentration of phosphate in water is crucial due to its role in eutrophication [11]. Nitrate fertilizers are increasingly used in agriculture, the quantity of nitrate leaching from fields into rivers and ground waters is increasing cumulatively. Once nitrate has entered the body it can be reduced to nitrite by bacteria in the stomach and further incorporated into carcinogenic N-nitrosamine compounds [12]. Nitrate is also an important analyte for environmental, food and human health monitoring and thus its detection and quantification is essential. Traditionally phosphates and nitrates were monitored manually by collecting and filtering samples which were later analyzed in a laboratory. Various detection strategies for phosphate include phosphate ion selective electrodes based on potentiometric techniques, indirect voltammetric detection based on the reaction of phosphate with various metals and associated complexes, and the development of sensors exploiting enzymatic reactions [13]. These methods are also used for the detection of other water pollutants, as described further in the chapter.

Current measurements of nutrients, such as phosphorous, ammonia and volatile fatty acids in water are mostly based on off-line monitoring and imply low frequency data sampling and delay between sampling and availability of the results. However, in past years there has been a growing interest in the use of on-line monitoring systems able to distinguish abnormal changes from normal variations [14, 15] for reasons of lower costs, faster response time and due to security concerns.

2 Current Technologies: Standard UV-Vis Optical Methods

The main method for phosphorus detection is using a photo sensor which measures the wavelength of a distinct color (e.g. blue or yellow) that results from a chemical reaction between phosphorus and special reagent [10]. The concentration of the resultant dye indicates the concentration of phosphorus in the sample. Fig. 1 shows conceptual diagram of the two standard UV-Vis optical methods, namely the molybdenum blue method and the vanadate / molydate yellow method [16]. In the molybdenum blue method, in an acidic medium, orthophosphate bonds with ammonium molybdenic to form phosphoric acid. With the aid of reducing agents this forms phosphorus molybdenum blue and photometric measurement then determines the dye intensity at 880 nm for the blue color. The vanadate / molydate yellow method is as follows: in acids, orthophosphate ions react with ammonium molybdate and ammonium vanadate to form yellow ammonium phosphoric vanadomolybdate, which can be analyzed at 380 nm using a photometric sensor [10].

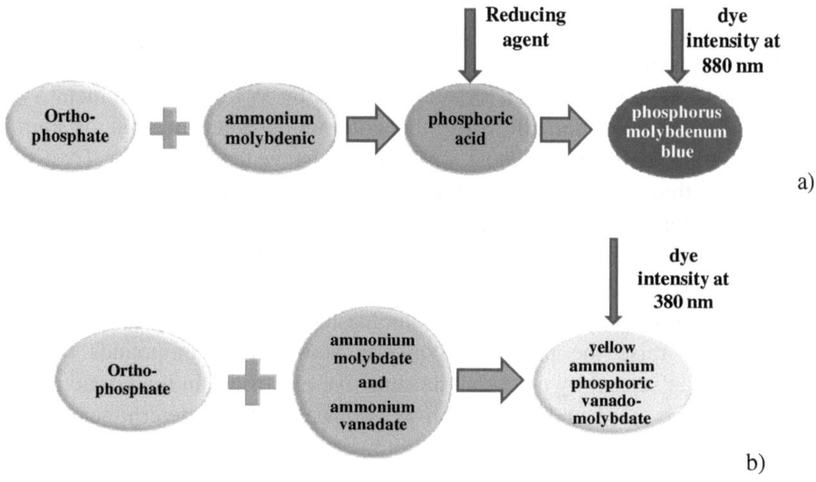

Fig. 1. Conceptual diagram of standard UV-Vis optical methods: a) Molybdenum blue method, b) Vanadate / molydate yellow method [16]

Most of the reported wastewater UV-Vis spectrometric applications are based on visual observation and direct comparison of the UV-Vis spectra. A few examples of commercially available systems are given below.

2.1 Examples of Commercially Available Systems

Hach Lange[1] is probably one of the most known and reputable suppliers of equipment and chemicals for various lab based measurements, and especially for screening of water parameters. They provide a broad spectrum of analytical techniques that are considered as a standard by many industrial and utilities companies.

Hache Lange are not alone however; EnviroTech Instruments offers, for example, a field-deployable nutrient analyzer system available in single or dual channel configurations that may be "daisy-chained" to provide up to six channels of simultaneous chemical analysis. Parameters include nitrate, phosphate, ammonia, silicate, chloride, iron and urea. Sampling pump control and a cellular modem interface for real-time / on-line data are built-in. AutoLAB 4 has a very low maintenance requirement and incorporates easy-load reagent storage for maximum up-time and onsite service.

The AutoLAB 4 system employs well established wet-chemistry to determine nutrient concentrations via standard analytical techniques. This allows direct comparison with laboratory data and avoids interferences and fouling problems intrinsic with optical systems. AutoLAB is a discrete (not flow-type) analyzer and so avoids problems with clogging and tube wear. The discrete analysis methods allow short sampling times, in-situ self-calibration and automatic maintenance.

[1] Full details can be found at `http://www.hach-lange.com/`

Beam attenuation from the color reaction is recorded. It detects nitrate at a range of 0-5 mg/L at 543 nm wavelength and phosphate in 0-0.8 mg/L range at 880 nm, both with 0.003 mg/L sensitivity.

TresCon on-line water analysis systems, offered by WTW Measurement Systems Inc., in Ft. Myers, Florida, allow for the continuous determination of ammonia, phosphate, nitrate, or nitrite in most water and wastewater processes. TresCon is a modular device that can be configured to monitor one, two, or three of the parameters mentioned simultaneously. To operate the TresCon system, a continuous supply of water to be analyzed is required. A flow rate of 2-3 liters/hour is sufficient. The analyzer continuously "sips" from this sample stream, and feeds the sample to the measurement modules installed. In the ammonia module, the sample is mixed with a basic reagent to raise the pH. This converts ammonium compounds to gaseous ammonia which is sensed by an ammonia-sensitive electrode. In the nitrate module, a UV light source is used and absorption at two wavelengths determines the nitrate concentration. Phosphate is determined photometrically: the sample is mixed with a molybdate-vanadate reagent. However, the data on sensitivity and range are not available.

Another example of a commercially available system for water quality measurement is the ChemScan UV-6100 Analyzer, manufactured by Applied Spectrometry Associates (ASA) Inc., Waukesha, WI. USA, which was considered by The City of Calgary's Sewer Divisions as the most economical system for 20-year operation. The UV-6100 analyzer system is an on-line UV spectrometer capable of measuring the concentration of multiple dissolved chemical constituents of an aqueous sample with a single analyzer. It works by transmitting UV light through the sample. A portion of the light is absorbed by the chemical constituents and the analyzer splits the resulting light into 256 individual wavelengths from 200 to 450 nm. The spectral signature is analyzed using Chemometrics, a pattern recognition technique, to calculate the concentration parameters.

Parameters that absorb light naturally, such as nitrate, can be analyzed without the addition of reagents. Parameters that do not have adequate natural light absorbing characteristics must be conditioned through the addition of chemical reagents. Orthophosphate is determined using a secondary chemical method. The orthophosphate is combined with ferric at a low pH in the sample conditioning unit. The analyzer measures the resulting combined spectra and uses the Chemometric algorithm to determine the orthophosphate concentration. The reported phosphate operating range is from 0.01 mg/L to 5.0 mg/L, while nitrate can be measured in a range from 0.5 mg/L to 20.0 mg/L.

2.2 Limitations of the Optical Methods of Detection

Sensors based on colorimetric UV spectral measurements are widely used in commercially available systems [17], but they all suffer from a number of limitations. In particular, sample handling is problematic, and the acquisition of a reference spectra and calibration process are necessary for samples of different origin. Although the methods for the determination of phosphate such as colorimetry or spectrophotometry are still predominant in a standard utilities operation [4, 17], they do not lend themselves particularly well to in-situ analysis due to the use of potentially toxic agents and lengthy preparation and analysis times. There are two main drawbacks:

- Linearity between the concentration of a compound and its absorbance performance differs from compound to compound. Therefore it is hard to identify a compound based on a single spectral wavelength. Only the ratio between different wavelengths can assist identification.
- Secondly, only a small fraction of the potential compound array in water absorbs light with wavelengths from 190 to 850 nm.

This means that UV-Vis spectroscopy must be combined with other techniques to cover a larger compound spectrum. IR, Raman and X-ray fluorescence spectroscopy are new techniques for on-line chemical water quality monitoring, but many challenges must be overcome before these techniques can be used routinely. An alternative approach is to develop fiber optic, potentiometric or electrochemical biosensors [18] as they have the potential to be used in-situ, and when combined with hand-held instrumentation [9], offer a high degree of selectivity and specificity and may be operated by lay personnel.

3 Mass Spectrometry for Water Micro-Pollutants Control

Surface water reservoirs and aquifers are exposed to contamination by thousands of micropollutants from industrial, pharmaceutical, agricultural and natural origins. Traditionally, the monitoring of contaminant levels was based mainly on gas chromatography mass spectrometry (GC-MS and GC-MS/MS) a method that involves multistep sample preparation and has low specificity. Nowadays, liquid chromatography with UV/fluorescence tandem mass spectrometry (LC-MS/MS) has become a routine analytical tool and a key technique for environmental analysis, allowing the detection of a wide range of polar and nonvolatile compounds [19]. The use of this method has increased the specificity and confidence of identification, while reducing sample preparation to a minimum. Thus, one of the major challenges of water analysis is to reduce sample preparation to achieve direct analysis by LC-MS/MS detection; that is, analysis without the need to concentrate ultra-trace-level contaminants in the water sample.

3.1 Atmospheric Pressure Ionization

Liquid chromatography mass spectrometers can be equipped with several mass analyzers, each of which provides unique features able to identify, quantify, and resolve ambiguities by selecting appropriate ionization and acquisition parameters [19]. Despite the enormous variability in the chemical features of water pollutants, most can be ionized by atmospheric pressure ionization (API) methods. The most common ionization method for residue analysis by LC-MS/MS is electrospray ionization (ESI), which is suitable for the analysis of most of the compounds that pollute water. The use of LC-ESI-MS/MS for the detection of 500 pesticides in food and water samples was tested by Alder et al [5].

Less polar compounds, which can be separated using reverse phase chromatography, require higher ionization energy. For these compounds, atmospheric pressure chemical ionization (APCI) is usually the method of choice for LC-MS/MS analysis [19]. For example, APCI was used for the analysis of organochlorine pesticides in surface waters [20] and for the detection and identification of rotenone in lake and river waters [21].

3.2 Liquid Chromatography/Quadrupole Time-of-Flight Mass Spectrometry

The inherent characteristics of time-of-flight (TOF) mass analyzers — high resolution, mass accuracy and good sensitivity in scanning wide mass ranges — make these analysers well-suited to the analysis of pesticide residues in water samples, mainly for qualitative purposes [22]. The hybrid quadrupole-TOF (QqTOF) permits the pre-selection of a precursor ion in the quadrupole filter and the measurement of the full-scan product ion spectra with high mass accuracy, which makes it one of the most valuable modern lab-based technique for elucidating unknown compounds in environmental waters [23]. For example, the QqTOF system was used for the detection of 22 triazole compounds including parent fungicides and metabolites in water with a limit of detection of 0.5 ppb [24].

3.3 Triple Quadrupole and QTrap Technologies

Triple quadrupole instruments (QqQ) operating in multiple reaction monitoring mode (MRM) achieve the best absolute sensitivity for targeted compounds. QqQ instruments exhibit excellent quantitation abilities owing to their wide linear dynamic range. For these reasons, and because of the relative simplicity of the MRM analysis, this technology has become the most common method for environmental analysis for contaminant residues by LC-MS/MS [19]. The MRM analysis introduces high specificity as it is based on the detection of both a parent ion and one of its known fragments. Another dimension of specificity can be introduced with the unique combination of a linear ion trap in the third quadrupole (QTrap) and an online information-dependent driven workflow that provides a full scan linear ion trap spectrum to confirm of the identity of compounds detected by MRM.

For increased specificity, the European Union regulatory authorities require the identification of a residual compound in water samples using one of three approaches: a high accuracy MS spectra with a resolution of >10000 (e.g. as obtained with QqTOF instruments); an MRM identification based on two different MRM transitions and their ratios (e.g. using QqQ instruments); or a single MRM transition combined with a full scan MS/MS spectrum (e.g. QTrap instruments) [19].

4 Fiber Optic Sensors for Real-Time Water Quality Monitoring

Fiber optic sensors are used in combination with the UV-Vis methods of water contaminants detection discussed above. Normally an optic fiber is suitably doped to produce luminescence when exposed to an excitation light source. Glass fibers are either doped with a rare earth metal or activated with a transition metal. Polymeric fibers are doped with a dye. The fibers have fast response and decay times and can achieve high efficiency through the design of appropriate delivery optics. Fiber optic systems are particularly suitable for harsh and difficult to reach places. The design and selection of the fiber determines the peak wavelength of the output illumination; options exist to span the UV–Vis–NIR spectrum. A detailed review of the recently reported fiber optics based systems provides in-depth analysis of these systems for various water contaminants monitoring [25]. Notably, the coating of the fiber determines the sensitivity and selectivity of the sensor and research is continuing to develop novel materials that would suit the need of a particular sensing task [26, 27].

5 Pollutant Monitoring Using Amperometric, Potentiometric and Conductimetric Sensors

Amperometric, potentiometric and conductometric sensing approaches are widely used in the measurement of pollution in water [28]. These sensors change their properties as a result of interaction with the component being measured. The species of interest are either oxidized or reduced at the working electrode causing a transfer of electrons, thus generating a measureable signal. This change can be recorded as a change in the output signal, i.e. output voltage, current, change in conductivity, capacitance or dielectric constant – whatever parameter gives the most pronounced sensor response [29].

For example, a portable amperometric three-electrode immunosensor for screening of polycyclic aromatic hydrocarbons (PAHs) in water was recently reported [28]. In particular, amperometric detection is based on the measurement of current when a potential is applied to the working and reference electrodes of the system.

Potentiometric detection is attractive since it possesses numerous advantages when considering the development of real-time sensing technologies, as the recording instrumentation is cost-effective and highly portable. Potentiometric DO detection is a better choice when dealing with fast changing large range of oxygen concentration, which cannot be accommodated by amperometric detection. A study on the performance of a Cu_2O-doped RuO_2 sensing electrode used in a solid-state potentiometric DO sensor in terms of both sensor capability and bio-fouling resistance for utilization in aqueous environments was recently conducted [30]. However, the acquisition of sufficient selectivity and sensitivity to operate at the concentrations below 0.1 ppm remains problematic for these sensors. Various potentiometric approaches to detect phosphate typically fall within one of five main categories, as illustrated in Fig. 2.

Although amperometric and conductimetric sensors are widely employed for various sensor systems, their use for monitoring of nitrates and phosphates in wastewater is in its infancy not least due to the limited research to find suitable materials that would provide not only desired sensitivity and selectivity, but also long-term stability and reusability of sensors.

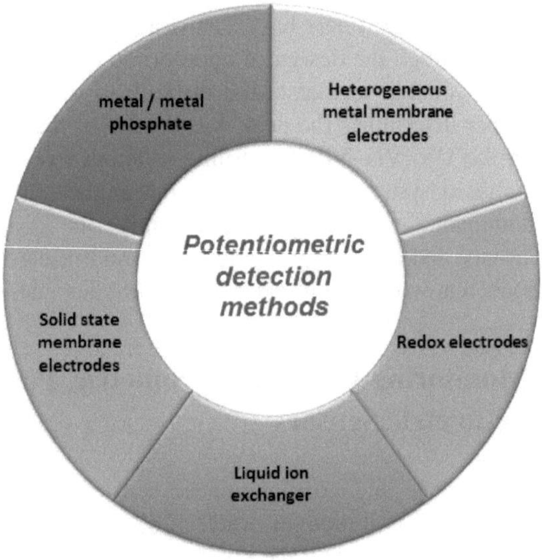

Fig. 2. Potentiometric detection methods

6 Biosensors

The presence of microorganisms in water is generally assessed with five indicators, such as total coliform, fecal coliform, fecal streptococcus, enterococcus, and Escherichia coli. To protect public health, microbiological standards have to be met at each individual treatment works and service reservoir. Notably, the presence of various microorganisms in water, including *salmonella, campylobacter, listeria, Bacillus cereus* and *Escherichia coli O157:h7* is a natural and unavoidable occurrence, but the level of these bacteria should be strictly monitored in real time. Infectious doses of these pathogens (as low as ~10 bacterial cells) increase the vulnerability of the elderly, infants, and people with immunological deficiencies or organ transplants [31]. In recent years, there has been greater recognition of the importance of biosecurity in relation to protection of the environment [32].

Examples of molecular methods used to identify specific microorganisms and to assess microbial community diversity using DNA sequences [33] are listed in Table 3. These methods are applicable for detecting and identifying microorganisms present in water (e.g. fecal indicators, pathogens) and examining community composition using microbial community DNA [34, 35].

Biosensors have been widely applied to a variety of analytical problems in medicine, food, process industries, security, defense and for environmental monitoring including water quality assessment. A biosensor is an analytical device which converts a biological response into an electrical signal. It consists of two main components: a bioreceptor or biorecognition element, which recognizes the target analyte and a transducer, for converting the recognition event into a measurable electrical signal [32]. A bioreceptor can be a tissue, microorganism, organelle, cell, enzyme, antibody, nucleic acid and biomimic etc. and the transduction may be optical, electrochemical, thermometric, piezoelectric, magnetic and micromechanical or combinations of one or more of the above techniques.

Table 3. Status of current genomic tools that can be used to assess water quality [36]

Tool	Description	Current status
Direct hybridization	Detection of genes, gene expression	Requires sequence knowledge and DNA to label for probe
PCR, RT-PCR	Rapid, sensitive detection of specific taxa, genes, or gene expression (RT-PCR)	In wide use, requires DNA sequence information
RAPD, RFLP, AFLP, AP-PCR, DGGE, T-RFLP, RFLP, ARDA	PCR-based, provides "fingerprint" of microbial community or the genome of microbial strains. Useful for description of community composition and source tracking	Widely used
Q-PCR, real-time PCR	Rapid, sensitive, quantitative detection of target	Use increasing rapidly, requires sequence knowledge
Fluorescent in situ hybridization (FISH)	Allows detection, visualization of individual cells	Widely used, requires actively growing cells.

Fig. 3 shows schematic diagram of a biosensor. The bioreceptor recognizes the target analyte and the corresponding biological responses are then converted into equivalent electrical signals by the transducer. The bioreceptor-protein interactions can be visualized with biosensors at μg/L levels (or even the sub-μg/L level). The amplifier in the biosensor responds to the small input signal from the transducer and delivers a large output signal that contains the essential waveform features of an input signal. The amplified signal is then processed by the signal processor where it can later be stored, displayed and analyzed.

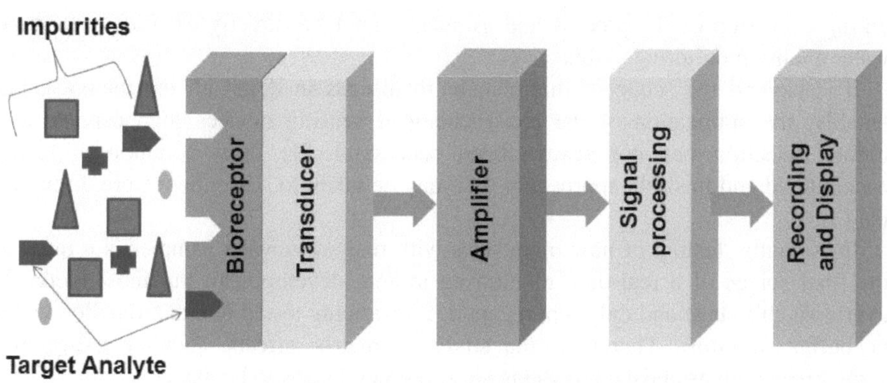

Fig. 3. Schematic diagram of a biosensor

Protein based biosensors have been developed for a series of chemical pollutants, but their sensitivity depends on the physical detection principle used and the affinity of the protein-analyte interaction. Their robustness is influenced by the selectivity of the interaction. Matrix effects can influence the binding between analyte and protein. Since water quality is variable, matrix effects are inevitable and result in a high rate of false positives and negatives, so regeneration or replacement of disposable reagents is needed between measurements.

Examples of analytes measured with biosensors are: propanil, atrazine, isoproturon, sulphamethizole, bisphenol A, estrone, 17ß-estradiol equivalents, nonylphenol, benzene, toluene, xylene, prometryn, trichloroetylene, ametryn, terbuthylazine, simazine, benzenesulfonamide, caffeine and so forth.

Biosensors for the determination of phosphate are normally based on mono- or multi-enzymatic reactions where phosphate acts as an inhibitor or substrate [12]. For example, an amperometric phosphate biosensor, based on a cobalt phthalocyanine screen-printed carbon electrode (CoPC-SPCE) was recently reported [11] to be successfully applied to the measurement of phosphate in pond water samples and a linear range of 2.5–130 μM with a limit of detection of 2 μM was obtained under optimal conditions, exhibiting a response time of ~13 s.

Also, a highly selective and sensitive monohydrogen phosphate membrane sensor based on a molybdenum bis(2-hydroxyanil) acetylacetonate complex (MAA) was reported in [37]. This sensor showed a linear dynamic range between 1.0×10^{-1} and 1.0×10^{-7} M, with a detection limit of 6.0×10^{-8} M (~6 ppb). The best performance was obtained with a membrane composition of 32% poly(vinyl chloride), 58% benzyl acetate, 2% hexadecyltrimethyl-ammonium bromide and 8% MAA. The sensor reportedly possesses additional advantages of short response time, very good selectivity towards a large number of organic and inorganic anions and is claimed to be suitable for at least 10 weeks usage without any considerable divergence in its slope and detection limit [37].

Nitrate concentrations are routinely determined using a method whereby nitrate is reduced to nitrite with a copper activated cadmium catalyst and the nitrite concentration is determined colorimetrically by its reaction with sulphanilamide and N-1-naphthylendiamine [12]. However, this method requires careful control of acidity during each step of the process and as such cannot be directly used as a real-time water quality monitoring system.

It is beyond the scope of this chapter to discuss in detail all the methods, but notably, the application of the electrochemical sensing devices to phosphate and nitrate detection has not always been successful. He most common problems encountered and possible approaches that may be taken to solve them were discussed in [13].

Importantly, testing of new biodevices with real wastewater samples is a must in the final stages of a real-time monitoring system development, but most literature overlooks this stage and only reports applications being tested in either distilled water or buffer solutions. Therefore, the study of matrix effects, stability issues and comparison with established methods are still crucial steps to be made.

7 Systems Based on Specifically-Sensitive Electrodes

7.1 Microelectrodes for Phosphate Detection

Due to the difficulty of miniaturizing traditional phosphate detection facilities and the problems with the classical colorimetric methods, a new approach, using phosphate sensitive materials to make the electrodes and detecting the concentration of phosphate by an electrochemical method has attracted increasing attention [38]. The phosphate ion sensitive electrode is one of the most important parts in the sensing system. Most reports on Co-based phosphate sensors have used a bulk Co-wire as the working electrode and used another isolated cell as the reference electrode. Potential response originates from a mixed potential due to slow oxidation of Co and simultaneous reduction of oxygen, and this response is directly based on the formation of $Co_3(PO_4)_2$ precipitate on the surface of Co electrodes [39].

Reportedly, a microelectrode with a tip size ~10 μm fabricated with cobalt wire was designed for in-situ and in vivo environmental analysis of orthophosphate ions (HPO_4^{2-} and $H_2PO_4^-$) that evaluates the wastewater phosphorus removal system and for biological applications [40]. Cobalt is known as a phosphate-sensitive electrode material, since oxidized surfaces of metallic cobalt could be used as a selective potentiometric sensor for phosphate [41]. General schematics of a phosphate microelectrode are illustrated in Fig. 4. It has a micro-sized tip, a cobalt wire enclosed in a lead glass micropipette, which also contains bismuth alloy and a copper wire for electrical connection. The detection range of the phosphate microelectrode was in the range of 10^{-5} to 10^{-1} M and the detection limit of the phosphate microelectrode was 7.5×10^{-5} M, while the response time during the evaluation was less than 1 min [40].

Electrochemical sensors offer a suitable platform for the development of microsystems for the remote detection and monitoring of pollution in waters. These sensors can be mass fabricated using the manufacturing technologies available to the microelectronics industry and their miniaturization improves their performance.

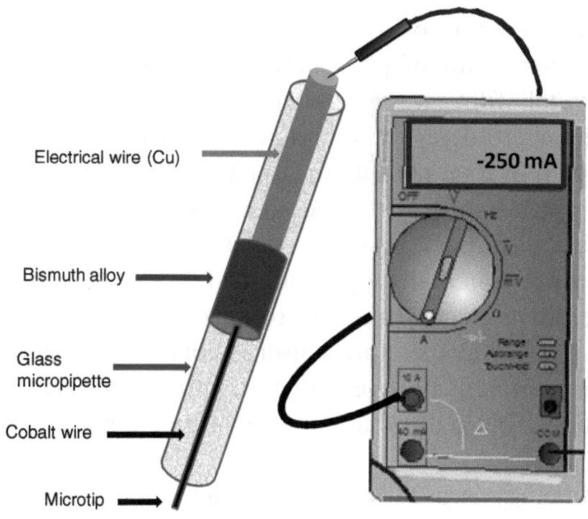

Fig. 4. General schematics of phosphate microelectrode [16]

7.2 MEMs Microelectrode Array Sensors

Microelectrode arrays have unique properties such as an improved mass-transport towards the microelectrode surface and a reduced ohmic drop, which both contribute to a higher sensitivity and hence, make them suitable for water-monitoring applications [42].

The major advantages of the micro-electro mechanical systems (MEMs) microelectrode array sensors (Fig. 5) include the ability to penetrate samples to perform measurements, small tip size for in situ measurements, array structure for higher robustness, and possibility of multi-analyte detection.

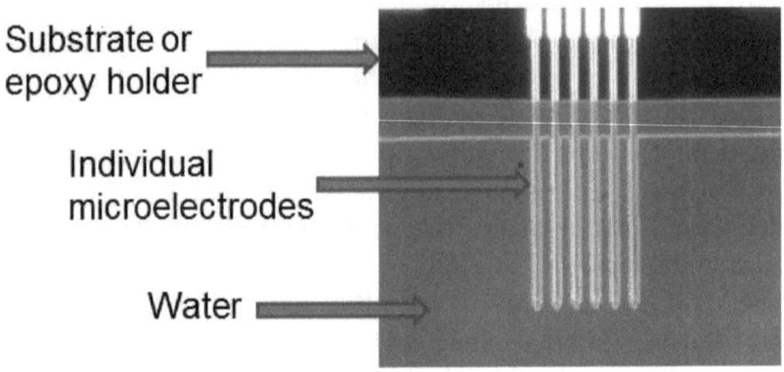

Fig. 5. Microelectrode arrays sampling water

For example, a cobalt-based MEMs microelectrode array sensors for direct measurement of phosphate in small environmental samples, such as microbial aggregates, has been introduced and applied for in situ measurement of phosphate within activated sludge floes in the enhanced biological phosphorus removal process as reported in [43]. These microelectrodes performed linearly and exhibited high sensitivity toward the phosphate ions in the range $10^{-5.0}$ to 10^{-25} M KH_2PO_4 solution at pH 7.5.

A similar MEMs type phosphorus microsensor with a microdigestion system (which converts various forms of phosphorus to phosphate) was reported in [38], where a cobalt metal film was introduced as the sensing material of the microelectrodes. The micro digestion chip is constructed with a silicon microfluidic device for the digestion of total phosphorus to phosphate. Three inlets are needed for digestion solution, water sample to be measured and cleaning solution that used to clean the channels and electrodes. Since the water sample to be measured needs to be digested at 120 °C for 30 min under a pressure of 1.14~1.2 MPa before flow onto the micro phosphate sensitive electrodes, a micro heating electrode is placed under the digestion well, and a micro temperature sensor is coupled with the heating electrode. However, the presence of a micro digestion bioreactor can be seen as both advantage for enhanced sensitivity, but a drawback for a real-time unmanned water monitoring system.

Microfluidic technology is used to minimize the volumes of reagent and sample required. The sensor is based on the yellow method for phosphate determination, using a simple colorimetric technique involving the formation of vanado-molybdo-phosphoric acid when a phosphate-containing sample is mixed with an acidic reagent containing ammonium molybdate and ammonium meta-vanadate. A yellow-colored compound is generated and the absorption of this compound is measured using a light emitting diode (LED) light source and a photodiode detector. The absorption is directly proportional to the phosphate concentration in the original sample.

However, due to the fragile nature of these sensors, they are capable of direct phosphate measurements only in small biological samples to enable in-situ analysis in many biological applications, but hardly for real-time long-term wastewater monitoring. A small volume of measured solution around the cobalt electrode is preferred, since the cobalt electrode deposited by sputtering [38] will dissolve when it is working as the sensing element. As a result, multiple long-term measurements with this system are a challenge.

7.3 Lab-on-Chip Sensors

Lab-on-chip and electrochemical sensing-based portable monitoring systems appear well suited to complement standard analytical methods for a number of environmental monitoring applications, including water quality monitoring. The concept of a lab-on-chip type system started from the integration of the various chemical operations involved in conventional analytical processes in a laboratory, such as sampling, preparation, mixing, reaction, and separation into a single unified system, requiring only a tiny volume of chemicals and sample and only a fraction of the time needed for the conventional approach.

Modern lab-on-chip is a complex system that combines amperometric/conductimetric sensor, microelectrodes and MEMs arrays, often along with microfluidics facilities. These techniques were discussed in previous sections.

The design, fabrication and characterization of four designs of microelectro-chemical systems on a silicon chip for the detection of pollutants in artificial seawater was reported in [42]. These systems were fabricated by photolithography and incorporate a Pt working microelectrode array, a Pt counter electrode and a Ag/AgCl reference electrode. They have been characterized by cyclic voltammetry of ferricyanide and behaved in good agreement with the theory. These systems were evaluated over 72 hours and showed good stability. Underpotential Deposition – Stripping Voltammetry experiments of measuring Cu^{2+} in artificial seawater have been carried out on an array of 35 micro-squares of 20 μm x 20 μm. The sensitivity achieved was (2.93±0.14) $\mu A\ cm^{-2}\ \mu M^{-1}$, with 1 μM being the lowest Cu^{2+} concentration measured. These devices are believed to provide the basis for the development into sensor systems for remote analysis applications [42].

For a comprehensive review of a recent advances in the lab-on-chip systems one may refer to Jang et al [44], who recently reported on their efforts towards a lab-on-chip sensor for environmental water monitoring and gave a detailed review of the methods employed.

8 Sensors Based on Sound and Electromagnetic Field Interaction

8.1 Ultrasonic Measurement

Ultrasonic technique is non-destructive and non-invasive method capable of rapid characterization of systems which are concentrated and optically opaque. Ultrasonic technique is sensitive to particles with radius between about 10 nm and 1000 mm. The propagation of ultrasound in a fluid is affected by its density, compressibility, temperature and composition. Low-power ultrasound and infrared waves which are emitted from equipment isolated from the target material are well-studied candidates for water and food analysis applications, as they do not introduce extraneous contamination nor alter the properties of the substance [45]. Ultrasonic sensor coupled with chemometric techniques for discrimination of different water samples was recently reported [46].

8.2 Electromagnetic Wave Sensors

The use of the electromagnetic waves for sensing purposes is an actively researched approach [47-49] with considerable potential for commercialization. One of the most successful attempts to use the electromagnetic sensing as a method for the detection of nitrates and contamination in natural water sources have been recently reported [48], where sensors based on the combination of planar meander and interdigitated electromagnetic structure were used. The meander sensors part consisted of two coils. The exciting coil was driven by high-frequency sinusoidal current to create a high-frequency magnetic field. To increase the magnetic field strength, the number of meander turns was increased by introducing several spiral rectangular turns toward the inner section. When this sensor was placed near a material, an electromagnetic field was induced on it, which modified the created field, and the sensing coil, placed above the exciting coil, was used to pick up the resultant field. The operating principle behind the interdigitated sensor is as follows [48]. The positive electrodes are driven by an AC electric potential (same source as the meander) to create an AC electric field between the positive and negative electrodes. These fields pass through the material under test; thus, material dielectric properties as well as the electrode and material property affect the capacitance and the conductance between the two electrodes. The meander type of coil was connected in series with the interdigitated coil, and an AC voltage was applied across the combination of the coils. The sensor response was determined by measuring the total impedance of the sensor at different frequencies between 10 kHz and 100 MHz [48].

8.3 Microwave Sensing

Microwave sensing is a novel and upcoming technology which has been successfully used as a sensing method for various industrial applications including water solution concentrations [50] and water level measurements [51], material moisture content [52, 53], for continuous process monitoring for biogas plants [54] and of course in the healthcare industry, for example for non-invasive real-time monitoring of glucose in diabetic patients [55, 56].

Microwave analysis (or microwave spectroscopy) can be applied to suit a broad range of requirements [57, 58] and has a number of advantages over competing technologies for wastewater sensing applications, including: 1) true real-time sensing, since an analyte material flows past or through the sensor and an instantaneous measurement is acquired without the need for significant pre-processing; and 2) direct sample measurement – a feature currently unavailable in many competing technologies.

The principle of sensing properties of materials, i.e. composition of water, using electromagnetic waves in the GHz frequency range is based on the fact that an object under test interacts with the electromagnetic waves in a unique manner, which can be specifically correlated with the properties of this material. Due to this interaction, the permittivity of the material changes and it manifests itself as a frequency change, attenuation or reflection of the signal. By considering how transmitted (S_{21}) and reflected (S_{11}) microwave powers vary at discrete frequency intervals, the change in the signal can be linked to the composition of the tested water sample.

A microwave resonator is made of a section of transmission line with open or shorted ends. Depending on the type of transmission line, the resonators can be for example coaxial, microstrip, stripline, slotline, or cavity resonators. When the resonator is used as a sensor, the object to be measured is brought into contact with at least some part of the electromagnetic field in the resonator. As a consequence, the resonant frequency and the quality factor will change in relation to the permittivity of the object. Due to the large variety of possible structures, sensors can be designed for measurement of almost any kind of object. The major advantage of the technique using microwave cavities is the capability to measure materials non-destructively, without contact from a short distance, using penetrating waves, without health hazards to personnel. Disadvantages are the usually high degree of specialization and the simultaneous existence of several variables affecting the microwave measurement (temperature, density, structure, etc.) in material measurements [59]. Initial results for in-situ monitoring of phosphorus levels in the treated wastewater [10, 49] validate the potential of microwave sensors for a real-time water quality monitoring, and this topic is discussed in greater depth elsewhere in this book.

9 Further Challenges and Possible Solutions in Developing Real-Time Water Monitoring Platform

This chapter reviewed the most common laboratory based methods of water pollutant detection, namely standard UV-Vis measurement methods, mass spectrometry, ion-sensitive electrodes and amperometric sensors. In addition, modern real-time monitoring approaches were examined, in particular fiber optic sensors, lab-on-chip sensors, biosensors and electromagnetic wave sensors, including microwave sensors. These are illustrated in Fig. 6.

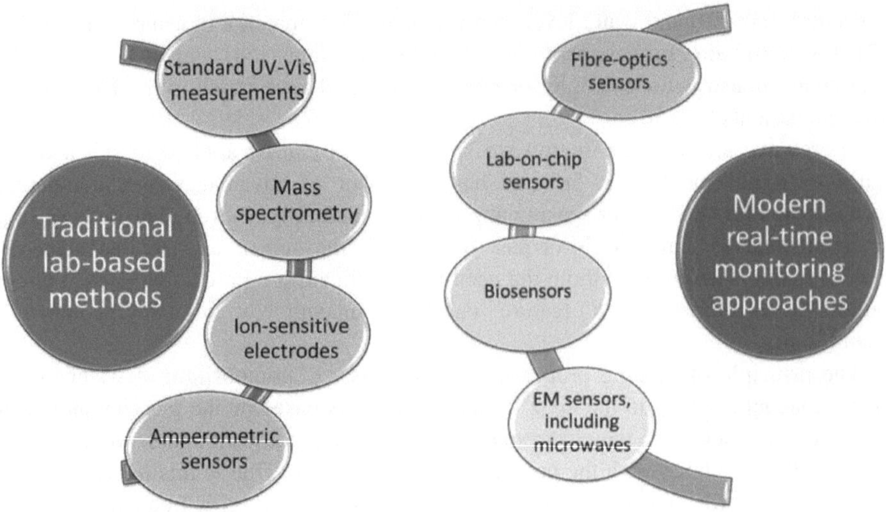

Fig. 6. Traditional lab-based and modern real-time approached to monitor wastewater

Even a brief overview of the traditional and state-of-the-art real time water monitoring methods given above reveals that it is not possible to achieve adequate simultaneous detection of different water parameters by using only one type of sensor. New instruments need to be introduced that are more robust, more sensitive and more accurate than before, to allow the detection of minute quantities of pollutants with high confidence and minimum sample preparation. Alternative solutions could be in merging various technologies into a single system that would employ the best available methods for the detection of specific water contaminants, providing overall superior sensitivity, selectivity and long-term stability, while at the same time enabling real-time wireless data collection for enhanced cost-effectiveness. This approach is commonly known as sensor fusion, which refers to the acquisition, processing and synergistic combination of information gathered by various knowledge sources and sensors to provide a better understanding of a phenomenon [60].

The feasible monitoring system might consist of a network of sensors deployed at key locations, capable of autonomous operation in the field for a year or more [61]. Such a system should be portable and would have to satisfy a broad range of requirements, including robustness and cost-effectiveness of the sensors; long battery life, which will depend on the sensors type and monitoring frequency [62]; data should be gathered in real-time from remote locations to a central server, where they will be processed, presented and communicated instantly, even by e-mail or SMS [61], if for example a set threshold of any water pollutant is exceeded. This will also help with mapping the spatial and temporal distribution of pollutants and may be of particular importance in identifying sources of water contamination [63, 64].

One such attempt, a multi-sensor system to monitor water quality parameters, namely pH, temperature, conductivity, turbidity and dissolved oxygen, is reported

[62]. Here a group of water quality sensors were integrated into a distributed communication network, through interfacing them with the PSoC Plug and Play system, with Zigbee telemetry, capable of transmitting the data to the SmartCoast server, which processed the data for transmission to the web based on the IEEE 1451 standard. However, this system suffered from bio-fouling within days of deployment and the sensors required regular maintenance.

Currently available multi-sensor systems give only an approximate estimation of the water composition. Artificial neural networks, which are usually a part of such systems, require specially developed software. Various response times of each system component are also an issue, i.e. sensors for various pollutants could present a problem in case of sudden changes in wastewater composition. Though these systems are promising, they are reliable for a short period only and need further research and development.

Real-time monitoring of wastewater quality remains an unresolved problem to the wastewater treatment industry [65]. Advanced autonomous platforms capable of performing complex analytical measurements at remote locations still require individual power, wireless communication, processor and electronic transducer units, along with regular maintenance visits. The above analysis of the available technologies shows that due to the complex pollutant matrix and generally hostile environment [66], the lack of accurate, cost-effective and robust sensors, the automation of wastewater treatment and monitoring systems remains a challenge. There exists a need for better on-line monitoring of water systems given that existing laboratory-based methods are too slow to develop operational response and do not provide a level of public health protection in real time [67].

When developing a real-time water monitoring system, one needs to take into account that:

- monitoring equipment should better fit practical utility needs and should be easy to operate and maintain;
- verification schemes should sufficiently match utility practices;
- available sensor technologies should link to water quality regulations;
- technologies and practices should be developed to manage the large quantities of data and translate these into information for operational processes.

To further preserve valuable water resources, a paradigm shift in water purification can be proposed by rethinking the distribution strategy at point-of-use and developing decentralized purification methodology based on region specific contaminants. This unique approach would assist in addressing the challenges in water purification so that clean water is available to everyone.

10 Conclusions

It is demonstrated that monitoring of water quality is currently still dominated by laboratory analysis of grab samples. Sensors are only available for a very limited number of parameters and frequently do not entirely meet the needs of the users.

Thus, the water quality control demands are not being met due to the infrastructure and maintenance costs of conventional sensing models. The interest in the use of field-based real-time measurements stems from a need to provide quick on-site assessments that could cover a greater geographical spread while negating much of the cost, time delay and issues of sample integrity associated with traditional laboratory-based methods. Sensors and on-line monitoring systems may have clear and multiple benefits for water utilities. Possible applications are intake protection, control of operations, security, and providing information to customers.

On the other hand, there are rapid advances in various sensors, integrated components and microelectronics technology, which promise to make comprehensive on-line water quality control an increasingly realistic alternative.

This chapter reviewed the most promising approaches for real-time water monitoring. It is suggested that since none of the methods alone can satisfy all the regulatory requirements, sensor fusion is required. Namely, multi-sensor platforms that utilize the best available methods combined into a single monitoring process are seen as the only way to achieve overall superior sensitivity, selectivity and long-term stability, while at the same time enabling real-time wireless data collection for enhanced cost-effectiveness. Also proposed is a paradigm shift in water purification by rethinking the distribution strategy at point-of-use and developing decentralized purification methodology based on region specific contaminants.

Acknowledgements. This work is financially supported by the European Community's Seventh Framework Programme through the FP7-PEOPLE-2010-IEF Marie-Curie Action project 275201, Water-Spotcheck.

References

[1] European Council, Directive 2000/60/EC of the European Parliament and of the Council of establishing a framework for Community action in the field of water policy. OJ L 327, 1–73 (2000)

[2] Schwarzenbach, R.P., Escher, B.I., Fenner, K., Hofstetter, T.B., Johnson, C.A., von Gunten, U., Wehrli, B.: The Challenge of Micropollutants in Aquatic Systems. Science 313, 1072–1077 (2006)

[3] Stuart, M., Lapworth, D., Crane, E., Hart, A.: Review of risk from potential emerging contaminants in UK groundwater. Science of the Total Environment 416, 1–21 (2012)

[4] Rodriguez-Mozaz, S., Lopez de Alda, M.J., Barceló, D.: Advantages and limitations of on-line solid phase extraction coupled to liquid chromatography–mass spectrometry technologies versus biosensors for monitoring of emerging contaminants in water. Journal of Chromatography A 1152, 97–115 (2007)

[5] Alder, L., Greulich, K., Kempe, G., Vieth, B.: Residue analysis of 500 high priority pesticides: better by GC-MS or LC-MS/MS? Mass Spectrometry Reviews 25, 838–865 (2006)

[6] Thomas, O., Théraulaz, F., Cerdà, V., Constant, D., Quevauviller, P.: Wastewater quality monitoring. TrAC Trends in Analytical Chemistry 16, 419–424 (1997)

[7] Heberer, T.: Occurrence, fate, and removal of pharmaceutical residues in the aquatic environment: a review of recent research data. Toxicology Letters 131, 5–17 (2002)
[8] Bourgeois, W., Burgess, J., Stuetz, R.: J. Chem. Techn. & Biotechnology 76, 337–348 (2001)
[9] Slater, C., Cleary, J., McGraw, C.M., Yerazunis, W.S., Lau, K.T., Diamond, D.: Autonomous field-deployable device for the measurement of phosphate in natural water, p. 67550L (2007)
[10] Al-Dasoqi, N., Mason, A., Alkhaddar, R., Al-Shamma'a, A.: Use of Sensors in Wastewater Quality Monitoring - a Review of Available Technologies. In: World Environmental and Water Resources Congress 2011: Bearing Knowledge for Sustainability, p. 354 (2011)
[11] Gilbert, L., Jenkins, A.T.A., Browning, S., Hart, J.P.: Development of an amperometric, screen-printed, single-enzyme phosphate ion biosensor and its application to the analysis of biomedical and environmental samples. Sensors and Actuators B: Chemical 160, 1322–1327 (2011)
[12] Amine, A., Palleschi, G.: Phosphate, Nitrate, and Sulfate Biosensors. Analytical Letters 37, 1–19 (2004)
[13] Villalba, M.M., McKeegan, K.J., Vaughan, D.H., Cardosi, M.F., Davis, J.: Bioelectroanalytical determination of phosphate: A review. Journal of Molecular Catalysis B: Enzymatic 59, 1–8 (2009)
[14] Allan, I.J., Vrana, B., Greenwood, R., Mills, G.A., Knutsson, J., Holmberg, A., Guigues, N., Fouillac, A.-M., Laschi, S.: Strategic monitoring for the European Water Framework Directive. TrAC Trends in Analytical Chemistry 25, 704–715 (2006)
[15] Allan, I.J., Vrana, B., Greenwood, R., Mills, G.A., Roig, B., Gonzalez, C.: A "toolbox" for biological and chemical monitoring requirements for the European Union's Water Framework Directive. Talanta 69, 302–322 (2006)
[16] Korostynska, O., Mason, A., Al-Shamma'a, A.I.: Monitoring of Nitrates and Phosphates in Wastewater: Current Technologies and Further Challenges. International Journal on Smart Sensing and Intelligent Systems 5, 149–176 (2012)
[17] Sten O, E.: The phosphate sensor. Biosensors and Bioelectronics 13, 981–994 (1998)
[18] Almeida, M.G., Silveira, C.M., Moura, J.J.G.: Biosensing nitrite using the system nitrite redutase/Nafion/methyl viologen—A voltammetric study. Biosensors and Bioelectronics 22, 2485–2492 (2007)
[19] Rosen, R.: Mass spectrometry for monitoring micropollutants in water. Current Opinion in Biotechnology 18, 246–251 (2007)
[20] Chusaksri, S., Sutthivaiyakit, S., Sutthivaiyakit, P.: Confirmatory determination of organochlorine pesticides in surface waters using LC/APCI/tandem mass spectrometry open diamond. Analytical and Bioanalytical Chemistry 384, 1236–1245 (2006)
[21] Donna, L.D., Mazzotti, F., Sindona, G., Tagarelli, A.: Assay of rotenone in river water by high-throughput tandem mass spectrometry and multiple-reaction monitoring methodology. Rapid Communications in Mass Spectrometry 19, 1575–1577 (2005)
[22] Pozo, O., Sancho, J.V., Ibanez, M., Hernandez, F.: Potential of liquid chromatography/time-of-flight mass spectrometry for the determination of pesticides and transformation products in water. Analytical and Bioanalytical Chemistry 386, 987–997 (2006)
[23] Lacorte, S., Fernandez-Alba, A.: Time of flight mass spectrometry applied to the liquid chromatographic analysis of pesticides in water and food. Mass Spectrometry Reviews 25, 866–880 (2006)

[24] Schermerhorn, P., Golden, P., Krynitsky, A.J., Leimkuehler, W.: Determination of 22 triazole compounds including parent fungicides and metabolites in apples, peaches, flour, and water by liquid chromatography/tandem mass spectrometry. Journal of AOAC International 88, 1491–1502 (2005)

[25] Capitán-Vallvey, L.F., Palma, A.J.: Recent developments in handheld and portable optosensing—A review. Analytica Chimica Acta 696, 27–46 (2011)

[26] Lyons, W.B., Ewald, H., Flanagan, C., Lewis, E.: A multi-point optical fibre sensor for condition monitoring in process water systems based on pattern recognition. Measurement 34, 301–312 (2003)

[27] McCue, R.P., Walsh, J.E., Walsh, F., Regan, F.: Modular fibre optic sensor for the detection of hydrocarbons in water. Sensors and Actuators, B: Chemical 114, 438–444 (2006)

[28] Ahmad, A., Paschero, A., Moore, E.: Amperometric immunosensors for screening of polycyclic aromatic hydrocarbons in water. In: 16th Conference in the Biennial Sensors and Their Applications, Cork, Ireland, September 12-14 (2011)

[29] Arshak, K., Korostynska, O.: Advanced materials and techniques for radiation dosimetry. Artech House (2006)

[30] Zhuiykov, S., Marney, D., Kats, E., Kalantar-Zadeh, K.: Potentiometric solid-state sensor for DO measurement in water using sub-micron CuO.4Ru3.4O7+RuO2 sensing electrode. Sensors and Actuators B: Chemical 153, 312–320 (2011)

[31] Van der Star, W.R.L., Abma, W.R., Blommers, D., Mulder, J.-W., Tokutomi, T., Strous, M., Picioreanu, C., van Loosdrecht, M.C.M.: Startup of reactors for anoxic ammonium oxidation: Experiences from the first full-scale anammox reactor in Rotterdam. Water Research 41, 4149–4163 (2007)

[32] Velusamy, V., Arshak, K., Korostynska, O., Oliwa, K., Adley, C.: An overview of foodborne pathogen detection: In the perspective of biosensors. Biotechnology Advances 28, 232–254 (2010)

[33] Devereux, R.: Environmental Monitoring and Assessment 116, 459–479 (2006)

[34] Campbell, G.A., Mutharasan, R.: Biosensors and Bioelectronics 22, 78–85 (2006)

[35] Arora, K., Chand, S., Malhotra, B.D.: Analytica Chimica Acta 568 (2006)

[36] Korostynska, O., Arshak, K., Velusamy, V., Arshak, A., Vaseashta, A.: Recent Advances in Point-of-Access Water Quality Monitoring. In: Vaseashta, A., Braman, E., Susmann, P. (eds.) Technological Innovations in Sensing and Detection of Chemical, Biological, Radiological, Nuclear Threats and Ecological Terrorism, pp. 261–268. Springer, Netherlands (2012)

[37] Ganjali, M.R., Norouzi, P., Ghomi, M., Salavati-Niasari, M.: Highly selective and sensitive monohydrogen phosphate membrane sensor based on molybdenum acetylacetonate. Analytica Chimica Acta 567, 196–201 (2006)

[38] Jianhua, T., Chao, B., Yang, L., Yin, B., Shanhong, X.: Design of a MEMS-Based Total Phosphorus Sensor with a Microdigestion System. In: 2010 4th International Conference on Bioinformatics and Biomedical Engineering (iCBBE), pp. 1–4 (2010)

[39] Lee, J.-H., Lim, T.-S., Seo, Y., Bishop, P.L., Papautsky, I.: Needle-type dissolved oxygen microelectrode array sensors for in situ measurements. Sensors and Actuators B: Chemical 128, 179–185 (2007)

[40] Lee, W.H., Seo, Y., Bishop, P.L.: Characteristics of a cobalt-based phosphate microelectrode for in situ monitoring of phosphate and its biological application. Sensors and Actuators B: Chemical 137, 121–128 (2009)

[41] Xiao, D., Yuan, H.-Y., Li, J., Yu, R.-Q.: Surface-Modified Cobalt-Based Sensor as a Phosphate-Sensitive Electrode. Analytical Chemistry 67, 288–291 (1995)

[42] Moujahid, W., Eichelmann-Daly, P., Strutwolf, J., Ogurtsov, V.I., Herzog, G., Arrigan, D.W.M.: Microelectrochemical Systems on Silicon Chips for the Detection of Pollutants in Seawater. Electroanalysis 23, 147–155 (2011)
[43] Lee, W.H., Lee, J.H., Bishop, P.L., Papautsky, I.: Biological Application of Micro-Electro Mechanical Systems Microelectrode Array Sensors for Direct Measurement of Phosphate in the Enhanced Biological Phosphorous Removal Process. Water Environment Research 81, 748–754 (2009)
[44] Jang, A., Zou, Z., Lee, K.K., Ahn, C.H., Bishop, P.L.: State-of-the-art lab chip sensors for environmental water monitoring. Measurement Science and Technology 22, 032001 (2011)
[45] Gan, T.H., Pallav, P., Hutchins, D.A.: Non-contact ultrasonic quality measurements of food products. Journal of Food Engineering 77, 239–247 (2006)
[46] Chouhan, R., Kiranmayee, A., Panchariya, P., Bhanu Prasad, P.: Acoustic signature based discrimination of drinking water. Presented at the ICST 2012: 6th International Conference on Sensing Technology, Kolkata, India (2012)
[47] Yunus, M.A.M., Mukhopadhyay, S., Punchihewa, A.: Application of independent component analysis for estimating nitrate contamination in natural water sources using planar electromagnetic sensor. In: 2011 Fifth International Conference on Sensing Technology (ICST), pp. 538–543 (2011)
[48] Yunus, M.A.M., Mukhopadhyay, S.C.: Novel Planar Electromagnetic Sensors for Detection of Nitrates and Contamination in Natural Water Sources. IEEE Sensors Journal 11, 1440–1447 (2011)
[49] Al-Dasoqi, N., Mason, A., Alkhaddar, R., Shaw, A., Al-Shamma'a, A.: Real-time non-destructive microwave sensor for nutrient monitoring in wastewater treatment. In: 16th Conference in the Biennial Sensors and Their Applications, Cork, Ireland, September 12-14 (2011)
[50] Kapilevich, B., Litvak, B.: Microwave sensor for accurate measurements of water solution concentrations. In: Asia-Pacific Microwave Conference, APMC 2007, pp. 1–4 (2007)
[51] Boon, J.D., Brubaker, J.M.: Acoustic-microwave water level sensor comparisons in an estuarine environment. In: OCEANS 2008, pp. 1–5 (2008)
[52] Jackson, B., Jayanthy, T.: A novel method for water impurity concentration using microstrip resonator sensor. In: Recent Advances in Space Technology Services and Climate Change (RSTSCC), pp. 376–379 (2010)
[53] Bernou, C., Rebière, D., Pistré, J.: Microwave sensors: a new sensing principle. Application to humidity detection. Sensors and Actuators B: Chemical 68, 88–93 (2000)
[54] Nacke, T., Barthel, A., Pflieger, C., Pliquett, U., Beckmann, D., Goller, A.: Continuous process monitoring for biogas plants using microwave sensors. In: 2010 12th Biennial Baltic Electronics Conference (BEC), pp. 239–242 (2010)
[55] Korostynska, O., Arshak, A., Creedon, P., Arshak, K., Wendling, L., Al-Shamma'a, A.I., O'Keeffe, S.: Glucose monitoring using electromagnetic waves and microsensor with interdigitated electrodes. In: IEEE Sensors Applications Symposium, SAS 2009, pp. 34–37 (2009)
[56] Mason, A., Wylie, S., Thomas, A., Keele, H., Shaw, A., Al-Shamma'a, A.: HEPA Filter Material Load Detection Using a Microwave Cavity Sensor. International Journal on Smart Sensing and Intelligent Systems 3, 16 (2010)
[57] Choi, J., Cho, J., Lee, Y., Yim, J., Kang, B., Oh, K., Jung, W., Kim, H., Cheon, C., Lee, H.: Microwave Detection of Metastasized Breast Cancer Cells in the Lymph Node; Potential Application for Sentinel Lymphadenectomy. Breast Cancer Research and Treatment 86, 107–115 (2004)

[58] Nyfors, E., Vainikainen, P.: Industrial microwave sensors. In: IEEE MTT-S International Microwave Symposium Digest, vol. 3, pp. 1009–1012 (1991)
[59] Nyfors, E., Vainikainen, P.: Industrial microwave sensors. Artech House, Norwood (1989)
[60] Varshney, P.K.: Multisensor data fusion. Electronics & Communications Engineering Journal 9, 245–253 (1997)
[61] O'Flynn, B., Regan, F., Lawlor, A., Wallace, J., Torres, J., O'Mathuna, C.: Experiences and recommendations in deploying a real-time, water quality monitoring system. Measurement Science and Technology 21, 124004 (2010)
[62] Regan, F., Lawlor, A., Flynn, B.O., Torres, J., Martinez-Catala, R., O'Mathuna, C., Wallace, J.: A demonstration of wireless sensing for long term monitoring of water quality. In: IEEE 34th Conference on Local Computer Networks, LCN 2009, pp. 819–825 (2009)
[63] O'Flynn, B., Martinez-Catala, R., Harte, S., O'Mathuna, C., Geary, J., Slater, C., Regan, F., Diamond, D., Murphy, H.: SmartCoast: A Wireless Sensor Network for water quality monitoring. In: 32nd IEEE Conference on Local Computer Networks, LCN 2007, Dublin, Ireland, October 15-18, pp. 815–816 (2007)
[64] Tschmelak, J., Proll, G., Riedt, J., Kaiser, J., Kraemmer, P., Bárzaga, L., Wilkinson, J.S., Hua, P., Hole, J.P., Nudd, R., Jackson, M., Abuknesha, R., Barceló, D., Rodriguez-Mozaz, S., López de Alda, M.J., Sacher, F., Stien, J., Slobodník, J., Oswald, P., Kozmenko, H., Korenková, E., Tóthová, L., Krascsenits, Z., Gauglitz, G.: Automated Water Analyser Computer Supported System (AWACSS): Part II: Intelligent, remote-controlled, cost-effective, on-line, water-monitoring measurement system. Biosensors and Bioelectronics 20, 1509–1519 (2005)
[65] Bourgeois, W., Burgess, J.E., Stuetz, R.M.: On-line monitoring of wastewater quality: a review. Journal of Chemical Technology & Biotechnology 76, 337–348 (2001)
[66] Qin, X., Gao, F., Chen, G.: Wastewater quality monitoring system using sensor fusion and machine learning techniques. Water Research 46, 1133–1144 (2012)
[67] Storey, M.V., van der Gaag, B., Burns, B.P.: Advances in on-line drinking water quality monitoring and early warning systems. Water Research 45, 741–747 (2011)

In Situ Phosphate Monitoring in Seawater: Today and Tomorrow

J. Jońca[1,*], M. Comtat[2], and V. Garçon[1]

[1] Laboratoire d'Etudes en Géophysique et Océanographie Spatiales, UMR 5566, 18 Avenue Edouard Belin, 31401 Toulouse Cedex 9, France
[2] Laboratoire de Génie Chimique, UMR 5503, Université Paul Sabatier, 118 Route de Narbonne, 31062 Toulouse Cedex, France
justyna.jonca@legos.obs-mip.fr

Abstract. Phosphorus is an important macronutrient and the accurate determination of phosphorous species (namely phosphate) in environmental matrices such as natural waters and soils is essential for understanding the biogeochemical cycling of this element, studying its role in ecosystem health and monitoring the compliance with legislation. This paper is focused on phosphate determination in seawater. Thus, the sources, occurrence and importance of phosphate together with several aspects regarding the analysis and terminology used in the determination of this element in the ocean are briefly described. Existing and future *in situ* analytical techniques for the determination of phosphate in seawater are presented. Today's *in situ* phosphate monitoring is dominated by different spectrophotometrical analyzers. Thus, a description of the basis, advantages and disadvantages of the different existing analyzers is provided. It seems that these techniques may be replaced in the near future by electrochemical sensors which provide excellent possibilities for phosphate determination with high precision, long lifetime, low detection limit and good reproducibility. Additionally, electrochemistry allows going further in miniaturization, provides a decrease in energy requirements and avoidance of additional reagents. Recently developed electrochemical methods for phosphate determination will lead to the first *in situ* autonomous sensor (ANESIS) which will fulfill all these expectations.

Keywords: Phosphate, nutrients analyzers, open ocean, sensor, electrochemistry, spectrophotometry.

1 Introduction

With 50 times more carbon dioxide than in the atmosphere, the oceans contain the largest reservoir of carbon actively circulating in the biosphere. In the long term, the ocean plays a dominant role in the natural regulation of CO_2 in the atmosphere via both physical and biological carbon pumps and thus exerts a powerful influence on

---

[*] Corresponding author.

S.C. Mukhopadhyay & A. Mason (Eds.): *Real-Time Water Quality Monitoring*, SSMI 4, pp. 25–44.
DOI: 10.1007/978-3-642-37006-9_2 © Springer-Verlag Berlin Heidelberg 2013

the climate [1]. A comprehensive and quantitative understanding of the way the ocean carbon cycle functions is fundamental to predict the consequences of rising levels of carbon dioxide and other greenhouse gases in the atmosphere. Phosphate is one of the main components of the biological carbon pump and its avaibility can impact primary production rates in the ocean as well as species distribution and ecosystem structure [2]. Thus, the availability of phosphates in marine systems can strongly influence the marine carbon cycle and the sequestation of atmospheric carbon dioxide. It is then critical to monitor properly phosphate concentration in the ocean, if we are to understand climate regulation.

1.1 Chemistry of Phosphate

From a chemical point of view, phosphates are the salts of phosphoric acid. In dilute aqueous solution, phosphate exists in four forms. In strongly-basic conditions, the orthophosphate ion (PO_4^{3-}) dominates, whereas in weakly-basic conditions, the hydrogen phosphate ion (HPO_4^{2-}) is prevalent. In weakly-acid conditions, the dihydrogen phosphate ion ($H_2PO_4^-$) is the most common. In strongly-acid conditions, aqueous phosphoric acid (H_3PO_4) is the main form. More precisely, considering the following three equilibrium reactions:

$$H_2PO_4 \Leftrightarrow H^+ + H_2PO_4^- \tag{1}$$

$$H_2PO_4^- \Leftrightarrow H^+ + HPO_4^{2-} \tag{2}$$

$$HPO_4^{2-} \Leftrightarrow H^+ + PO_4^{3-} \tag{3}$$

the corresponding constants at 25°C (in mol/L) are:

$$K_{a1} = \frac{[H^+][H_2PO_4^-]}{[H_3PO_4]} \approx 7.5 \cdot 10^{-3}, pK_{a1} = 2.12 \tag{4}$$

$$K_{a2} = \frac{[H^+][HPO_4^{2-}]}{[H_2PO_4^-]} \approx 6.2 \cdot 10^{-8}, pK_2 = 7.21 \tag{5}$$

$$K_{a3} = \frac{[H^+][PO_4^{3-}]}{[HPO_4^{2-}]} \approx 2.14 \cdot 10^{-13}, pK_{a3} = 12.67 \tag{6}$$

1.2 The Cycle of Phosphate in the Ocean

The marine phosphorus cycle is presented on Fig. 1 [2] with a special attention to different phosphorus fluxes. The phosphorus content on both oceans and lands is very small. The main reservoir of phosphorus on land is rocks formed in ancient geological epochs. These rocks gradually ventilate and release phosphorus compounds into marine ecosystems. Continental weathering is the primary source of phosphorus to the oceanic phosphorus cycle [3]. Most of this phosphorus is delivered via rivers with a smaller portion delivered via dust deposition [4]. In recent times, anthropogenic sources of phosphorus have become a large fraction of the phosphorus delivered to the marine environment, effectively doubling the pre-anthropogenic flux. An excess of phosphate causes uncontrolled growth of algae leading to eutrophication or over fertilization of receiving waters [5]. The rapid growth of aquatic vegetation can cause the death and decay of vegetation and aquatic life due to the decrease in dissolved oxygen levels [6, 7]. The main anthropogenic sources of phosphorus are detergents and sewage. The primary sink for phosphorus in the marine environment is the loss to sediments. Much of the particulate flux from rivers is lost to sediments on the continental shelves, and a smaller portion is lost to deep-sea sediments [8]. In surface waters, certain phosphate forms (PO_4^{3-}) are taken up by phytoplankton during photosynthesis. Subsequently, the phosphate travels up through the food chain to zooplankton, fish and other top marine organisms [9]. Much of organic-P is converted back to PO_4^{3-} in surface waters as phytoplankton die but some of it finds its way to the deep ocean (via downwelling and sinking of organic matter) where it is remineralized back to inorganic-P. It may return to the surface waters (via upwelling) or be bound by the widely prevalent cations (Al^{3+}, Ca^{2+}, Fe^{2+}, Fe^{3+}) and stored as minerals for a long time in the rocks and sediments. In that case, only geological processes can bring them back to cycle.

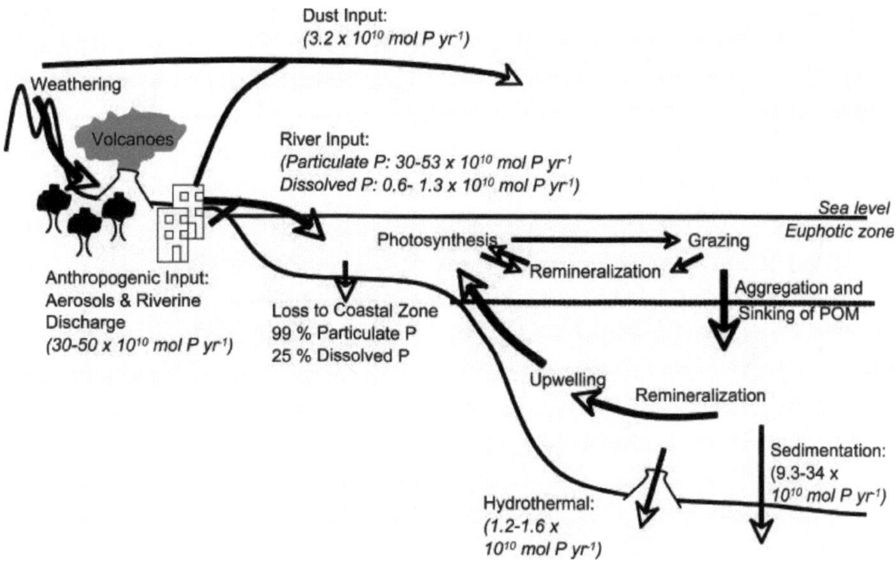

Fig. 1. The marine phosphorus cycle [2]

Vertical sections for phosphate concentrations in sea water are presented on Fig. 2 for the Atlantic, Pacific and Indian Oceans.

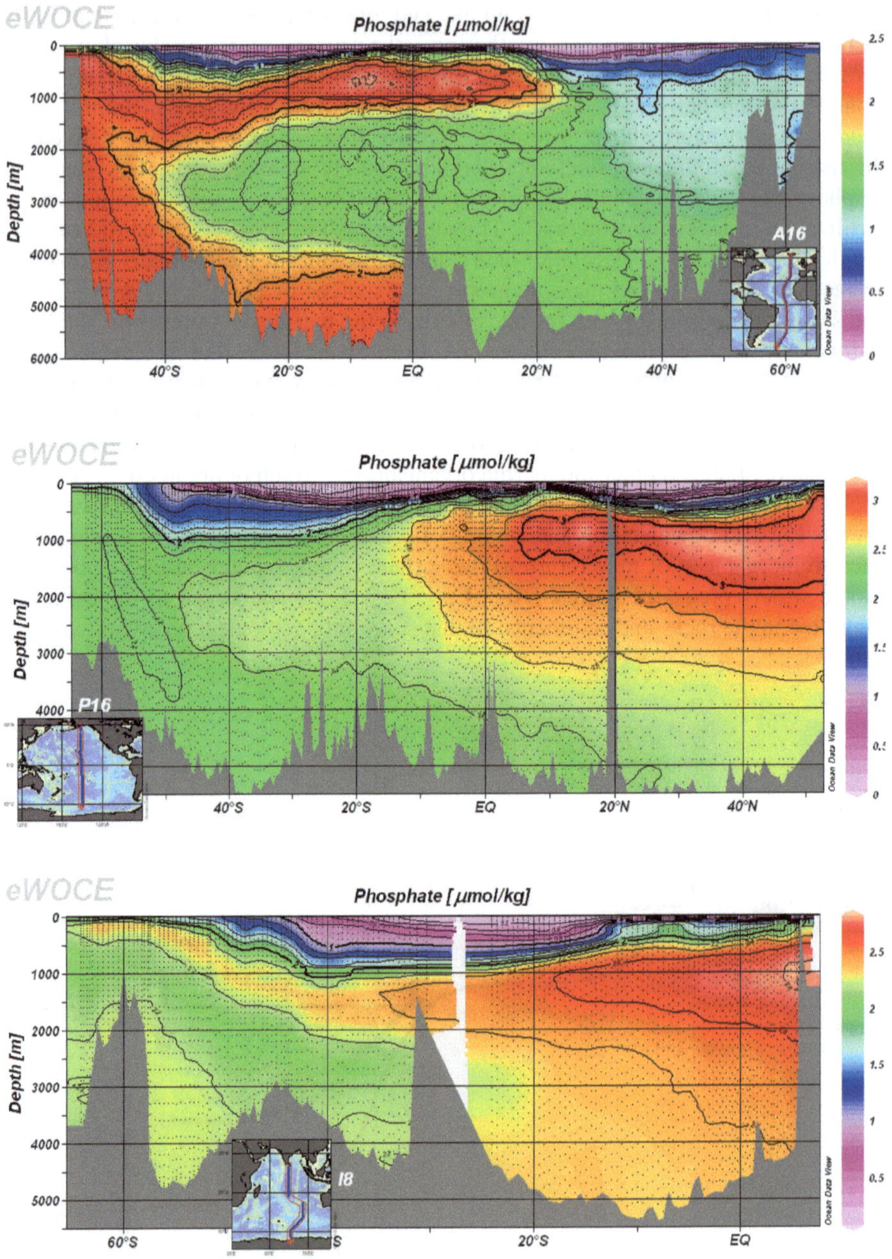

Fig. 2. Vertical section of phosphate concentrations obtained during the WOCE A16, P16 and Indian Ocean I8 expeditions [10]

As phytoplankton and other organisms die, PO_4^{3-} is regenerated in the water column. A maximum regeneration occurs near 1000 m, which is the same depth as the oxygen minimum layer. The maximum values of phosphate concentration in the Atlantic are about 2.5 μM, while in the Pacific values around 3.25 μM occur. The higher values observed in the Pacific (and Indian) oceans as compared to the Atlantic are due to the age of the water masses (older in the Pacific thus accumulating more oxidized plant material). Generally, knowledge of the phosphate concentration in the ocean surface allows us to deduce information on the biological activity in the ocean. Secondly, phosphate is one of the chemical tracers (along with temperature, salinity, and other nutrients concentration) which allows us to analyse water mass mixing and determine the origin of the water masses.

Phosphate concentration and ratio of nitrate over phosphate concentration (NO_3^-/PO_4^{3-}) can inform us about the biogeochemical processes in the ocean where microbial processes coexist: denitrification, nitrification, anammox, nitrogen fixation [11]. These parameters were used to explain unexpected phosphate profiles obtained during the Pelagico 1011-12-BIC OLAYA cruise aboard R/V Josa Olaya Balandra off Peru during austral spring 2010 (November-December 2010) when an intense upwelling occurred whithin this Oxygen Minimum Zone region. At depths ranging from 150-200m a relative minimum in phosphate concentrations was observed. Two cases were noticed: phosphate minimum associated with nitrate minimum and phosphate minimum associated with nitrate maximum. The first case is characteristic of the predominance of denitrification/anammox processes (which consume nitrate over nitrification/remineralization) associated with N_2-fixation (phosphate consumption). The second would be indicative of predominance of remineralization/ nitrification processes (production of nitrate over denitrification/anammox) associated also with N_2-fixation (phosphate consumption, but no nitrate consumption) [12, 13].

1.3 Sensors and Platforms for Long Term Monitoring of the Ocean

Much of what we currently know about the phosphate (and other nutrients) content in the ocean comes from very expensive and time consuming ship-based sampling. An alternative is to obtain autonomous observations supported by spectrophotometric analyzers based on wet chemistry but their use for phosphate analysis is not very much widespread [14]. The development of multi-disciplinary oceanic observatories for long-term monitoring will help to increase the observing capacities in the constantly changing marine environment. This monitoring requires an *in situ* miniaturized autonomous instrumentation able to achieve excellent figures of merit: long lifetime, high precision, low detection limit, fast response time, good reproducibility, resistance to biofouling and high pressure, provision of stable long-term operation, and low energy consumption. Within the past years, sensor technologies for oxygen, chlorophyll, particles and nitrate have been refined [15]. These sensors are capable of deployment on long-endurance missions on autonomous platforms such as profiling floats and gliders.

The ARGO profiling floats are good example of autonomous platforms. ARGOs drift at a fixed parking depth (1000 or 2000m for instance below sea level) for 9 days and then they come up to the surface. On their way up, they measure temperature and salinity (and sometimes nitrate and oxygen concentrations) through the water column. Once they reach the surface, they transmit recorded data to land via satellite and then the cycle is repeated (Fig. 3). The buoyancy of an ARGO float is controlled to make it rise or sink. This is done by pumping oil between internal and external bladders. There is about 3000 of these floats in the ocean [16] equipped with typical physical sensors whereas the same floats equipped with biogeochemical sensors are in number of about 300. The major challenge will be to increase the number of biogeochemical sensors adapted on ARGO floats and other autonomous platforms.

Many sensors detect analytes directly, without addition of liquid reagents. Among them, the most often used are oxygen and pH sensors. Clark-type sensors are essential to study primary production and metabolism of organisms. They are also essential for oxygen measurement of special marine ecosystems like those found in the Oxygen Minimum Zones. There are many examples of applications of these sensors on CTD profilers or floats. pH sensors are essential for measurements of growing acidification of the oceans and its impact on the carbon pump and life of marine organisms [17]. However, there are many species undetectable by direct measurement (phosphate, silicate, etc.). Analyzers have been developed to measure *in situ* these elements but these methods require a set of elements for measuring samples where the detection is done after a chemical reaction (pump, valves, sensors, electronics, etc.).

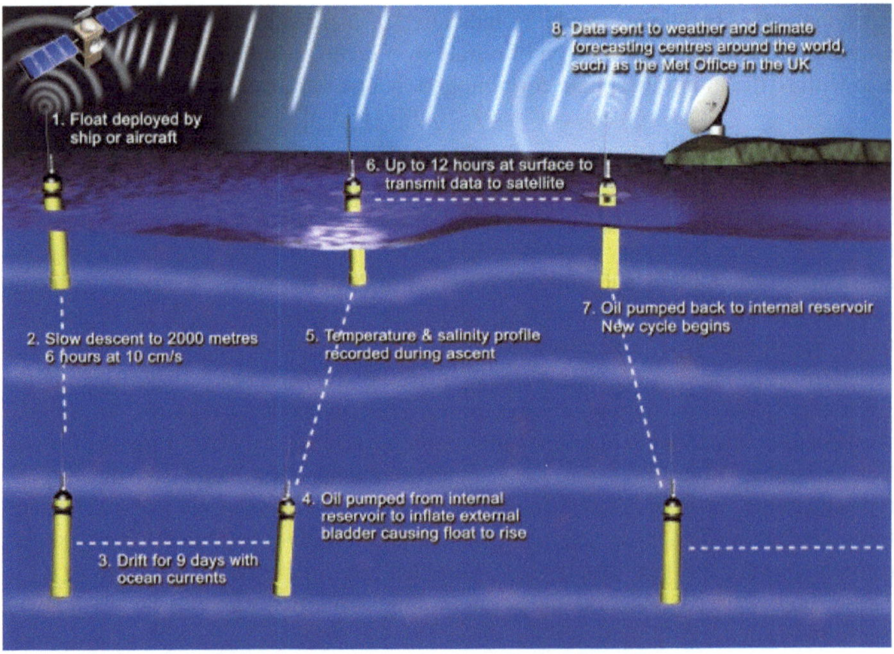

Fig. 3. Operational cycle of an ARGO float [18]

Within a large number of analytical methods, electrochemistry provides promising reagentless methods to go further in miniaturization, decrease in response time and energy requirements. In aquatic systems, electrochemical methods are used routinely for monitoring of pH by potentiometry, dissolved oxygen by amperometry [19] trace metals and speciation by voltammetry [20, 21], conductivity and therefore salinity by impedimetry. Electrochemistry offers also a wide range of possibilities for achieving an excellent phosphate determination in seawater but no autonomous electrochemical sensor for *in situ* phosphate determination exists nowadays. Generally, the use of electrochemical *in situ* sensors in the ocean is not well developed but the opportunities provided by some teams show an increasing interest in this research field.

2 Today's In Situ Analyzers for Phosphate Monitoring in Seawater

Phosphate analysis in the ocean is still based generally on traditional shipboard sampling techniques with subsequent analysis either on board or in a shore-based facility. The *in situ* analyzers use wet chemistry like ANAIS or NAS 3-X. They are able to work about 1-2 months without human intervention and are capable for simultaneous determination of phosphate, silicate, nitrate and nitrite. Such analyzers provide also good accuracy but require addition of reagents and release waste to the ocean. The drawback of these techniques is also their large size and weight.

2.1 Principles of Spectrophotometric Phosphate Measurements

The analysis is performed using the method of Murphy and Riley [22] adapted by Strickland and Parsons [23]. Molybdic acid is formed by conversion of ammonium heptamolybdate in acidic medium (reaction 7). The phosphate in reaction with molybdic acid forms a phosphomolybdate complex which absorbs yellow in the UV (reaction 8):

$$(NH_4)_6Mo_7O_{24}, 4H_2O + 6H^+ \rightarrow 7(MoO_3, H_2O) + 6NH_4^+ \quad (7)$$

$$PO_4^{3-} + 12MoO_3 + 3H^+ \rightarrow H_3PMo_{12}O_{40} \quad (8)$$

$$PMo^{VI}{}_{12}O_{40}{}^{3-} + 4e \rightarrow PMo^{V}{}_4Mo^{VI}{}_8O_{40}{}^{7-} \quad (9)$$

The phosphomolybdate complex (the oxidation state of molybdenum is VI) is reduced using ascorbic acid (reaction 9). This gives a blue complex with molybdenum in an oxidation state of V and VI. This complex has an absorption maximum at 882 nm. The measurement is also possible at 715 nm but with a loss of sensitivity. The concentration can be found by measurements of absorbance which is directly proportional to the concentration as described by Lambert-Beer's law:

$$I = I_0 \, e^{-\alpha l C} \quad (10)$$

$$A_\lambda = -\log_{10} \frac{I}{I_0} = \varepsilon_\lambda \, l \, C \tag{11}$$

With I_0-intensity of excitation light, I-intensity of light output, I / I_0-transmittance of the solution, A-absorbance or optical density at a wavelength λ, ε-molar extinction coefficient (in mol^{-1} cm^{-1}); l-optical path length traversed in the solution, it corresponds to the thickness of the cell used (cm); C-molar concentration of the solution (mol L^{-1}), α-absorption coefficient (m^2 mol^{-1} or m^3 mol^{-1} cm^{-1}).

The absorption measurements required the use of spectrophotometers. Today, spectrophotometers with simple measurements using cuvettes are still performed but methods coupled with flow-injection apparatus are preferable.

Arsenate and silicate form the same kind of complex which can be a source of interferences. However, normally arsenate concentrations in the open ocean are very small as compared to phosphate and do not exceed 0.03 μM in seawater. Thus, the influence of arsenate on phosphate detection is negligible. From the other hand, silicate concentration is much higher in the ocean and without special treatment it can cause a significant interference. To avoid silicate interferences, the ratio of protons over molybdates H^+/Mo and pH of solution should be in the range of 60-80 and 0.4-0.9, respectively [24].

Some fractions of particulate phosphorus (included in phytoplankton or minerals) react like dissolved phosphate. For this reason, the procedure of analysis of rich coastal waters must be preceded with the removal of particles before analysis, by filtration or centrifugation [25]. Another problem during phosphate sampling is associated with adsorption of phosphate on plastic and glass bottles (especially in brackish waters) and contamination in contact with the skin [25].

2.2 ANAIS – Autonomous Nutrient Analyzer In Situ

The spectrophotometrical *in situ* analyzer ANAIS developed in our team allowed simultaneous measurement of nitrate, phosphate and silicate. This instrument was suitable for the autonomous vehicle YOYO, profiler developed by the LODYC (Laboratory of Dynamic Oceanography and Climatology, now LOCEAN laboratory) in the framework of the European MAST III "YOYO 2001: The Ocean Odyssey". The YOYO vehicle was profiling along a cable stretched between 1000 meters and the surface with a range of 400 up-down cycles with the same principles of motion used by the ARGO floats (change of buoyancy, see Fig 3). This vehicle was also equipped with several other bio-optical, biological, physical and chemical sensors. This instrument was autonomous and capable of *in situ* measurements in pressure between 0 and 1000 m depth [26, 27].

The ANAIS analyzer represented on Fig. 4 is a set of:

1) Three chemical sensors (nitrate, silicate, phosphate) which contain:

- A manifold where the chemical reaction takes place at a constant temperature maintained between 20 ° and 25 ° C by a heating resistor integrated into the manifold,

- A colorimeter, integrated into the manifold with characteristic wavelength for the studied analyte,
- Two clamping plates sealing the manifold and the fixing of pumps (pumps solenoids Lee Co.).

2) A container in which the three sensors mentioned above are fixed.
3) A set of bags for storing reagents and standard solutions.
4) An electronic card for the sensors' control and data storage (1 card for each salt), receiving control commands transmitted by a central brain outside (located within the YOYO body) and data transmission to the same central brain (one card called master card).

Fig. 4. ANAIS applied on YOYO vehicle (left), and the container with reagent bags and spectrophotometer (right)

The device offers an analytical measurement reproducibility of about 1%, and very good calibration results similar to those obtained with a Technicon colorimetric chain (conventional laboratory equipment).

A series of tests under real conditions, as part of the YOYO 2001 EC project milestones, was performed with ANAIS along offshore Blanes (Spain) and offshore Banyuls sur Mer in the Mediterranean Sea from 1998-2001 [28]. The results show that the standard deviation on ANAIS results is much lower than that achieved on the Niskin samples. This analyzer (only nitrate) was tested in 2003 off the shore of Argentina as part of the CLIVAR programme. The scientific objectives were articulated along two axes: 1) to monitor the Malvinas Current transport at 41 °S with a line of current meters on moorings under a track of the satellite altimeter JASON, 2) to study variations and characteristics of Antarctic Intermediate Water in the Argentine Basin [28, 29].

2.3 *Other Spectrophotometric Analyzers*

Many adaptations of classic colorimetric methods for the measurement of phosphate (silicate and nitrate, nitrite) have been developed for *in situ* analysis. Many of these instruments are now commercially available (Fig. 5). The basic principle is an assembly of pumps, valves, manifolds and colorimetric detectors that can be combined to measure multiple analytes simultaneously. The calibration has to be done regulary and thus, bags with standards are needed. Characterization of some phosphate analyzers is presented in Table 1. The characterization is based on a few parameters like: range of measured chemicals, detection limit for phosphates, size and weight, and application of domain analyzer.

Both commercial and experimental analyzers are characterized by a good precision (1-3%) and high sensitivity. Detection limit is in the range of 10 to 100 nM. The analyzers can measure simultaneously all salts (phosphate, nitrate, nitrite, silicate). Some of them have a possibility of additional measurement of iron or sulphur (CHEMINI, SubChemPak Analyzer). Cycle-PO_4 analyzer is dedicated to phosphate measurements. Most of the analyzers are not resistant to high pressure (except ANAIS and CHEMINI) and can not go deeper than 10-200 m. Most of the analyzers can work autonomously for a few months, but require addition of reagents which can be unstable (ascorbic acid). They are usually large and heavy.

Although some analyzers were deployed in natural waters, there are not many examples of phosphate monitoring using these analyzers. Nitrate is the most commonly measured nutrient. For example, the *SubChemPak Analyzer* was installed on a winch-deployable, electronic profiling package. The profiling package also included a Sea Bird Electronics Model 25 Sea Logger CTD with modular sensors for the measurement of conductivity, temperature, pressure, dissolved oxygen, pH, chlorophyll fluorescence, and light transmission and irradiance.

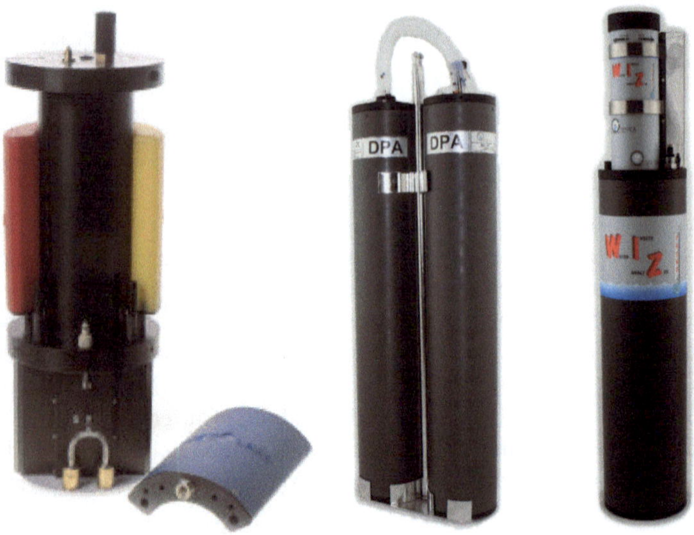

Fig. 5. Commercially available *in situ* nutrients analyzers: Wet Labs Cycle-PO_4 (left) [30], Systea Deep-sea Nutrient Analyzer – DPA (center) [31], Systea Water *In situ* Analyzer – WIZ (right) [32]

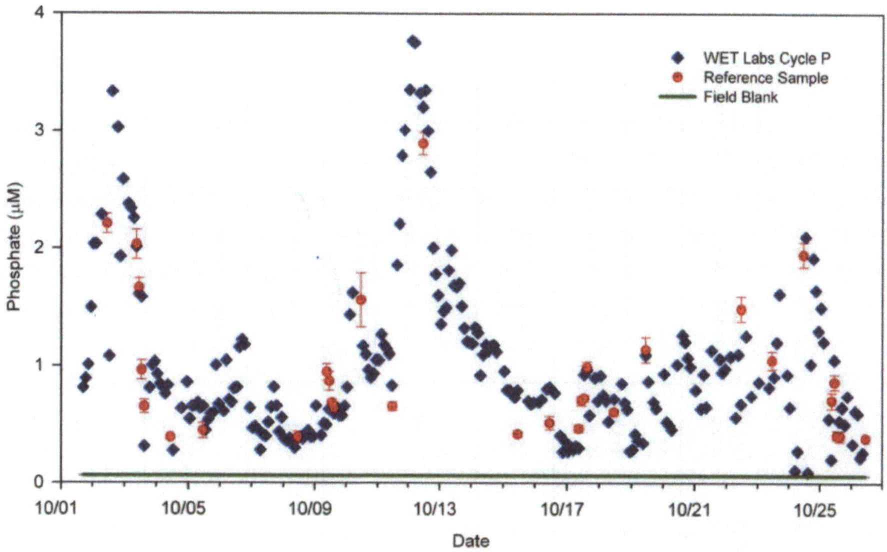

Fig. 6. Time series comparison of WET Labs Cycle-PO_4 measured phosphate concentrations against laboratory measured reference samples and field trip blanks for the Clinton River, MI moored deployment test [30]

The Wet Labs Cycle-PO_4 was successfully used at three different fixed mooring stations. The field tests were performed during 4 weeks at a fixed depth in coastal freshwaters (Clinton River), brackish waters (Chesapeake Bay), and seawater (Resurrection Bay). Phosphate concentrations at Chesapeake Bay were found in the range of 0.039-0.092 µmol L^{-1}, despite a significant contribution of river water to the site. Phosphate concentrations were higher at the Resurrection Bay (even 0.701 µmol L^{-1}) and the highest in the Clinton River (0.09-3.77 µmol L^{-1}) (Fig. 6). The same analyzer was used on CHARM (**CHA**nnel Island **R**elocatable **M**ooring) mooring at Hawai Ocean Time Series (HOTS) station in February-March 2004 at depth of 22 m. The phosphate there varied from 0.1 to 0.9 µmol L^{-1}.

3 Electrochemical Sensors – The Future of Phosphate Monitoring?

As an alternative to spectrophotometry, methods based on electrochemistry have also been proposed. Amperometric procedures have been reported for the determination of phosphate as phosphomolybdate complex [38-40]. Phosphate has been also determined by using voltammetric methods with carbon paste electrode [41], gold microdisk electrode [42] and glassy carbon electrode [43]. Different modifications were proposed to increase the sensitivity or eliminate the interferences from other coexisting molecules. Applications of these methods are proposed but only

Table 1. In situ analyzers for phosphate detection in seawater, detection limits only for phosphate

Name	***Laboratory/ Society***	***Detected chemicals***	***Detection limit***	***Size, mm***	***Weight in air, kg***	***Depth max. m***	***Applications***	***Literature***
ANAIS (Autonomous Nutrient Analyzer In Situ	*LEGOS*	$Si(OH)_4$, PO_4^{3-}, NO_3^-+NO_2^-,	*100 nmol* L^{-1}	*320(d)x1000*	*50*	*1000*	*Open ocean, coastal waters*	*[26, 27]*
NAS – 3X MicroLab EcoLab	*EnviroTech LLC*	$Si(OH)_4$, PO_4^{3-}, NO_3^-+ NO_2^-,	*50 nmol* L^{-1}	*246(d)x799*	*27*	*250*	*Bays, coastal waters*	*[33]*
DPA/NPA (Deep – sea/ Nutrient Probe Analyer	*Systea*	$Si(OH)_4$, PO_4^{3-}, NO_3^-+NO_2^-,	*1 ppb (10 nmol* L^{-1}*)*	*320(d)1220*	*36*	*10-4500*	*Open ocean, coastal waters*	*[31)*
APP 4004	*Me Grisard Gmbh*	$Si(OH)_4$, PO_4^{3-}, NO_3^-+NO_2^-,	*np*	-	*8*	*7*	*Bays, coastal waters*	*[34]*
SubChemPak Analyzer	*SubChemSystems, Inc. /Wet Labs*	$Si(OH)_4$, PO_4^{3-}, NO_3^-+NO_2^-, *Fe*	*20-50 nmol* L^{-1}	*12(d)x635*	*6*	*200*	*Bays, costal waters*	*[35]*
CHEMINI (Chemical Miniutarized Analyzer)	*IFREMER*	$Si(OH)_4$, PO_4^{3-}, *Fe*, H_2S, NO_3^-+ NO_2	*np*	*120x140+ 140(d)x264*	*4*	*6000*	*Coastal waters, hydrothermal vents*	*[36, 37]*
Cycle – PO_4	*Wet Labs*	PO_4^{3-},	*75 nmol* L^{-1}	*560x180*	*6.8*	*200*	*Bays, coastal waters*	*[30]*

a few of them were used for natural analysis of water in laboratory conditions [39, 40, 44]. In spite of many advantages of voltammetric or amperometric methods, some serious problems may appear. In particular, electrode surface fouling by absorbing compounds, or oxidation by oxygen are the most important issues. Additionally, these electrochemical methods still require addition of acid and molybdate ions in order to convert electroinactive phosphate species into electroactive phosphomolybdate complex.

The potentiometric analysis of phosphate requires application of ion-selective electrodes (ISE). In these conditions, the water sample does not have to be pretreated. Numerous membranes are being used [45-49]. Some of them are dedicated to environmental water analysis but so far no *in situ* application of potentiometric sensors in seawater is described. Several issues affect potentiometric methods: the electrode response is often slow, interferences from other species are very common, and miniaturization is tricky especially because the potential is unstable when the electrode approaches micrometric dimensions. Finally, electrodes require very often calibration, sometimes before and after the measurement of interest.

Some teams work on electrochemical phosphate biosensors [50], but so far no *in situ* application is described in the literature. Their development is complicated and requires use of enzymes which are sensitive to the environment changes (temperature, ionic force) [51] and to the presence of heavy metals and pesticides which may inhibit enzyme activity [52].

3.1 *ANESIS – Autonomous Nutrients Electrochemical Sensor In Situ*

Recently, a totally new electrochemical method was developed in our group for determination of silicate [53, 54] and phosphate [55, 56] without addition of any liquid reagents. The method is based on the oxidation of molybdenum in seawater in order to form silico- or phosphomolybdate complex electrochemically detectable either by means of voltammetry or amperometry. Both methods were tested *in situ* offshore Peru (phosphate) and Drake Passage (silicate) (Fig. 7 and 8).

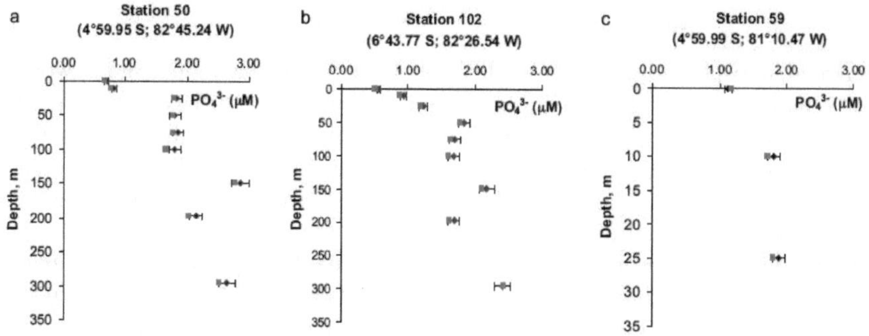

Fig. 7. Comparison of both methods, amperometry (black) and colorimetry (grey) at CTD stations 50 (a), 102 (b), 59 (c). Station 59 is situated very close to the coast and stations 50 and 102 more off-shore [55].

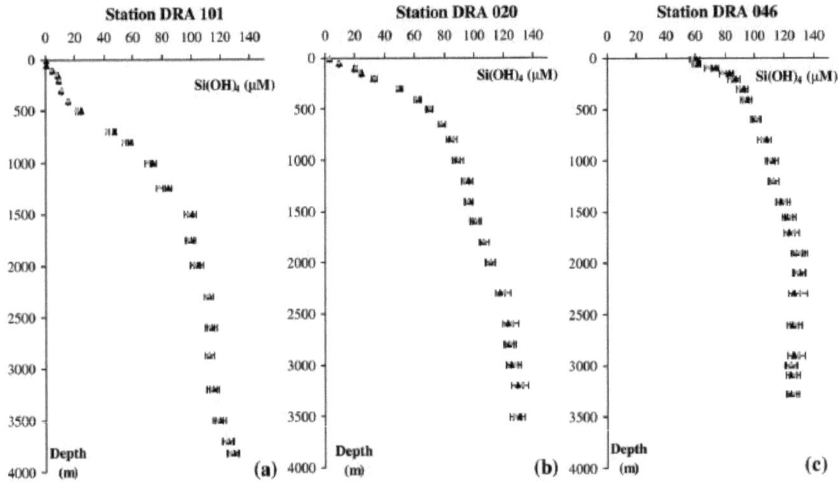

Fig. 8. Comparison of the two methods, voltammetry in black and colorimetry in grey at stations (a) DRA 101, (b) DRA 020 and (c) DRA 046 [53]

The results show excellent agreement with colorimetric classical analysis [53, 55]. The deviation between both methods is 4.9% for phosphate and 3% for silicate concentrations higher than 10 µM. The description below corresponds, however, only to the phosphate sensor.

The electrochemical determination of the non electroactive phosphate is based on the formation of a complex with molybdate. The complex is formed by the reaction with K_2MoO_4 in an acidic solution to form a Keggin anion [55, 56].

$$PO_4^{3-} + 12\ MoO_4^{2-} + 27\ H^+ \rightarrow H_3P(Mo_{12}O_{40}) + 12\ H_2O \tag{12}$$

In order to eliminate the addition of molybdate and protons, a reagentless method was developed. Molybdate salts and protons are produced as the product of molybdenum oxidation in the reaction cell due to the following reaction [55-57].

$$Mo + 4\ H_2O \rightarrow MoO_4^{2-} + 8\ H^+ + 6e^- \tag{13}$$

The created phosphomolybdate complex can then be detected on a gold working electrode by means of amperometry with the detection limit of 0.12 µM. The method described above has an obvious problem of cross interference in the ocean samples which contain both silicate and phosphate. A few methods to avoid these interferences are tested using differences in electrochemistry of the two complexes, influence of pH on complex formation and differences in kinetics of complex formation by choosing an appropriate ratio of protons over molybdates of 70 [55]. The latter one was used during the Pelagico 1011-12 cruise aboard the R/V OLAYA offshore Peru. The results show great potential in the developed method for phosphate detection but it still required extra addition of acid to achieve an appropriate ratio of protons over molybdates (70). According to reaction (13), the ratio of protons over molybdate is 8 during the oxidation of molybdenum. At the beginning, we achieved an appropriate ratio of 70 by addition of acid. Recently, we developed a method where the desired ratio of protons over molybdate is achieved without addition of any reagents. This method is based on the use of specific membrane technology and a special design of the electrochemical cell made by GIS laboratory (Groupe d'Intrumentation Scientifique, Observatoire Midi-Pyrenées, Toulouse) [56]. The cell consists of 3 compartments. In the first one (1 mL), a primary molybdenum electrode is oxidized

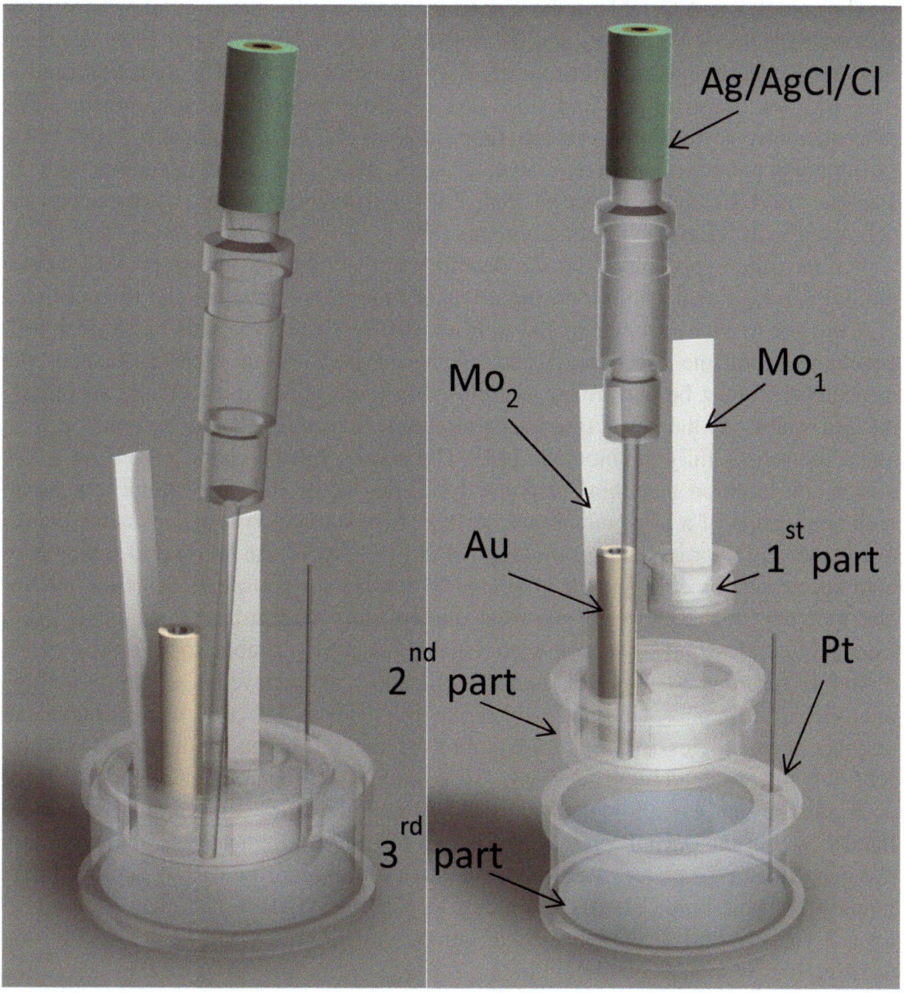

Fig. 9. Schematics of the electrochemical cell with application of membrane technology (Ag/AgCl/Cl-reference electrode, Au-gold working electrode, Pt-platinum counter electrode, Mo_1-first molybdenum electrode, Mo_2-second molybdenum electrode, 1st part-first compartment of the cell with the proton exchange membrane, 2nd part-second compartment of the cell with the non-proton exchange membrane, 3rd part-third compartment of the cell, simulating the open ocean) [56]

and thanks to a proton exchange membrane only protons can pass through to the second compartment and thus acidify the media to pH 1. In the second compartment (5 mL), a secondary molybdenum electrode is oxidized during a short time to produce 1.5 mM of molybdates achieving a ratio of protons over molybdates of 70. The third compartment is in contact with previous parts of the cell by a non-proton exchange membrane where a platinum cathode is immersed to avoid the reduction of protons formed during the two previous oxidations of molybdenum (Fig. 9).

The complex is detectable in the second compartment by means of amperometry or differential pulse voltammetry and the detection time is about 15 min. The detection time can be much shorter by decreasing the volume of the cell. The detection limit is 0.11 μM for amperometry and 0.19 μM for differential pulse voltammetry. The reproducibility tests show average precisions of 5.7% (amperometry) and 3.8% (differential pulse voltammetry). Results show also quite good deviation with an average of 4.3% (amperometry) and 5.0% (differential pulse voltammetry) as compared with colorimetric measurements [56].

For the future phosphate sensor development, it will be necessary to continue monitoring the signal stability during a longer time (a few months) and to establish a protocol for *in situ* cleaning of the gold electrode. Next steps will be to deal with development of the mechanical and electronic parts of the sensor. Indeed, the measurement must be performed in a closed volume implying to sample accurately the sea water. At the same time, a calibrationless method should be developed to make the sensor fully autonomous [58]. The sensor will be characterized by a low power consumption and short response time due to the reduced volume combined with application of a large membrane surface. Miniaturization is also possible thanks to advances in microsystem technology (MST) [57]. MST can be used to miniaturize both spectrophotometric and electrochemical sensors for phosphate detection but for the moment this technology is used mostly for reagent-based protocols [60]. Currently, a close collaboration with LAAS laboratory (Laboratoire d'Analyse et d'Architecture des Systèmes, Toulouse) will allow us to develop a "lab-on-a-chip" microdevices for phosphate and silicate electrochemical detection using silicon and polymer technologies [61]. The adaptation of the sensor on autonomous vehicles profiling the water column (ARGO, AUV) would be the next step. Finally fixed eulerian moorings can allow the acquisition of long data time series at a fixed station. Indeed, within the AMOP project (Activities of research dedicated to the Minimum Oxygen Zone in the eastern Pacific), a shallow (200m) mooring will be deployed next January 2013 offshore Callao, Peru, along 12°S at the IMARPE historical station N°5 for a 3 years period. It is planned that our phosphate sensor will be adapted on the mooring line as soon as possible.

4 Summary

Monitoring of oceanic biogeochemical cycles at proper spatial and temporal scales requires miniaturized *in situ* sensors with excellent figures of merit. Development of the marine sensors is even more complicated due to harsh and difficult access to the oceanic environment. In spite of the large number of requirements, some very good sensors actually exist and new ones are currently being developed.

Phosphate monitoring is based on wet chemical analyzers. Analyzers have good accuracy, reliability and can measure a few analytes at the same time. However, phosphate detection is based on addition of liquid reagents to the sample and thus analyzers contain reagents and standards bags which make them large and heavy. There is also the issue of stability of the reagents and of high energy consumption of the analyzers. Actually, many teams work on electrochemical sensors for phosphate detection. They are few voltammetric and amperometric sensors for phosphate detection but none of them has been applied *in situ*. Moreover, they still require addition of liquid reagents to transform electroinactive phosphate species into an active phosphomolybdate complex.

Recently, a new electrochemical possibility to detect phosphate is developed in our team. The method is based on oxidation of molybdenum in order to convert phosphate ions into an electroactive phosphomolybdate complex. The method requires application of an electrochemical cell with use of membrane technology. The method is still under investigation but already, the method was validated for natural seawater samples and the results show excellent comparison with classical colorimetric method. If needed, the sensor can be optimized for other applications such as natural freshwaters or waste waters.

Acknowledgments. Justyna Jońca is supported by a Marie Curie PhD grant within SENSEnet ITN (EC Framework Programme 7, grant agreement No 237868).

References

[1] Millero, F.J.: The Marine Inorganic Carbon Cycle. Chemical Reviews 107, 308–341 (2007)
[2] Paytan, A., McLaughlin, K.: The Oceanic phosphorus cycle. Chemical Reviews 107, 563–576 (2007)
[3] Delaney, M.L.: Phosphorus accumulation in marine sediments and the oceanic phosphorus cycle. Global Biogeochemical Cycles 12, 563–572 (1998)
[4] Ridame, C., Guieu, C.: Saharian input of phosphate to the oligotrophic water of the open western Mediterranean Sea. Limnology and Oceanography 47, 856–869 (2002)
[5] Beman, J.M., et al.: Agricultural runoff fuels large phytoplankton blooms in vulnerable areas of the ocean. Nature 434, 211–214 (2005)
[6] Cai, W.J., et al.: Acidification of subsurface coastal waters enhanced by eutrophication. Nature Geosciences 4, 766–770 (2011)
[7] Gilbert, D., et al.: Evidence for greater oxygen decline rates in the coastal ocean than in the open ocean. Biogeosciences 7, 2283–2296 (2010)
[8] Wheat, C.G., et al.: Phosphate removal by oceanic hydrothermal processes: An update of the phosphorus budget in the oceans. Geochimica et Cosmochimica Acta 60, 3593–3608 (1996)
[9] Chisholm, S.W.: Stirring times in the Southern Ocean. Nature 407, 685–687 (2000)
[10] eWOCE gallery tables, `http://www.ewoce.org/gallery/eWOCE_Tables.html`

[11] Lam, P., Kuypers, M.M.M.: Microbial nitrogen cycling processes in Oxygen Minimum Zones. Annual Review of Marine Science 3, 317–345 (2011)

[12] Jońca, J.: Electrochemical methods for autonomous monitoring of chemicals (oxygen and phosphate) in seawater: Application to the Oxygen Minimum Zone. PhD Thesis, Université Paul Sabatier, Toulouse (2012)

[13] Giraud, M., et al.: Phosphate minimum in the Oxygen Minimum Zone off Peru (in preparation)

[14] Adornato, L., et al.: In situ nutrient sensors for ocean observing systems. In: Hall, J., Harrison, D.E., Stammer, D. (eds.) Proceedings of OceanObs 2009: Sustained Ocean Observations and information for Society, Venice, Italy, September 21-25, vol. 2, ESA Publication WPP-306,, doi:10.5270/OceanObs09.cwp.01

[15] Johnson, K.S.: Observing biogeochemical cycles at global scales with profiling floats and gliders. Oceanography 22, 216–225 (2009)

[16] Roemmich, D., Gilson, J.: The 2004-2008 mean and annual cycle of temperature, salinity, and steric height in the global ocean from the Argo Program. Progress in Oceanography 82, 81–100 (2009)

[17] Hofman, G.E., et al.: High-Frequency Dynamics of Ocean pH: A Multi-Ecosystem Comparison. Plos ONE 6, e28983 (2011)

[18] The ARGO programme, `http://research.metoffice.gov.uk/research/ocean/argo/index.html`

[19] Revsbech, N.P., et al.: Determination of ultra-low oxygen concentrations in oxygen minimum zones by the STOX sensor. Limnology and Oceanography: Methods 7, 371–381 (2009)

[20] Luther III, G.W., et al.: Sulfur speciation monitored in situ with solid state gold amalgam voltammetric microelectrodes: polysulfides as a special case in sediments, microbial mats and hydrothermal vent waters. Journal of Environmental Monitoring 3, 61–66 (2001)

[21] Tercier-Waeber, M.L., et al.: Multi Physical–Chemical profiler for real-time in situ monitoring of trace metal speciation and master variables: Development, validation and field applications. Marine Chemistry 97, 216–235 (2005)

[22] Murphy, J., Riley, J.P.: A modified simple solution method for the determination of phosphate in natural waters. Analytica Chimica Acta 27, 31–36 (1962)

[23] Strickland, J.D.H., Parsons, T.R.: A practical handbook of seawater analysis, 2nd edn. Bulletin, vol. 167. Fisheries Research Board of Canada, Ottawa (1972)

[24] Drummond, L., Maher, W.: Determination of phosphorus in aqueous solution via formation of the phosphoantimonylmolybdenum blue complex Re-examination of optimum conditions for the analysis of phosphate. Analytica Chimica Acta 302, 69–74 (1995)

[25] Kérouel, R., Aminot, A.: A procedure to reduce pre-analytical contamination during the analysis of dissolved nutrient elements (fingerprints, atmosphere, particulates) in seawater. Marine Environmental Research 22, 19–32 (1987)

[26] Vuillemin, R., et al.: ANAIS: Autonomous Nutrient Analyzer In Situ. Sea Technology 40, 75–78 (1999)

[27] Thouron, D., et al.: An Autonomous Nutrient Analyzer for Oceanic Long-term *in situ* biogeochemical monitoring. Analytical Chemistry 75, 2601–2609 (2003)

[28] Thouron, D., et al.: Monitoring des nitrates en milieu côtier. Le capteur autonome ANAIS en Baie de Banuyls/Mer. Atelier Expérimentation Instrumentation, AEI (2006)

[29] Thouron, D., Garçon, V.: ANAIS: Analyseur de sels nutritifs autonome *in situ*: 1995-2005: développement, validation, fiabilisation, Avenir? Rapport d'audit LEGOS, Toulouse, p. 45. INSU/CNRS (2006)
[30] Cycle-PO4 in situ dissolved phosphate analyser, WET Labs brochure, http://www.planet-ocean.co.uk/PDF/CYCLE-PO4.pdf
[31] DPA PRO, http://www.systea.it
[32] Vuillemin, R., et al.: Continuous Nutrient automated monitoring on the Mediterranean Sea using in situ flow analyser (2009), http://research.metoffice.gov.uk/research/ocean/argo/index.html
[33] NAS-3X-The Original *In-situ* Nutrient Analyzer, http://www.n-virotech.com
[34] Underwater APP 4004, http://www.me-grisard.de
[35] The SubChemPak Analyzer, http://www.subchem.com/prod01.htm
[36] CHEMINI: CHEmical MINIaturised analyzer, http://www.ifremer.fr/dtmsi/programmes/chemini/index.htm
[37] Vuillemin, R., et al.: CHEMINI: CHEmical MINIaturised analyse: A new generation of *in situ* chemical analysers for marine applications. In: EGU, European Geoscience Union Geophysical Research Abstract, EGU2007-A-06213 (2007)
[38] Harden, S.M., Nonidez, W.K.: Determination of orthophosphate by flow injection analysis with amperometric detection. Analytical Chemistry 56, 2218–2223 (1984)
[39] Quintana, J.C., et al.: Investigation of amperometric detection of phosphate. Application in seawater and cyanobacterial biofilm samples. Talanta 63, 567–574 (2004)
[40] Udnan, Y., et al.: Evaluation of on-line preconcentration and flow-injection amperometry for phosphate determination in fresh and marine waters. Talanta 66, 461–466 (2005)
[41] Guanghan, L., et al.: Studies on 1:12 phosphomolybdic heteropoly anion film modified carbon paste electrode. Talanta 49, 511–515 (1999)
[42] Carpenter, N.G., et al.: Microelectrode procedures for the determination of silicate and phosphate in waters-Fundamental Studies. Electroanalysis 9(17), 1311–1317 (1997)
[43] Fogg, A.G., Bsebsu, N.K.: Differental-pulse voltammetric determination of phosphate as molybdovanadophosphate at a glassy carbon electrode and assessement of eluents for the flow injection voltammetric determination of phosphate, silicate, arsenate and germanate. Analyst 106, 1288–1295 (1981)
[44] Matsunaga, K., et al.: Differential-pulse anodic voltammetric determination of dissolved and adsorbed phosphate in turbid natural waters. Analytica Chimica Acta 185, 355–358 (1986)
[45] Ganjali, M.Z., et al.: Highly selective and sensitive monohydrogen phosphate membrane sensor based on molybdenum acetylacetonate. Analytica Chimica Acta 567, 196–201 (2006)
[46] Jain, A.K., et al.: A newly synthesized macrocyclic dithiooxamine receptor for phosphate sensing. Talanta 69, 1007–1012 (2006)
[47] Kivlehan, F., et al.: Potentiometric evaluation of calix (4)arene anion receptors in membrane electrodes: Phosphate detection. Analytica Chimica Acta 585, 154–160 (2007)
[48] Zou, Z., et al.: A disposable on-chip phosphate sensor with planar cobalt microelectrodes on polymer substrate. Biosensors and Bioelectronics 22, 1902–1907 (2007)
[49] Ejhieh, A.N., Neda, M.: Application of a new potentiometric method for determination of phosphate based on a surfactant-modified zeolite carbon-paste electrode (SMZ-CPE). Analytica Chimica Acta 658, 68–74 (2010)
[50] Lawal, A.T., Adeloju, S.B.: Comparison of enzyme immobilisation methods for potentiometric phosphate biosensors. Biosensors and Bioelectronics 25, 406–410 (2009)

[51] Mousty, C., et al.: Trienzymatic biosensor for the determination of inorganic phosphate. Analytica Chimica Acta 443, 1–8 (2001)
[52] Kaniewska, et al.: Enantioselective inhibition of immobilized acetylcholinesterase in biosensor determination of pesticides. Central European Journal of Chemistry (2012) (accepted)
[53] Lacombe, M., et al.: Silicate determination in sea water: Toward a reagentless electrochemical method. Marine Chemistry 106, 489–497 (2007)
[54] Lacombe, M., et al.: Silicate electrochemical measurements in seawater: Chemical and analytical aspects towards a reagentless sensor. Talanta 77, 744–750 (2008)
[55] Jońca, J., et al.: Phosphate determination in seawater: Towards an autonomous electrochemical method. Talanta 87, 161–167 (2011)
[56] Jońca, J., et al.: Reagentless and silicate interference free electrochemical phosphate determination in seawater (2012) (submitted)
[57] Jońca, J., et al.: Electrochemical behavior of isopoly- and heteropolyoxomolybdates formed during anodic oxidation of molybdenum in sweater. International Journal of Electrochemical Sciences 7, 7325–7348 (2012)
[58] Giraud, W., et al.: Reagentless and calibrationless silicates measurement in oceanic water. Talanta 97, 157–162 (2012)
[59] West, J., et al.: Micro total analysis systems: Latest achievements. Analytical Chemistry 80, 4403–4419 (2008)
[60] McGraw, C.M., et al.: Autonomous microfluidic system for phosphate detection. Talanta 71, 1180–1185
[61] Vanhove, E., et al.: Final capping passivation for long-life microelectrodes in real fluids. Lab on a Chip (2012) (submitted)

Fluorescence and Phosphorescence Chemical Sensors Applied to Water Samples

G.A. Ibañez and G.M. Escandar*

Instituto de Química Rosario (CONICET-UNR),
Facultad de Ciencias Bioquímicas y Farmacéuticas,
Universidad Nacional de Rosario, Suipacha 531, 2000 Rosario, Argentina
escandar@iquir-conicet.gov.ar

Abstract. Due to the outstanding characteristics of both fluorescence and phosphorescence signals, they are widely applied to the construction of chemical sensors. The development of these sensors for the detection and quantification of compounds of environmental concern is an active research topic, daily enriched with the contribution of new works in the area. This chapter provides fundamentals and practical aspects of this type of sensors, focusing in those works devoted to quantify both organic and inorganic pollutants in environmental waters. The chapter is organized upon the basic functional units of a chemical sensor. A few examples of strategies of evaluation involving multivariate chemometric analysis are also discussed.

Keywords: Fluorescence, phosphorescence, chemical sensors, pollutants, natural waters.

1 Introduction

The definition of a chemical sensor is ambiguous, and different concepts have been proposed in the scientific literature. Many researchers of the sensor community indicate that sensing refers to a continuous process (i.e., continuously recording) of detection, and agree with the Cambrigde definition which states: "C*hemical sensor is a miniaturized device that can deliver real-time and on-line information on the presence of specific compounds or ions even in complex samples"* [1,2]. Similarly, Gauglitz considers that a complete sensor system involves a sensitive layer, transduction (recognition) principles, the signal processing and the strategies (software) of evaluation [3], and Janata suggests that the term "chemical sensor" should be restricted to those devices which acquire information *continuously*, while sensing systems which obtain information in *discrete steps* should be named as "analytical assays" [4]. These specifications are made because the term "sensor" is often used to refer to a probe or indicator (molecule or nanoparticle) that communicates the presence of an analyte *via* modulation of an analytical signal [5].

* Corresponding author.

S.C. Mukhopadhyay & A. Mason (Eds.): *Real-Time Water Quality Monitoring*, SSMI 4, pp. 45–64.
DOI: 10.1007/978-3-642-37006-9_3 © Springer-Verlag Berlin Heidelberg 2013

Valcarcel and Luque de Castro [6] define a chemical sensor as a device providing information about the system under study through the integration of a transducer (optical, electroanalytical, thermal, etc.) with a microzone where a chemical reaction takes place. This device can result in either a "probe" [able to sense the analyte in batch mode (single-shot) or in a flow-system (continuous recording)], or in a flow-through sensor (continuous recording). Finally, according to IUPAC, "*A chemical sensor is a device that transforms chemical information, ranging from the concentration of a specific sample component to total composition analysis, into an analytically useful signal. The chemical information mentioned above may originate from a chemical reaction of the analyte or from a physical property of the system investigated*" [7]. Fig.1 summarizes some configurations and sensing modes.

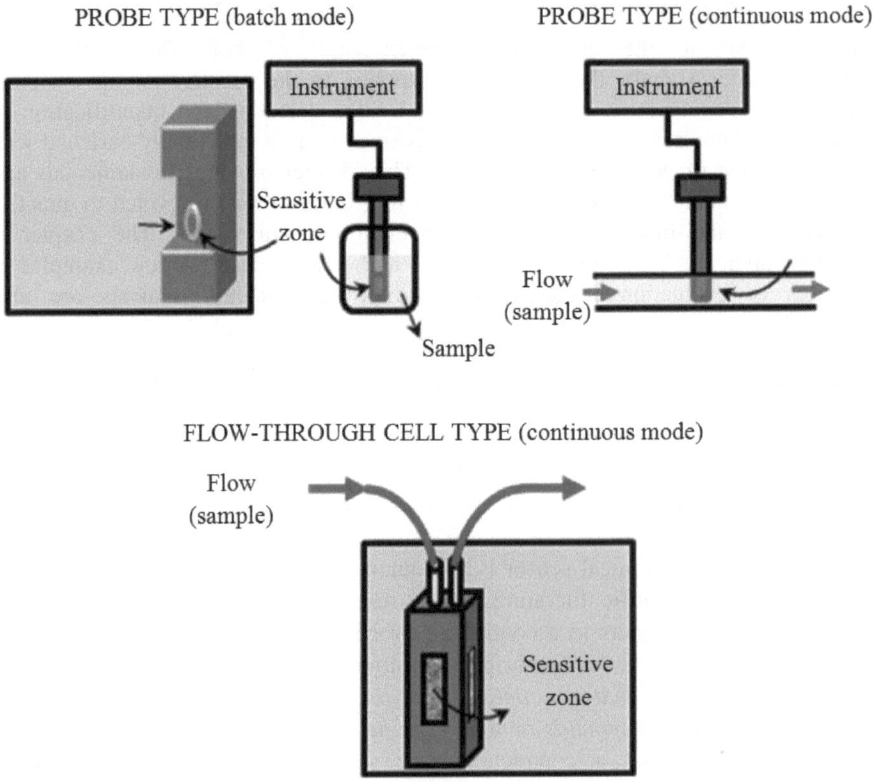

Fig. 1. Schematic representation of different configurations and optosensing modes

The dissimilarities in sensor definitions makes it difficult to provide a classification that covers all involved aspects (e.g. type of measured signal, external shape, sensing scheme, platform employed, etc.). Generally, the designation of a sensor is related to both the area of investigation where the sensor will be applied and to the concept of sensor adopted by the analyst. Fortunately, all researchers agree with a classification based on the transduction mechanism which categorizes sensors in electrochemical, optical (absorbance, reflectance, chemoluminescence, luminescence, refractive index, optothermal effect and light scattering signals), electrical, mass sensitive, magnetic, thermometric and based on other physical properties [7]. In addition, several characteristics may be indicated: probes or flow-through sensors [8], portable/handheld or not [9,10], optical fibers-based and planar waveguide systems, direct (the measured signal corresponds to an intrinsic optical property of the analyte) or reagent-mediated (an auxiliary reagent is needed to detect the analyte presence). In many cases, the immobilization approaches in both direct and reagent-mediated sensors are indicated in the name of the sensor (*e.g.* sol-gel based sensors, polymer based sensors, molecularly-imprinted polymer (MIP) sensors, etc.). Classifications can be complementary and it would be desirable to avoid a rigid scheme, trying to cover most principles involved in the nature of the sensor under study. IUPAC considers that chemical sensors contain two basic functional units: a *receptor* part and a *transducer* part. In the receptor part the chemical information is transformed into a form of energy, which is measured by the transducer, where the energy is transformed into a useful analytical signal [7].

Due to the many advantages of optical sensors, such as sensitivity, versatility, possibility of on-line detection and miniaturization, among others, they are of great utility in the environmental area, and numerous publications devoted to their application for the determination of compounds of environmental concern have been discussed in the literature [11-15].

This chapter deals with luminescent sensors and probes reported in recent years for the determination of both organic and inorganic pollutants, focusing the attention mainly in the above mentioned basic functional units, which have suffered significant changes over the years. However, a few examples of new strategies of evaluation will be treated. Neither sensing systems, which only involve measurements of signals produced by a chemical reaction in solution, nor bio-sensors are included in this chapter. The organization adopted is shown in Fig. 2, and includes both probes and complete chemical sensor systems. In this figure, planar waveguide systems [2] have been omitted as transduction platforms because most of them are based on biological reactions [16], which are beyond the scope of this chapter.

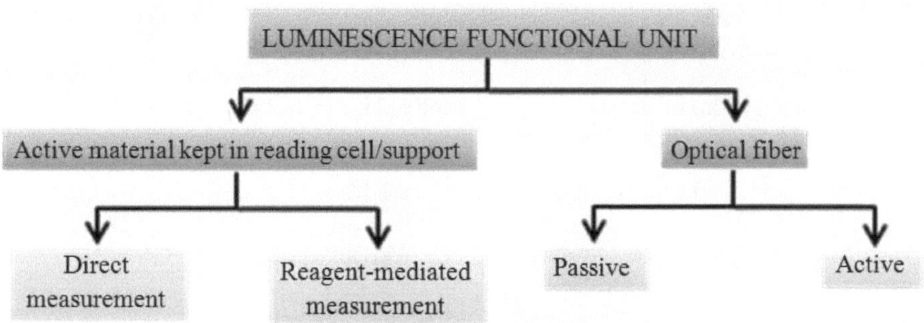

Fig. 2. Organization of the chapter

2 Active Material Kept in Reading Cell/Support

2.1 *Direct Measurement*

In this case, the emitted luminescent signal is directly measured after the analyte is retained in a suitable solid support [6,17-19]. In probe configurations, the analyte is retained by adsorption, binding or entrapment, either on a plane surface or in microbeads. The deposit of the analyte in plane surfaces can be carried out by different procedures. One possibility is spotting a few microliters of the sample in the support with the aid of a micropipette (Fig. 3A). A convenient mode of both retaining and simultaneously preconcentrating the analyte is to perform a solid-phase extraction procedure (Fig. 3B). In this case, the active support is a disk loaded into a stainless steel filter holder, and placed into a syringe. Through manual or mechanic positive pressure, the sample solution is forced through the membrane and analytes are retained on its surface.

After immobilization, plane surfaces are placed in an appropriate reading cell/support. Solid-sample holder accessories are commercially available and are adequate for thin films, powders, pellets and fibers. However, there are other experimental approaches which can be implemented for reading from solid supports. For example, if the sample is spotted in a sheet of paper or membrane, the surface can be located in a conventional fluorescence cell in a correct position to measure either reflection or transmission of the fluorescence signal (Fig. 4A). The layer can also be located between two quartz sheets and then in a chamber for spectrofluorimetric measurements (Fig. 4B). Disks used in solid-phase extraction can be placed in laboratory constructed holders, or in capped cells with nitrogen bubbling, for example, for phosphorescence measurements (Fig. 4C).

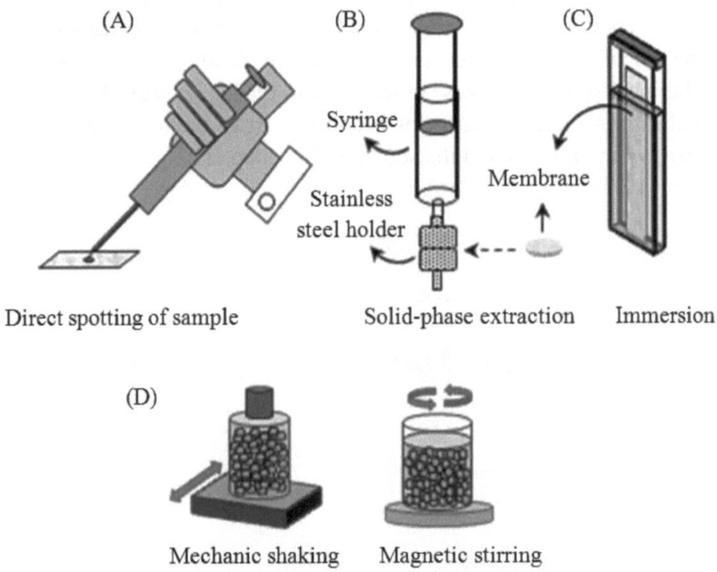

Fig. 3. Some experimental modes for immobilize the analyte in probe configurations. (A-C) plane surfaces, (D) microbeads.

As will be shown below, plane supports can also be set in laboratory-constructed special holders for performing determinations in flow-systems.

When particulate materials are used as probes, equilibrium is established between microbead active sites and the analyte (Fig. 3C). Beads are then transferred to a specially designed spectrofluorimetric holding. On the other hand, in flow-through optosensors, microbeads are contained in a flow-cell. In simpler configurations, a carrier solution transports the sample containing the analyte in a continuous flow system. When the analyte reaches the solid support filling the flow-cell, its on-line immobilization is produced and its luminescence signal is measured.

Filter-papers were common plane-supports used in the first works on solid-phase luminescence (SPL) and were gradually replaced by other materials, such as C18-disks, nylon membranes and polymeric membranes, with improved sensor qualities. Nylon membranes showed to be adequate film supports for the phosphorescence quantification of thiabendazole in natural waters [20]. Based on a solid-phase extraction on a nylon membrane, benzo[*a*]pyrene was determined through both fluorescent and phosphorescent measurements at sub-parts-per-trillion (sub-ng L^{-1}) levels in spiked tap, underground and mineral water samples of different origins [21]. Besides, polymeric materials, ion-exchange resins, bonded-phase silicagel [22] and nylon powder, among others, are employed in manufacturing microbeads and filler particles. Salinas Castillo *et al* presented the phosphorescence characterization of polycyclic aromatic hydrocarbons (PAHs) using non-ionic resins. The study allowed them to develop a flow-through phosphorescence optosensor for the selective determination of benzo(*a*)pyrene [23].

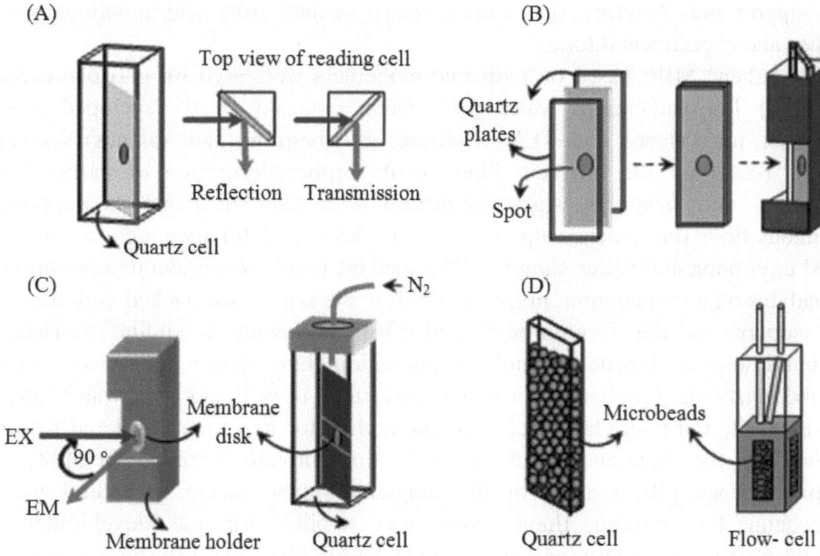

Fig. 4. (A-B) Fluorescence reading configurations for planar spotted surfaces. (C) Disks holder configurations. (D) Microbead holders for probe and flow-through optosensors.

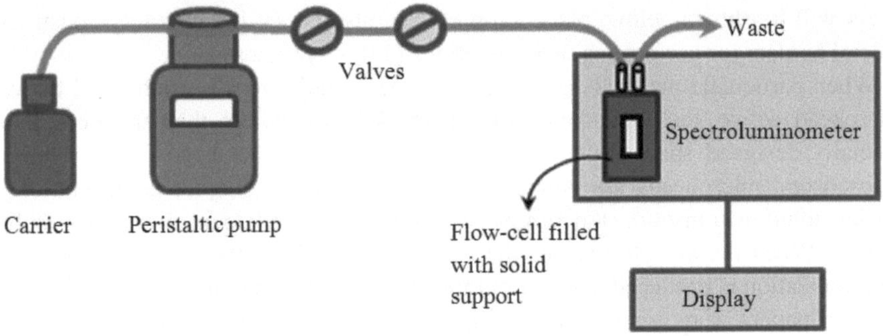

Fig. 5. Scheme of a flow-through luminescence sensor

MIPs are materials containing cavities capable of binding the analyte through a selective recognition, and are being increasingly used as solid-supports, especially when determinations are carried out in complex matrices where selectivity is mandatory [24-27]. MIPs are used in flow-through optosensor and in fiber-optic configurations (see below). Microbeads, powders and MIPs are adequate materials to pack a spectrofluorimetric cell coupled to a flow-system, resulting in a flow-through optosensor (Fig. 5). Flow-injection analysis (FIA) [17,28], sequential-injection analysis (SIA) [29], multi-syringe flow-injection analysis (MSFIA) [30], bead-injection spectroscopy (BIS) [31], and multi-commutation [32] are techniques usually employed for performing this on-line procedure. In comparison with batch methodologies, most flow-through optosensors allow the on-line regeneration of the solid-support and, therefore, they can be reused in successive determinations, saving supplies and experimental time.

Fluoranthene MIPs based on iodinated monomers were used for phosphorescence sensing of fluoranthene in water [33]. Sainz Gonzalo *et al* developed a MIP optosensor for xylenes and TEXs (toluene, ethylbenzene and xylenes) screening [34,35]. Based on the different kinetics of sorption/desorption of carbendazim, carbofuran, and benomyl when interacting with C_{18} silica gel as support, a continuous-flow fluorescence optosensor was developed for their determination in spiked environmental water samples. The method involves a previous separation of the analytes on a minicolumn, placed just before the sensor and packed with the same solid support as the flow-through cell [36]. Following a similar concept of sorption/desorption kinetics, a multi-commutation fluorescence optosensor analysis for fuberidazole and *o*-phenylphenol was performed in well and river water samples [37]. Piccirilli and Escandar developed and applied for first time powdered nylon as support for the implementation of both flow-through fluorescence [38] and phosphorescence [39] sensors of the fungicide thiabendazole in natural waters. Photochemically induced fluorescence was applied for the development of optosensors for the determination of imidacloprid [40], metsulfuron methyl [41], linuron [42] and thiabendazole and metsulfuron methyl [43] in environmental waters. 1-Naphthaleneacetic acid was investigated in water through a flow-through fluorescence sensor [44], and the same research group analyzed 2-naphthoxyacetic acid through both fluorescence and phosphorescence optosensors [45]. Tetracycline, oxytetracycline, chlortetracycline and doxycycline were detected in water using a phosphorescence flow-through optosensor and amberlite XAD-4 as solid support [46]. Recently, tetracycline was also determined using a fluorescence flow-through optosensor based on sephadex G-50 [47].

2.2 *Reagent-Mediated Measurement*

When an analyte does not have native luminescence properties, sensing can be achieved by monitoring the optical signal of an auxiliar reagent (fluorophore, dye, enzyme, etc), whose response is modulated by the presence of the analyte. The reagent is generally immobilized (by adsorption, covalent bond, ionic attachment or encapsulated) in a matrix which is permeable to the analyte. The matrix can be a thin polymer membrane, controlled-pore glass or microbeads [9,48]. Many of the configurations discussed in Section 2.1 can be implemented in reagent-mediated sensors.

Table 1. Luminescence sensors/probes for inorganic species in water samples

Analyte	Signal	Method	Remark	LOD	Sample	Ref
Cu (II)	RET	Disposable sensor	TP/Zincon/PVC matrix	3.9×10^{-8} M 2.5 ppb	River, well, spring, swimming pool water	[63]
	QF	Probe	LY immobilised on AE particles	1×10^{-8} M, 0.63 ppb	Tap water	[49]
Al(III) and Be(II)	F	BIS-FIA	Sephadex QAE A-25/morin	0.024 ppb Al(III), 10 ppb Be(II)	Tap and river water	[50]
V(V)	F	BIS-FIA	Sephadex QAE A-25/Alizarin Red S	0.450 ppb	Tap and well water	[51]
Hg(II)	F	FIA and batch	Rhodamine B derivative in PVC membrane	8.3×10^{-11} M	Pond water	[52]
	QF	OM, probe	L/PVC membrane/ KTpClPB		Well water	[53]
	QF	OM, flow-cell	H2tpp/PVC membrane	4.0×10^{-8} M	Tap water	[54]
	QF	OM, flow-cell	Porphyrin/PVC membrane	8.0×10^{-9} M	River, tap, waste water	[55]
	QF	OM, flow-cell-bifurcated OF	H_3(tpfc)/PVC membrane		River and tap water	[56]
	QF	OM, flow-cell-bifurcated OF, HCl (R)	DTPP/PVC membrane		River water, industrial wastewater	[57]
	QF	OM, probe	EHT/PVC membrane	1.8×10^{-10} M, 36 ppb	Tap water	[58]

Table 1. (*continued*)

	F	Syringe procedure, probe	Rhodamine 6G derivative and nylon membrane	0.4 ppb	Mineral, underground river water	[64]
	QF	Probe, EDTA (R)	Mesoporous silica /1,8-naphthalimide	200 ppb	Drinking, seawater	[65]
	QF	Probe	Hacid/LDH film *via* EPD method	6.3×10^{-8} M	Spiked, tap, lake water	[66]
	F	Probe	Rhodamine 6G/ QDs-silica NPs	0.520 ppb, 2.59×10^{-9} M	Tap, river water	[76]
	SF	Probe	MPA stabilized CdTe/ dOB/QDs	4.2×10^{-9} M	Tap, lake water	[81]
Hg(II) and Cu(II)	PLQ	Probe, GSH (R)	QDs-multilayer films	Lineal range: $(0.05\text{-}5)\times10^{-7}$ M for Hg^{2+}, $(0.01\text{-}1)\times10^{-6}$ M for Cu^{2+}	Artificial water sample	[82]
Zn(II)	QF	Probe, EDTA (R)	MCB/PVC membrane	2.5×10^{-8} M, 1.6 ppb	Tap water	[59]
Ni(II)	QF	Probe, EDTA (R)	PVC/KTpClPB/TTBB		Wastewater	[60]
Pb(II)	F	Probe, HCl-thiourea (R)	Triazolo-thiadiazin derivative/PVC	2.2×10^{-8} M	Tap and river water	[61]
	QF	Probe, dithiotritol (R)	Fluoroionophore L2/ PVC/ KTpClPB	2.0×10^{-7} M	Tap water	[62]
Be(II)	QF	Probe, F^- (R)	Beryllon II/MgAl-LDH	4.2×10^{-9} M	Lake and tap water	[67]
CN^-	Increase of F with Cu(II)	Probe	Boronic acid-linked hydrazide/fluorescein encapsulated in AMP/Gd^{3+}CNPs	20×10^{-6} M (naked-eye) 4.03×10^{-6} M (instrumental)	Drinking water	[68]
Cr(VI)	IFE	Probe	Luminescence NPs ($NaYF_4$:Yb^{3+}, Er^{3+})	2.4×10^{-8} M	Tap, river and waste water	[69]
Cd(II)	PL in NIR	Probe	CdTe/CdS QDs	Lineal range: $(0.1\text{-}2)\times10^{-6}$ M	Tap and river water	[80]
As(III)	QF	SIA	AsH_3 diffused across PTFE membrane and interacted with CdS-MAA QDs	70 ppb	Ground water	[83]
Cu(II) and CN^-	Cu^{2+}: QF; CN^-: F	FIA, EDTA (R for Cu^{2+}), Cu(II) (R for CN^-)	Imidazole microspheres/functionalized CPs	Cu(II): 1 ppb CN^-: 8 ppb	Tap and mineral water	[86]
Ag(I)	F	Distal OF bundle	PVC microspheres onto the distal end of an OF bundle		Pond water	[89]

Table 1. *(continued)*

Abbreviations (other are defined in the text): AE, anion exchange; AMP/Gd^{3+} CNPs: nucleotide/lanthanide coordination nanoparticles; Beryllon II, 2-(3,6-disulfo-8-hydroxynaphthylazo)-1,8-dihydroxynaphthalene-3,6-disulfonate; CPs, conjugated polymers; DFC, diphenylcarbazone; dOB, chemically denatured ovalbumin; DTPP, 5-*p*-[[4-(10'-15',20'triphenyl-5'-porphinato)phenyloxyl]-1-butyloxyl]-phenyl-10,15,20 triphenylporphine; EDTA, ethylenediaminetetraacetic acid; EHT, 4-ethyl- 5-hydroxy-5,6-di-pyridin-2-yl-4,5-dihydro-2H-[1,2,4]triazine-3-thione; EPD: electrophoretic deposition method; F, fluorescence; GSH, glutathione; Hacid, 1-amino-8-naphthol-3,6-disulfonic acid; H2tpp, 5,10,15,20-tetraphenylporphyrin; H_3(tpfc), 5,10,15-tris(pentafluorophenyl) corrole; KTpClPB: potassium tetrakis(*p*-chlorophenyl) borate; IFE: inner filter effect; LDH, layered double hydroxide; LY: Lucifer yellow; L, 1-(dansylamidopropyl)-1-aza-4,10-dithia-7-oxacyclododecane; L2, (5,8-bis((5'-chloro-8'-hydroxy-7'-quinolinyl) methyl)-2,11-dithia-5,8-diaza-2,6-pyridinophane); LOD, limit of detection; MAA, mercaptoacetic acid; MCB, 1-methyl-1-phenyl-3-[1-hydroxyimino-2-(succinimido)ethyl] cyclobutane; morin: 2',3,4',5,7-pentahydroxyflavone; NIR, near infrared; NPs, nanoparticles; OF, optical fiber; OM, optode membrane; MPA, 3-mercaptopropionic acid; PL, photoluminescence; QF, quenching of fluorescence; QPL, quenching of photoluminescence; R, regenerating reagent; RET, radiationless resonance energy transfer; SIA, sequential injection analysis; TP, porphyrazine 2,7,12,17-tetra-*tert*-butyl-5,10,15,20-tetraaza-21H,23H-porphine; TTBB, 2,5-thiophenylbis(5-*tert*-butyl-1,3-benzexazole).

This approach is widely used for the construction of probes and sensors devoted to determine metal cations and anions, which are not naturally fluorescent, through the signals produced by their interaction with auxiliary reagents. Table 1 summarizes reviewed works involving probes and sensors for ionic compounds found in environmental waters.

Different fluorophors (lucifer yellow, morin, alizarin red S) immobilized in ionic exchanger particles were used to determine Cu(II) [49], Al(III) and Be(II) [50] and V(V) [51]. Based on either the enhancement or quenching of fluorescence, polyvinyl chloride (PVC) membranes containing several organic molecules have been used to determine Hg(II) [52-58], Zn(II) [59], Ni(II) [60], Pb(II) [61,62], and Cu(II) [63] in natural waters.

A PVC membrane probe based on a Rhodamine B derivative (RND) was recently described for the determination of Hg(II) ion in environmental water samples and fish [52]. The method was developed in batch and also in a flow-system. For bulk measurements, RND-PVC-coated glass slides were placed in diagonal positions into a standard quartz cell, which was filled with a sample solution. The flow-through cell consisted of two separate teflon blocks, an optical glass window, a sample chamber containing the sensing membrane coated on the glass slide, a silicon gasket and screws for clamping the blocks, as is shown in Fig. 6. The measuring cell was mounted in the spectrofluorimeter and connected with a peristaltic pump. In this case, the conditioning, measurement and regeneration steps were implemented by passing carrier, Hg(II) ion and regenerating solutions, respectively, through the flow cell. In both batch and flow-through configurations the emission was detected at a 90° angle relative to the incident light.

The viability of determining Hg(II) ion residues in real water samples was demonstrated through a probe based on the mercury-promoted ring opening of the spirolactam moiety of a rhodamine 6G spirocyclic phenylthiosemicarbazide derivative retained in commercial nylon membranes [64].

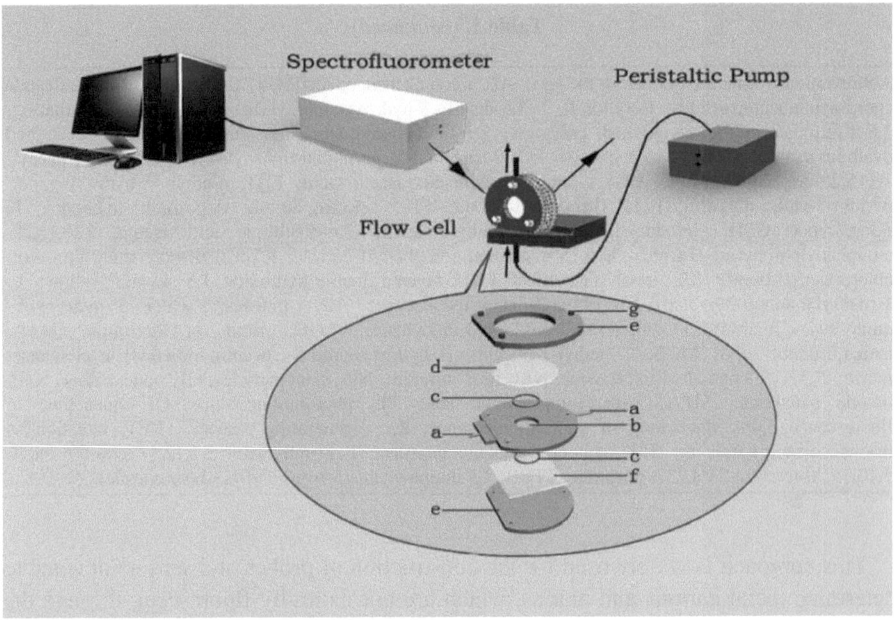

Fig. 6. Schematic diagram of the flow-through cell: (a) Teflon tube for sample in and out; (b) sample chamber; (c) gasket; (d) optical glass window; (e) Teflon cover; (f) sensing membrane coated on the glass slide; (g) fixing screws. (Reprinted with permission from [52]. Copyright 2012 Elsevier).

For monitoring Hg(II) in domestic and sea waters through fluorescence quenching, Meng *et al* developed an hybrid mesoporous silica functionalized with 1,8 naphtalimide-base receptor [65], and Sun *et al* obtained a layered double hydroxide film with intercalated 1-amino-8-naphthol-3,6-disulfonic acid *via* an electrophoretic deposition method [66]. The assembly of Beryllon II and MgAl–layered double hydroxide (LDH) nanosheets on quartz using the layer-by-layer deposition technique was used to quantify Be(II) in lake and tape water [67]. Kulchat *et al* proposed a method to determine CN^- in presence of Cu^{2+} based on a sensor molecule, containing boronic acid-linked hydrazide and fluorescein moieties. This molecule was encapsulated in nucleotide/lanthanide coordination nanoparticles to avoid hydrolysis in water [68]. The inner filter effect of luminescent nanoparticles ($NaYF_4$:Yb^{3+}, Er^{3+}) produced by the Cr(III)-diphenylcarbazone complex was exploited to assess Cr(VI) in tap, river and waste waters [69]. Azab *et al* developed a fluorescent sensing tool for screening primary aliphatic amines in water and soil samples [70]. This sensor is based on a pyrilum dye embedded into a polymeric cocktail, which is deposited on the bottom of wells in microtiter plates.

Because of their superior optical properties, quatum dots (QDs) are increasingly employed for sensing purposes [71,72]. Wang *et al* determined phosphorimetrically pentachlorophenol using surface molecular imprinting on Mn-doped ZnS QDs [73]. Stringer *et al* quantified 2,4 dinitrotoluene and 2,4,6-trinitrotoluene through fluorescence quenching of QD labeled microparticles [74]. The latter compound also was determined through the phosphorescence quenching of Mn-doped ZnS QDs [75]. The enhancement of emission signal of rhodamine 6G grafted on silica nanoparticles-QDs was proposed to determine Hg(II) [76]. A low limit of detection (2.4×10^{-12} M) was achieved in the determination of 1,4–dihydroxybenzene using the fluorescence resonance energy quenching of mercaptosuccinic acid–capped CdTe QDs immobilized on silica particles [77]. Fullerene in spiked river water was analyzed using CdSe/ZnS QDs coated by *p-ter*butylcalix[8]arene. The method allows the detection of 5 μg L^{-1} of fullerene in a simple and fast way [78]. QDs-MIP composite nanospheres were successfully applied to direct fluorescence quantification of the organophosphate insecticide diazinon in water. This method selectively detects down to 50 ng mL^{-1} of diazinon and it provides a strategy to obtain inorganic-organic nanocomposites with potential applications in environmental analysis [79]. Cd(II) and Hg(II) were quantified in diverse types of waters (tap, river, lake) using CdTe/CdS [80] and CdTe-chemically denatured ovalbumin QDs [81], respectively. In addition, Hg (II) and Cu(II) were quantified by fluorescence quenching of QD-multilayer films [82]. Butwong *et al* proposed a method to quantify arsenic in ground water which consists in the measurement of the fluorescence quenching of mercaptoacetic-acid capped CdS QDs after on-line arsine generation with BH_4Na and HCl through a SIA system [83].

The so-called “sol-gel” process is a technique widely used for the production of materials, starting from a colloidal solution that acts as the precursor (e.g. metal alkoxides) which undergoes hydrolysis and polycondensation reactions to produce a porous matrix in which the reagent is encapsulated and into which the analyte molecules can diffuse. Materials obtained from this process have outstanding mechanical and thermal stability and good optical characteristics and, therefore, are very suitable for the development of optical sensors [84]. A sol-gel technique was applied to develop luminescent Eu(II) transparent films deposited on glass slides, in order to obtain sensor devices capable of monitoring transition metal ions in aqueous solution [85]. Álvarez Diaz *et al* proposed an optosensor based on imidazole-functionalized colpolymer microspheres, obtained by a sol-gel method, to quantify Cu(II) and cyanide ion [86]. A sol-gel membrane modified with 2-hydroxy-1-naphthaldehydene-8-aminoquinoline was developed as a model for detecting Pb(II) [87].Trace levels of anthracene, phenanthrene and pyrene were detected in tap and mineral water through the enhancement of fluorescence of hydrophilic QDs (modified CdSe/ZnS) entrapped into sol-gel membranes [88].

3 Fiber-Optic Sensor Platforms

Optical fibers are usually employed as platforms in optical sensors, especially because they allow optical spectroscopy to be performed on sites inaccessible to conventional spectroscopy. McDonagh *et al* [2] summarized common configurations used for fiber-optic platforms (Fig. 7). In turn, fibers can only act to transport the optical signal to and from the sensing environment (passive fiber-optic system) or be doped with a selective indicator, having thus an active participation in the detection (active fiber-optic system). In the first case, the intrinsic optical property of the analyte (e.g. fluorescence emission) is measured. In the second one, the signal of an immobilized indicator is monitored.

3.1 Passive Fiber-Optic System

A typical picture of a passive fiber-optic fluorescence sensor comprises a sensitive material that fluoresces upon interaction with the analyte of interest and a fiber bundle (as shown in Fig. 7d). In the fiber bundle, the central fiber delivers the excitation light to the external sensor element and the outer fibers are used to collect and transport the emitted fluorescence to a suitable detector.

A fluorescent optical microsensor array for rapid sensing of trace levels of Ag^{+} in pond water was fabricated using plasticized PVC-based fluorescent microspheres deposited onto the etched distal end of an optical fiber bundle [89].

The bifurcated fiber bundle can be used to deliver excitation light with one fiber and to collect the excited fluorescence with the second. This latter configuration has been used for the fluorescence-based detection of 2,6-dinitrophenol (A selective optical chemical sensor for 2,6-dinitrophenol based on fluorescence quenching of a novel functional polymer). Bifurcated fibers were also employed to quantify Hg(II) ion through fluorescence quenching [56,57].

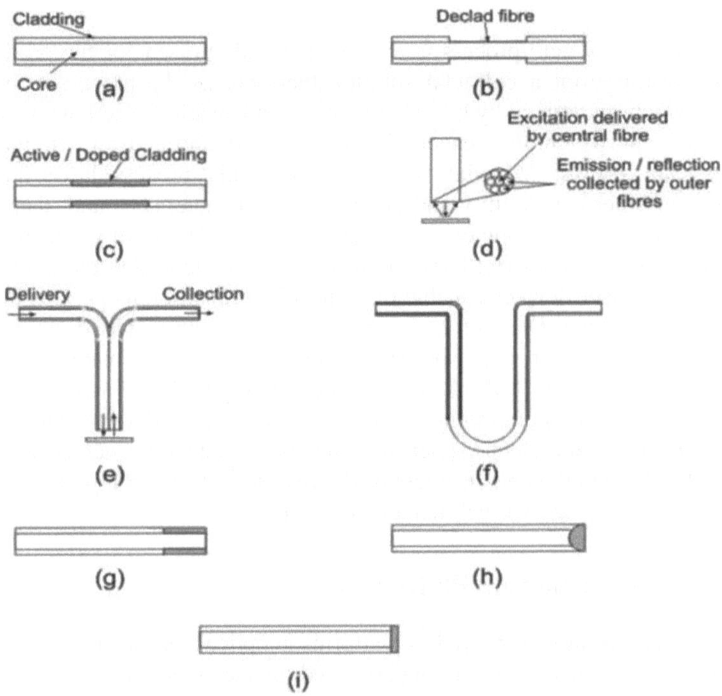

Fig. 7. Configurations for fiber-optical chemical sensors: (a) unmodified; (b) declad; (c) active or doped cladding; (d) fiber bundle; (e) bifurcated fiber bundle; (f) U-bend (shown declad); (g) tip with active cladding; (h) etched tip; (i) modified end-face. (Reprinted with permission from [2]. Copyright 2008 American Chemical Society).

A MIP using Al(III) ion as the template molecule and 8-hydroxyquinoline sulfonic acid as the the ligand to form a fluorescence complex was characterised for the implementation of an optosensor for this metal ion [90]. A self-designed flow-cell included a glass tube filled with the MIP fitted into a black box (to avoid interference from stray light) and a fibre optic cable attached at the top of the box.

3.2 Active Fiber-Optic System

In reagent-based optical sensor systems, fluorescent indicator molecules are added to the fiber platform. In most active systems, the sensor material is included in a solid matrix in the form of a monolith or a thin film. The matrix encapsulates the reagent such that it is accessible to the analyte. In their review, McDonagh *et al* give a detailed explanation of the different configurations which can be achieved in these types of sensors [2].

Two commonly used immobilization matrices are polymeric and sol-gel materials. The most widely used polymers are polystyrene (PS), polyvinyl chloride (PVC), polymethyl methacrylate (PMMA), polydimethyl siloxanes (PDMS), and polytetrafluoroethylenes (PTFE) and ethyl cellulose. Besides, sol-gel materials are very adequate to be coupled to a fiber-optic system. The process can be controlled in order to obtain materials of desired porosity and polarity [91]. The auxiliary reagent can be either entrapped or covalently bounded to the sol-gel matrix. Fiber-optic sensors based on sol-gel films have been intensely explored as pH, gas, ionic species and solvent sensors [84].

Lieberzeit *et al* highlighted some features that imprinting offers for sensor design and showed the ability and variability of this technological platform for the detection of PAHs (naphthalene and pyrene) among other compounds [92].

The chemical agents EA2192, VX, sarín and soman were determined at part-per-trillon levels in multiple water matrices using lanthanide-sensitized luminescence and MIP containing europium polymerized onto a fiber-optic [93]. A plasticized PVC–curcumin moiety membrane was used in a bifurcated optical fiber based flow-optosensor for the determination of 2,6-dinitrophenol through fluorescence quenching measurements [94]. Employing similar bifurcated optical fiber systems but different supports (β-CD polymer and pyrene/dimethyl-β-CD complex immobilized in plasticized PVC membranes), the same authors determined bisphenol A through fluorescence [95] and fluorescence quenching [96] signals, respectively.

4 Strategies of Evaluation Involving Multivariate-Chemometric Calibration

Because of the complexity of environmental samples, luminescence optonsensors, like most of sensors based on spectroscopic properties, suffer lack of selectivity due to the probability that similar compounds show similar luminescence properties. A current strategy for resolving the spectral overlapping in complex environmental matrices and, thus, ensuring interference-free quantitation is the coupling to multivariate calibration methods [97]. Specifically useful are second-order calibration methods which achieve the so called 'second-order advantage'. These methods allow the quantitation of analytes even in the presence of unexpected sample components [98].

The simultaneous determination of the systemic fungicides carbendazim and thiabendazole in natural waters was achieved using fluorescence excitation-emission matrices obtained after the extraction of the analytes over a C18-membrane surface [99]. The applied algorithm, partial least-squares with residual bilinearization (PLS/RBL), was capable of both correctly predicting the concentration of the studied compounds in the presence of unsuspected species and overcoming the inner-filter effect. Valero Navarro *et al* determined two naphthylamines in drinking waters coupling second-order calibration to a MIP-fluorescence optosensing system [100]. Using a flow-through optosensor based on C18-bonded phase and second-order algorithms, thiabendazole and fuberidazole where quantified in the presence of carbaryl, carbendazim and 1-naphthylacetic acid [101]. A flow-through spectrofluorimetric sensor, also based on C18-bonded phase as substrate in the flow cell and a first-order calibration approach, was proposed for the resolution α-naphthol, o-phenylphenol and thiabendazole in environmental water samples [102]. First derivative emission spectra of the analytes recorded during the process of retention-elution were used to provide multivariate data which were processed with the partial least-squares (PLS) algorithm. Benzo[*a*]pyrene and dibenz[*a,h*]anthracene were fluorimetrically determined in the presence of the remaining EPA-PAH priority pollutants using a probe based on nylon membranes and second-order calibration as strategy of evaluation [103]. The same research group determined six heavy-PAHs in the presence of another 10 interfering PAHs, applying second-order multivariate calibration to the data obtained with a flow-through optosensor (sensor phase: silica gel C18) interfaced to a fast-scanning spectrofluorimeter [104]. Very recently, microporous nylon membrane was employed as a fluorimetric probe for the quantification of the herbicide bentazone in natural waters. Second-order calibration using parallel factor analysis (PARAFAC) allowed the determination at low levels in a highly interfering medium [105].

5 Conclusions

Chemical sensors based on luminescence signals are relevant in environmental measurements due to their multiple advantages, highlighting their high sensitivity and usefulness for the determination of contaminants at very low concentrations. Fortunately, most optosensors are compatible with green-chemistry principles, and this quality makes them even more attractive. Bulky sensors are being gradually replaced by miniaturized devices, and flow methodologies are preferable than those carried out in batch. However, independently of their final configurations, new approaches for the development of novel probes and sensors for the determination of contaminants in the environment are continuously reported.

Acknowledgments. Universidad Nacional de Rosario and CONICET (Consejo Nacional de Investigaciones Científicas y Técnicas, Project PIP 1950) are gratefully acknowledged for financial support.

References

[1] Wolfbeis, O.S.: Fiber-optic chemical sensors and biosensors. Analytical Chemistry 78, 3859–3874 (2006)
[2] McDonagh, C., et al.: Optical chemical sensors. Chemical Review 108, 400–422 (2008)
[3] Gauglitz, G.: Direct optical sensors: principles and selected applications. Analytical and Bioanalytical Chemistry 381, 141–155 (2005)
[4] Janata, J.: Introduction: Modern topics in chemical sensing. Chemical Reviews 108, 327–328 (2008)
[5] Powe, A.M., et al.: Molecular fluorescence, phosphorescence, and chemiluminescence spectrometry. Analytical Chemistry 82, 4865–4894 (2010)
[6] Valcárcel, M., Luque de Castro, M.D.: Integration of reaction (retention) and spectroscopic detection in continuous-flow systems. Analyst 115, 699–703 (1990)
[7] Hulanicki, A., et al.: Chemical sensors detinitions and classification. Pure & Applied Chemistry 63, 1247–1250 (1991)
[8] Fernández Córdoba, M.L., Ortega Barrales, P.: The potential of flow-base optosensing devices for pesticides assessment. In: Stoytcheva, M. (ed.) Pesticides in the Modern World-Trends in Pesticides Analysis, ch. 20, pp. 489–514. InTech (2011)
[9] Capitán Vallvey, L.F., Palma, A.J.: Recent Developments in Handheld and Portable Optosensing-A Review. Analytica Chimica Acta 696, 27–46 (2011)
[10] Miró, M., Hansen, E.H.: Recent advances and future prospects of mesofluidic Lab-on-a-Valve platforms in analytical sciences - A critical review. Analytica Chimica Acta (2012) (in press)
[11] Vukovic, J., et al.: Characterization of disposable optical sensors for heavy metal determination. Talanta 94, 123–132 (2012)
[12] Ibañez, G.A., Escandar, G.M.: Luminescence Sensors Applied to Water Analysis of Organic Pollutants-An Update. Sensors 11, 11081–11102 (2011)
[13] Bosch Ojeda, C., Sánchez Rojas, F.: Recent development in optical chemical sensors coupling with flow injection analysis. Sensors 6, 1245–1307 (2006)
[14] Wang, X., et al.: Flow-based luminescence-sensing methods for environmental water analysis. Trends in Analytical Chemistry 28, 75–87 (2009)
[15] Aragay, G., et al.: Recent trends in macro-, micro-, and nanomaterial-based tools and strategies for heavy-metal detection. Chemical Reviews 111, 3433–3458 (2011)
[16] Sapsford, K., et al.: Planar waveguides for fluorescence biosensors. In: Ligler, F.S., Taitt, C.R. (eds.) Optical Biosensors: Today and Tomorrow, 2nd edn., ch. 3, pp. 139–184. Elsevier (2008)
[17] Molina Díaz, A., et al.: Solid-phase spectroscopy from the point of view of green analytical chemistry. Trends in Analytical Chemistry 29, 654–662 (2010)
[18] Hurtubise, R.J.: Solid-Surface Luminescence Spectrometry. Analytical Chemistry 61, 889A–895A (1989)
[19] Matsuoka, S., Yoshimura, K.: Recent trends in solid phase spectrometry: 2003-2009. A review. Analytica Chimica Acta 664, 1–18 (2010)

[20] Correa, R.A., Escandar, G.M.: A new analytical application of nylon- induced room-temperature phosphorescence: determination of thiabendazole in water samples. Analytica Chimica Acta 571, 58–65 (2006)
[21] Bortolato, S.A., et al.: A novel application of nylon membranes to the luminescent determination of benzo[a]pyrene at ultra trace levels in water samples. Analytica Chimica Acta 613, 218–227 (2008)
[22] Molina Díaz, A., et al.: The potential of flow-through optosensors in pharmaceutical analysis. Journal of Pharmaceutical and Biomedical Analysis 28, 399–419 (2002)
[23] Salinas Castillo, A., et al.: Solid-surface phosphorescence characterization of polycyclic aromatic hydrocarbons and selective determination of benzo[a]pyrene in water samples. Analytica Chimica Acta 550, 53–60 (2005)
[24] Henry, O.Y.F., et al.: Optical interrogation of molecular imprinted polymers and development of MIP sensors: A review. Analytical and Bioanalytical Chemistry 382, 947–956 (2005)
[25] Beltran, A., et al.: Molecularly-imprinted polymers: useful sorbents for selective extractions. Trends in Analytical Chemistry 29, 1363–1375 (2010)
[26] Caro, E., et al.: Application of molecularly imprinted polymers to solid-phase extraction of compounds from environmental and biological samples. Trends in Analytical Chemistry 25, 143–154 (2006)
[27] Medina Castillo, A.L., et al.: Novel strategy to design magnetic, molecular imprinted polymers with well-controlled structure for the application in optical sensors. Macromolecules 43, 55–61 (2010)
[28] Fernández Sánchez, J.F., et al.: Fluorescence optosensors based on different transducers for the determination of polycyclic aromatic hydrocarbons in water. Analytical and Bioanalytical Chemistry 377, 614–623 (2003)
[29] Mesquita, R.B.R., Rangel, A.O.S.S.: A review on sequential injection methods for water analysis. Analytica Chimica Acta 648, 7–22 (2009)
[30] Miró, M., et al.: Potentials of multisyrenge flow injection analysis for chemiluminescence detection. Analytica Chimica Acta 541, 57–68 (2005)
[31] Ruzicka, J., Scampavia, L.: From flow injection to bead injection. Analytical Chemistry 71, 257A–263A (1999)
[32] Llorent Martínez, E.J., et al.: Implementation of multicommutation principle with flow-through multioptosensors. Analytica Chimica Acta 545, 113–118 (2005)
[33] Sánchez Barragán, I., et al.: Fluoranthene molecularly imprinted polymers based on iodinated monomers for selective room-temperature phosphorescence optosensing of fluoranthene in water. Analytical Chemistry 77, 7005–7011 (2005)
[34] Sainz Gonzalo, F.J., et al.: The development of a screening molecularly imprinted polymer optosensor for detecting xylenes in water samples. Microchemical Journal 99, 278–282 (2011)
[35] Sainz Gonzalo, F.J., et al.: Synthesis and characterization of a molecularly imprinted polymer optosensor for TEXs-screening in drinking water. Biosensors and Bioelectronics 26, 3331–3338 (2011)
[36] Llorent Martínez, E.J., et al.: Flow-through fluorescence-based optosensor with on-line solid-phase separation for the simultaneous determination of a ternary pesticide mixture. Journal of the Association of Official Analytical Chemists International 88, 860–865 (2005)
[37] Llorent Martínez, E.J., et al.: A multi-commutated flow-through multi-optosensing: a tool for environmental analysis. Spectroscopy Letters 39, 619–629 (2006)

[38] Piccirilli, G.N., Escandar, G.M.: A novel flow-through fluorescence optosensor for the determination of thiabendazole. Analytica Chimica Acta 601, 196–203 (2007)
[39] Piccirilli, G.N., Escandar, G.M.: Flow injection analysis with online nylon powder extraction for room temperature phosphorescence determination of thiabendazole. Analytica Chimica Acta 646, 90–96 (2009)
[40] López Flores, J., et al.: Development of a photochemically induced fluorescence-based optosensor for the determination of imidacloprid in peppers and environmental waters. Talanta 72, 991–997 (2007)
[41] López Flores, J., et al.: Flow-through optosensing device implemented with photochemically-induced fluorescence for the rapid and simple screening of metsulfuron methyl in environmental waters. Journal of Environmental Monitoring 11, 1080–1085 (2009)
[42] Piccirilli, G.N., et al.: Flow-through photochemically induced fluorescence optosensor for the determination of linuron. Talanta 77, 852–857 (2008)
[43] López Flores, J., et al.: Simultaneous flow-injection solid-phase fluorometric determination of thiabendazole and metsulfuron methyl using photochemical derivatization. Analytical Science 25, 681–686 (2009)
[44] Fernández Argüelles, M.T., et al.: Flow-through optosensing of 1-naphthaleneacetic acid in water and apples by heavy atom induced–room temperature phosphorescence measurements. Talanta 66, 696–702 (2005)
[45] Casado Terrones, S., et al.: The development and comparison of a fluorescence and a phosphorescence optosensors for determining the plant growth regulator 2-naphthoxyacetic acid. Sensors and Actuators B 107, 929–935 (2005)
[46] Traviesa Alvarez, J.M., et al.: Direct screening of tetracyclines on water and bovine milk using room temperature phosphorescence detection. Analytica Chimica Acta 589, 51–58 (2007)
[47] Shen, L.M., et al.: A novel flow-through fluorescence optosensor for the sensitive determination of tetracycline. Talanta 85, 1285–1290 (2011)
[48] Bosch Ojeda, C., Sánchez Rojas, F.: Recent development in optical chemical sensors coupling with flow injection analysis. Sensors 6, 1245–1307 (2006)
[49] Mayr, T., Werner, T.: Highly selective optical sensing of copper(II) ions based on fluorescence quenching of immobilised Lucifer Yellow. Analyst 127, 248–252 (2002)
[50] Ruedas Rama, M.J., et al.: Implementation of flow-through multi-sensors with bead injection spectroscopy: fluorimetric renewable surface biparameter sensor for determination of berillium and aluminum. Talanta 62, 879–886 (2004)
[51] Ruedas Rama, M.J., et al.: A flow-injection renewable surface sensor for the fluorimetric determination of vanadium(V) with Alizarin Red S. Talanta 66, 1333–1339 (2005)
[52] Ling, L., et al.: An optical sensor for mercuric ion based on immobilization of Rhodamine B derivative in PVC membrane. Talanta 91, 65–71 (2012)
[53] Shamsipur, M., et al.: Novel fluorimetric bulk optode membrane based on a dansylamidopropyl pendant arm derivative of 1-aza-4,10-dithia-7-oxacyclododecane ([12]aneNS_2O) for selective subnanomolar detection of Hg(II) ions. Analytica Chimica Acta 533, 17–24 (2005)
[54] Chan, W.H., et al.: Development of a mercury ion-selective optical sensor based on fluorescence quenching of 5,10,15,20-tetraphenylporphyrin. Analytica Chimica Acta 444, 261–269 (2001)

[55] Yang, Y., et al.: An optical sensor for mercury ion based on the fluorescence quenching of tetra(p-dimethylaminophenyl)porphyrin. Analytica Chimica Acta 636, 261–269 (2009)

[56] He, C.L., et al.: A fluorescent chemical sensor for Hg(II) based on a corrole derivative in a PVC matrix. Talanta 70, 364–369 (2006)

[57] Zhang, X.B., et al.: An optical fiber chemical sensor for mercury ions based on a porphyrin dimer. Analytical Chemistry 74, 821–825 (2002)

[58] Aksuner, N., et al.: A sensitive and selective fluorescent sensor for the determination of mercury(II) based on a novel triazine-thione derivative. Dyes and Pigments 88, 143–148 (2011)

[59] Aksuner, N., et al.: Highly sensitive sensing of zinc(II) by development and characterization of a PVC-based fluorescent chemical sensor. Spectrochimica Acta Part A 78, 1133–1138 (2011)

[60] Shamsipur, M., et al.: Development of a new fluorimetric bulk optode membrane based on 2,5-thiophenylbis(5-tert-butyl-1,3-benzexazole) for nickel(II) ions. Analytica Chimica Acta 501, 55–60 (2004)

[61] Aksuner, N.: Development of a new fluorescent sensor based on a triazolo-thiadiazin derivative immobilized in polyvinyl chloride membrane for sensitive detection of lead(II) ions. Sensors and Actuators B 157, 162–168 (2011)

[62] Shamsipur, M., et al.: Novel fluorimetric bulk optode membrane based on 5,8-bis((5'-chloro-8'-hydroxy-7'-quinolinyl)methyl)-2,11-dithia-5,8-diaza-2,6-pyridinophane for selective detection of lead(II) ions. Talanta 80, 2023–2033 (2010)

[63] Cano Raya, C., et al.: Fluorescence resonance energy transfer disposable sensor for copper(II). Analytica Chimica Acta 555, 299–307 (2006)

[64] Lozano, V., et al.: A novel nylon membrane-rhodamine 6G spirocyclic phenylthiosemicarbazide derivative system as a fluorimetric probe for mercury(II) ion. Analytical Methods 4, 2002–2008 (2012)

[65] Meng, Q., et al.: A hybrid mesoporous material functionalized by 1,8-naphthalimide-base receptor and the application as chemosensor and absorbent for Hg^{2+} in water. Talanta 84, 53–59 (2011)

[66] Sun, Z., et al.: An optical sensor based on H-acid/layered double hydroxide composite film for the selective detection of mercury ion. Analytica Chimica Acta 702, 95–101 (2011)

[67] Ji, X., et al.: A ratiometric fluorescence sensor for Be^{2+} based on Beryllon II/layered double hydroxide ultrathin films. Analytica Chimica Acta 728, 77–85 (2012)

[68] Kulchat, S., et al.: Self-assembled coordination nanoparticles from nucleotides and lanthanide ions with doped-boronic acid-fluorescein for detection of cyanide in the presence of Cu^{2+} in water. Talanta 89, 264–269 (2012)

[69] Chen, H., Ren, J.: Sensitive Determination of Chromium (VI) Based on the Inner Filter Effect of Upconversion Luminescent Nanoparticles ($NaYF_4$:Yb^{3+}, Er^{3+}). Talanta (2012) (in press)

[70] Azab, H.A., et al.: Reactivity of a luminescent "off-on" pyrylium dye towards various classes of amines and its use in a fluorescence sensor microtiter plate for environmental samples. Journal of Photochemistry and Photobiology A 243, 41–46 (2012)

[71] Costa Fernandez, J.M., et al.: The use of luminescent quantum dots for optical sensing. Trends in Analytical Chemistry 25, 207–218 (2006)

[72] Frigerio, C., et al.: Application of quantum dots as analytical tools in automated chemical analysis: A review. Analytica Chimica Acta 735, 9–22 (2012)

[73] Wang, H.F., et al.: Surface molecular imprinting on Mn-doped ZnS quantum dots for room-temperature phosphorescence optosensing of pentachlorophenol in water. Analytical Chemistry 81, 1615–1621 (2009)
[74] Stringer, R.C., et al.: Detection of nitroaromatic explosives using a fluorescent-labeled imprinted polymer. Analytical Chemistry 82, 4015–4019 (2010)
[75] Zou, W.S., et al.: Room-temperature phosphorescence chemosensor and rayleigh scattering chemodosimeter dual-recognition probe for 2,4,6-trinitrotoluene based on manganese-doped ZnS quantum dots. Analytical Chemistry 83, 30–37 (2011)
[76] Liu, H., et al.: Rhodamine-based ratiometric fluorescence sensing for the detection of mercury(II) in aqueous solution. Talanta 81, 433–437 (2010)
[77] Sui, B., et al.: Ultrasensitive determination of 1,4-dihydroxybenzene based on fluorescence resonance energy quenching of luminescent quantum dots modified on surface of silica nanoparticles. Talanta 85, 1609–1613 (2011)
[78] Carrillo Carrión, C., et al.: Calix[8]arene coated CdSe/ZnS quantum dots as C60-nanosensor. Analytical Chemistry 83, 8093–8100 (2011)
[79] Zhao, Y., et al.: Composite QDs@MIP nanospheres for specific recognition and direct fluorescent quantification of pesticides in aqueous media. Analytical Chemistry 84, 386–395 (2012)
[80] Gui, R., et al.: A near-infrared-emitting CdTe/CdS core/shell quantum dots-based OFF–ON fluorescence sensor for highly selective and sensitive detection of Cd^{2+}. Talanta 94, 257–262 (2012)
[81] Wang, Y.Q., et al.: Highly sensitive synchronous fluorescence determination of mercury(II) based on the denatured ovalbumin coated CdTeQDs. Talanta (2012) (in press)
[82] Ma, Q., et al.: Synchronous determination of mercury(II) and copper(II) based on quantum dots-multilayer film. Analytica Chimica Acta 701, 60–65 (2011)
[83] Butwong, N., et al.: Determination of arsenic based on quenching of CdS quantum dots fluorescence using the gas-diffusion flow injection method. Talanta 85, 1063–1069 (2011)
[84] Jerónimo, P.C.A., et al.: Optical sensors and biosensors based on sol–gel films. Talanta 72, 13–27 (2007)
[85] Barja, B.C., et al.: Luminescent Eu(III) hybrid sensors for in situ copper detection. Biosensors and Actuators B 158, 214–222 (2011)
[86] Álvarez Diaz, A., et al.: Conjugated polymer microspheres for turn-off/turn-on fluorescence optosensing of inorganic ions in aqueous media. Analytical Chemistry 83, 2712–2718 (2011)
[87] Guo, L., et al.: An organically modified sol-gel membrane for detection of lead ion by using 2-hydroxy-1-naphthaldehydene-8-aminoquinoline as fluorescence probe. Sensors and Actuators B 130, 789–794 (2008)
[88] Duong, H.D., et al.: Amplification of fluorescence emission of CdSe/ZnS QDs entrapped in a sol–gel matrix, a new approach for detection of trace level of PAHs. Sensors and Actuators B 157, 139–145 (2011)
[89] Wygladacz, K., et al.: Fiber-optic microsensor array based on fluorescent bulk optode microspheres for the trace analysis of silver ions. Analytical Chemistry 77, 4706–4712 (2005)
[90] Ng, S.M., Narayanaswamy, R.: Fluorescence sensor using a molecularly imprinted polymer as a recognition receptor for the detection of aluminium ions in aqueous media. Analytical and Bioanalytical Chemistry 386, 1235–1244 (2006)

[91] Wright, J.D., Sommerdijk, N.A.: Sol-gel Materials: Chemistry and Application. In: O'Brien, P., Phillips, D., Roberts, S. (eds.) CRC Press (2000)
[92] Lieberzeit, P.A., et al.: Imprinting as a versatile platform for sensitive materials-nanopatterning of the polymer bulk and surfaces. Sensors and Actuators B 111-112, 259–263 (2005)
[93] Jenkins, A.L., Bae, S.Y.: Molecularly imprinted polymers for chemical agent detection in multiple water matrices. Analytica Chimica Acta 542, 32–37 (2005)
[94] Wang, X., et al.: Selective optical chemical sensor for 2,6-dinitrophenol based on fluorescence quenching of a novel functional polymer. Talanta 70, 160–168 (2006)
[95] Wang, X., et al.: A reversible fluorescence sensor based on insoluble b-cyclodextrin polymer for direct determination of bisphenol A (BPA). Sensors and Actuators B 114, 565–572 (2006)
[96] Wang, X., et al.: Selective determination of bisphenol A (BPA) in water by a reversible fluorescence sensor using pyrene/dimethyl b-cyclodextrin complex. Analytica Chimica Acta 556, 313–318 (2006)
[97] Ballesteros Gómez, A., Rubio, S.: Recent advances in environmental analysis. Analytical Chemistry 83, 4579–4613 (2011)
[98] Escandar, G.M., et al.: Second- and third-order multivariate calibration: data, algorithms and applications. Trends in Analytical Chemistry 26, 752–765 (2007)
[99] Piccirilli, G.N., Escandar, G.M.: Partial least-squares with residual bilinearization for the spectrofluorimetric determination of pesticides. A solution of the problems of inner-filter effects and matrix interferents. Analyst 131, 1012–1020 (2006)
[100] Valero Navarro, A., et al.: Chemometric-assisted MIP-optosensing system for the simultaneous determination of monoamine naphthalenes in drinking waters. Talanta 78, 57–65 (2009)
[101] Piccirilli, G.N., Escandar, G.M.: Second-order advantage with excitation-emission fluorescence spectroscopy and a flow-through optosensing device. Simultaneous determination of thiabendazole and fuberidazole in the presence of uncalibrated interferences. Analyst 135, 1299–1308 (2010)
[102] Domínguez Vidal, A., et al.: Environmental water samples analysis of pesticides by means of chemometrics combined with fluorimetric multioptosensing. Journal of Fluorescence 17, 271–277 (2007)
[103] Bortolato, S.A., et al.: Chemometrics-assisted excitation-emission fluorescence spectroscopy on nylon membranes. simultaneous determination of benzo[a]pyrene and dibenz[a,h]anthracene at parts-per-trillion levels in the presence of the remaining EPA PAH priority pollutants as interferences. Analytical Chemistry 80, 8276–8286 (2008)
[104] Bortolato, S.A., et al.: Chemometrics assisted fluorimetry for the rapid and selective determination of heavy polycyclic aromatic hydrocarbons in contaminated river waters and activated sludges. Environmental Sciences and Technology 45, 1513–1520 (2011)
[105] Chiarandini, J.P., Escandar, G.M.: Nylon membrane as a fluorimetric probe for the herbicide bentazone. Analytical and Bioanalytical Chemistry 402, 2221–2225 (2012)

Monitoring Wastewater Treatment Using Voltammetric Electronic Tongues

J. Soto*, I. Campos, and R. Martínez-Máñez

Centro de Reconocimiento Molecular y Desarrollo Tecnológico,
Unidad Mixta Universidad Politécnica de Valencia-Universidad de Valencia,
Departamento de Química, Universidad Politécnica de Valencia, Spain
juansoto@qim.upv.es

Abstract. In the early 1990s, the design and preparation of new control and quantification systems called Electronic Noses and Tongues became popular. The common denominator of these is the use of an array comprising a significant number of sensors (potentiometric, voltammetric, mass, optical, etc.) which, when conveniently grouped and using chemometric techniques, allow us to obtain qualitative, and even quantitative information on certain samples. One of the most interesting aspects of these systems is their capacity as classification techniques to distinguish and group products or complex mixtures based on their physicochemical similarity.

Keywords: Electronic tongue, wastewaster, voltammetric, chemometric, water monitoring.

1 Introduction: State of the Art

1.1 Reusing and Recycling Treated Wastewater for Irrigation

Since the most water demand is associated with agriculture, a simple way to compensate structural water deficit is to reuse treated wastewater. Working in this direction leads to better protection and helps conserve the environment. Environmental conservation promotes tourism and is a policy followed by advanced regions such as Israel, Japan and California and in some Mediterranean countries. Moreover, it has been demonstrated that purified waste water provides a large number of nutrients and trace elements to soil. These compounds may have a positive impact on the nutritional status of crops irrigated with this water [1,2].

The main problem of recycling treated water is the potential threat to public health, soil and water when reuse is not carefully done. This is precisely where research into recycling treated water must lie in order to adapt usages to effluent characteristics. To achieve reasonable levels of assurance in treated wastewater quality terms, it is necessary to control the content of the pollutants in the water obtained. However, the

* Corresponding author.

S.C. Mukhopadhyay & A. Mason (Eds.): *Real-Time Water Quality Monitoring*, SSMI 4, pp. 65–103.
DOI: 10.1007/978-3-642-37006-9_4 © Springer-Verlag Berlin Heidelberg 2013

effect that the use of low quality water may have on the surrounding environment should be investigated, and special emphasis must be placed on ground water, which may be contaminated with toxic components [3].

1.2 Quality Control of Treated Water and Wastewater

Water quality control is required by law and there is generally a number of parameters (physicochemical and biological in nature) to be controlled, which usually depend on the subsequent application of treated water. Moreover, each country's regulations may require the control of additional parameters. However, due to the high economic cost that the determination of certain chemicals in water involves, normally very few parameters are controlled or are only partially determined in time.

One key aspect in the design and operation of a Wastewater Treatment Plant (WWTP) is the knowledge of the most important pollutants in the water to be treated. There are several methods for this type of characterization, but the most widely used ones are physicochemical [4,5] and respirometric [6,7]. Both methods require much effort, and laboratory analytical results are not immediately available (e.g., biodegradable organic matter after 5 days). To optimize WWTP aeration and to reduce pumping costs, it is necessary to know the concentration of certain pollutants (for instance, ammonium, nitrate, phosphate, BOD, etc.) throughout the treatment process. This information may be available when using online analyzers. However, these entail making significant investments and usually imply high maintenance costs. Given these drawbacks, control in a WWTP typically relies on the information obtained from low-cost sensors (pH, redox, conductivity) in order to roughly estimate variations in the concentration of pollutants. Based on these indirect measures and on operators' experiences, decisions are made to amend water treatment processes. With a view to overcoming these problems, a number of new sensors and measurement technologies are being developed today. Current research focuses on lowering costs, improving response times and searching for a better relationship between water composition and the information obtained [8].

1.3 The Electronic Tongue

In 1994, Gardner and Bartlett [9] defined the Electronic Nose concept, which was an instrument made up of a group of sensors with partial specificity which, when coupled with pattern recognition techniques, was capable of recognizing simple or complex odors in gaseous samples. In this field, for instance, several studies have demonstrated the utility of Electronic Noses for quality control of water purification [10, 11].

Electronic Tongues are related to Electronic Noses, and they are also formed by an array of sensors capable of being used in liquid samples [12, 13]. These sensor systems have been used successfully for the characterization of food [14], and there are several patents and commercial available equipment based on the use of Noses, Tongues, and even Electronic Eyes, for certain applications.

In Potentiometric Electronic Tongues (PETs), the information obtained by the sensing array consists of a set of values of electrical potential versus time, obtained for a set of electrodes versus a reference electrode [15]. PETs have been widely used, but there have been relatively few works conducted to characterize water samples. Among these studies, the implementation of PETs to classify natural waters has been reported [16, 17]. PETs have also been applied to the detection and quantification of certain ions in water. In particular a potentiometric array, containing several chalcogenide glass sensors [18], has illustrated this system's ability to detect metal ions, chloride and sulphate in natural waters. In another work, an array of ISFET-based sensors, containing different membranes sensitive to distinct ions, combined with conductivity and redox potential measurement [19], allowed the classification of different types of bottled waters and the quantification of the presence of some ions, such as Ca^{2+}, Na^{+}, K^{+} and Cl^{-}. A comprehensive review on the use of Potentiometric Tongues applied to water, food and medicine, using different sensor arrays (ISE, ISFET, LAPS, etc.) combined with flow injection analyses, has been recently provided by Batrov et al [20].

1.3.1 Challenges for Electronic Tongues

Probably some of the most important groups of potentiometric sensors are the so-called ion selective electrodes (ISE), which are able to measure a potential dependent on the concentration of the target ions present in the solution. The oldest, best-known and most widely used electrode is the glass electrode pH sensor. Besides, there is currently a variety of useful commercial ISEs available for the direct determination of the activity of certain ions in water samples (i.e., Ca^{2+}, Cl^{-}, F^{-}, etc.). These ISE electrodes can be integrated into Electronic Tongues in order to be applied to the problem of determining wastewater pollution or to monitor the evolution of water during the purifying process. Following this idea, a PET which uses all-solid-state ISE's with a solid electrical contact made from a conductive epoxy composite has been reported. With this array, monitoring remote pollution in rivers and lakes has been achieved [21]. Besides, ammonium, Na^{+}, K^{+} Cd^{2+}, Pb^{2+}, Cu^{2+} and Zn^{2+} were quantified using this electronic tongue. The working device was tested at Rio Salado (Mexico) with fine results.

The most important problem of this type of sensor, regardless of its working principle (ISE, liquid membrane, all-solid state, ISFET, etc.), is the fouling of membranes or sensitive surfaces by chemicals or biological agents. In fact, it is generally accepted that basically two situations lead to the deactivation of potentiometric electrodes:

(1) Presence of substances that can be absorbed and transported through the membrane, which lead to fouling of the internal electrode solution.
(2) Physical or chemical adsorption of different species on the sensor surface, which diminishes the active surface and deactivates its recognition capability. For instance in ISE electrodes, it is quite typical that absorption on the membrane of lipophilic compounds (i.e., fats, fatty acids, alkanes, aromatics, etc.) progressively leads to a change in membrane response.

There are many pollutants in wastewater, which include a large amount of organic and inorganic compounds (both dissolved and suspended) and a number of living organisms. This set of elements pollutes different sensors, and the fouling rate depends mainly on the nature and concentration of the different elements present in the system. In the particular case of ISE's, membrane pollution and fouling processes are also accompanied by a leaching of the internal reference electrolyte electrode due to a diffusion concentration gradient.

Another potential inconvenience in PETs is the need to use a Reference Electrode, which also has the same pollution, fouling and leaching problems as those discussed above. Yet another problem is the pollution of the filling solution with compounds from the test solution (halophilic algae penetration sometimes occurs). Moreover, when the reference electrode is immersed in wastewater for a long time, the porous membrane electrode can often become clogged [22]. Furthermore, it is not recommended to immerse the reference electrodes in environments containing chemical species which can displace chloride from the reference electrode. For example, it is undesirable to dip the reference electrode into wastewater containing sulphide.

These are some of the reasons that make many WWTPs so difficult to monitor by potentiometric electrodes; maintenance to regenerate the electrode, and cleaning and calibrating sensor systems should be performed daily.

1.3.2 Voltammetric Electronic Tongues

In Voltammetric Electronic Tongues (VETs), the information obtained from the sensor array consists of a series of electric current values as a function of time. Data is obtained by measuring the current due to the application of a potential pulse sequence on an array of multiplexed electrodes. The first pulse VET was developed and studied by Winquist et al in 1997 [23]. The authors built a device consisting of an array of four noble metals (Pt, Au, Rh and Ir) as working electrodes, embedded in epoxy resin inside a stainless steel cylinder, which was used as an auxiliary electrode. By applying a pulse sequence in this VET, different intensity versus time curves for each electrode is obtained. Information about a given sample is obtained when applying chemometric techniques to current intensity values. One of the most important advantages of this electronic tongue reported by Winquist is that it is formed by a group of noble metal microelectrodes, which are physically and chemically robust sensing systems that do not require high maintenance costs.

Winquist et al successfully applied this VET to monitor the efficiency of the tap water filtration process [24], and to also detect and quantify surfactants during cleaning stages in automatic washing machines [25]. Finally this series of pioneering works of Winquist concluded that a mechanical [26] or electrochemical [27] polishing of electrodes surfaces is required to ensure the reproducibility of the results between different measurement sequences. The electrode must be polished to remove contamination from the electrode surface, which is caused by adsorption of products or by oxidation of the metal surface. It is also known that the initial surface contamination rate due to adsorption is directly proportional to the adsorbate concentration. With a view to slowing down the electrode fouling rate, methods in which the sample is diluted prior to the study can be applied. One result of this is, for instance, the implementation of FIA techniques with VETs. As an example of such hybridisation, we found some papers describing the application of an electronic tongue-FIA system to study a wastewater paper mill [28]. In this work, the authors showed that it was possible to not only discern the factory that generated the samples, but to also estimate the chemical oxygen demand value by means of statistical PCA and PLS analyses.

Although there are many applications employing VETs, their ability to identify, classify and quantify has been somewhat limited due to the fact that the electrodes used were essentially only noble metals. Noble metal electrodes are able to directly detect species, and show reversible or irreversible redox processes in solution (Faradic systems). It was also found that these electrodes indirectly respond to the concentration of ionic or neutral species with no redox actives because their presence brings about changes in current pulse voltammetric relaxation. These changes are associated with viscosity, the electrical resistance of the solution and the capacitance of the electrical double layer (non-Faradic processes). A recent work has shown that in VETs the ability to classify and quantify significantly increases when arraying noble metals in combination with electrodes showing some chemical reactivity, such as silver and copper or non-noble metal electrodes like nickel or cobalt [29]. In fact, this combination of metals has proved useful to detect and quantify anions. This may be a consequence of the chemical reactions occurring between the anion present in the solution and the metal cations generated by electrode surface oxidation. It has been reported that the electronic tongues built with noble and non-noble metals are able to detect and quantify (among others) the following ions: NO_3^-, NO_2^-, NH_4^+, $CO_3^=$, Cl^-, CN^-, PO_4^{-3} [30]. The formation of complex species can also play an important role in the detection of ammonia, amines or nitrites, among others (e.g., via the formation of Co(II)-NO_2^-, or Cu(II)-NH_4^+ complexes). Recently, a comparative study of some parameters typically used to track the wastewater purification process in a pilot plant has been conducted. The results obtained by classical analysis methods for COD, BOD, nitrates, phosphates, ammonia, etc., have been compared with those obtained using a VET with satisfactory results [30]. The results of this study suggest that a control system for a WWTS based on this technology can be adapted to recognize abnormal operating conditions so, in theory, it can anticipate the appearance of possible contingences to allow corrective action. The main potential application of electronic tongues focuses on providing virtually instantaneous values of important parameters, whose acquisition by traditional routes may involve waiting between a few hours and several days.

1.3.3 Pulse Train Operation

One of the features upon which electronic tongue performance is based is dependence on the existence of cross selectivity between the different sensors forming the array. From this point of view, one metallic electrode is a fully configurable sensor whose response depends on both the redox activity of the studied system and the applied potential of the electrode. By way of example, let us consider a system consisting of three redox active species, A, B and C. Let us assume that the respective oxidation potentials are +0.3V, +0.7 V and that compound C reduced to-0.4 V. If we apply the working electrode to a potential of -0.5V, a cathodic current proportional to the C concentration in the test solution is obtained. If the applied potential jumps to 0 volts later, when the system's steady state is reached, there is no current flow. Let us now assume that the potential jumps to +0.5V. Under these conditions, an anodic current associated with the oxidation of A is observed, which is proportional to its concentration. Finally, we then force a jump to +0.8 V. Under these conditions, the simultaneous oxidation of A and B takes place. If instead of the three compounds A, B and C, the system consists of two more species, D and E, whose potentials are clearly differentiated, we can add only two electrode potential steps to discern the presence of these species.

The sequence of potentials that we have applied is called voltammetric pulse sequence. The number of pulses forming the train and the selected potential values (pulse train design) is a particular parameter that depends on the test solution to be monitored. It is evident that pulse train design is very important for the detection and quantification of the components forming a system. Moreover, and as discussed later, the electric potential at which a given redox process occurs is determined by surface adsorption, which depends on the molecular structure of the adsorbate, the nature of the electrode surface, etc. This means that a redox-active substance itself may present different voltammetric behaviors when studied in distinct electrode types. This, and the fact that electrode selectivity can be modulated as a function of the electric potential applied, makes the pulse voltammetry an ideal system for building electronic tongues and for applying chemometric techniques.

1.3.4 Electrochemical Instrumentation

The basic instrumentation required to implement a pulse VET, basically consists of the following modules:

(1) The first modular element is the potentiostat, which must ensure the stability of the applied potential and should have a good response time to quickly jump potential steps. It needs to be noted that it is sometimes necessary to work by applying pulses whose duration is only a few milliseconds.
(2) The Pulse Train Generator is the second modular element. This element must be fully configurable and capable of information exchange with the potentiostat and the computer. The modular element must allow the application of different wave pulses and the triangular wave techniques conventionally used in cyclic voltammetry in order to characterize the electrochemical behavior of the solution.
(3) Finally, there must be a multiplexing module to sequentially apply the pulse train to the array of electrodes forming the VET.

The potentiostatic module should allow work with two or three electrodes. Later, the causes of this particular requirement of the module are explained. It must also be able to apply and control any electric potential value between approx. +3 and -3 volts to be able to work with aqueous or organic solvents.

VETs characteristics include the tremendous speed of data collection and processing, and it can be considered a non-destructive method of analysis. To gain an idea of how fast this system works, we can take an Electronic Tongue with eight electrodes. By applying a pulse array with 20 potential steps with duration of 50 milliseconds each, the total time to measure one sample is approx. 8 seconds. Typically, the information collected while applying the pulse array ranges from a few hundred to a few thousand data. This vast amount of information requires the use of chemometric tools to study the electronic tongue response.

2 Chemometrics Multivariate Analysis

Due to the huge amount of data generated with the electronic tongue [31,32], multivariate analysis is essential. In particular, the use of this analysis and its combination with sensor arrays has been shown to be a very powerful analytical tool. The literature [33] offers a number of works that use different data analysis techniques. Of all these techniques, the most widely used are Principal Component Analysis (PCA), Partial Least Square Analysis (PLS) and Neural Networks Analysis (NNA).

2.1 *Principal Component Analysis (PCA)*

PCA is a well-established dimensionality-reduction technique. Many applications using this procedure (e.g., data compression, image processing, visualization, exploratory data analysis, pattern recognition, time series prediction, and multivariate process monitoring) have been reported [34-37]. PCA is an orthogonal lineal transformation that transforms data into a new coordinate system so that the greatest variance by any projection of the data comes to lie on the first coordinate (called the first principal component), the second greatest variance on the second coordinate, and so on. The goal of this is to map raw data vector $x = [x_1, x_2,...,x_m]^T$ onto vectors $z = [z_1,z_2,...,z_m]^T$, where $k < m$ [38]. Vector x can be represented as a linear combination of a set of m orthonormal vectors u_i as in (1).

$$x = \sum_{i=1}^{m} \mathrm{z}_1 \, u_\mathrm{i} \tag{1}$$

Where coefficients z_i can be found from (2).

$$z_i = u_i^T x \tag{2}$$

This corresponds to a rotation of the coordinate system from the original x to a new set of coordinates given by z. T reduces the dimensions of the data by which a subset ($k < m$) of the basis vectors u_i is kept. The remaining coefficients are replaced, and constants b_i and each vector x are then approximated as in (3).

$$\tilde{x} = \sum_{i=1}^{M} z_1 u_i + \sum_{i=M+1}^{d} b_i u_i \tag{3}$$

The basis vector u_i, also called principal components, are the eigenvectors of the covariance matrix of the data set. Coefficients b_i and the principal components should be chosen so that the best approximation of original vector x is obtained on average. However, the reduction of dimensionality from m to k causes an approximation error. The sum of the squares of the errors [38] over the whole data set is minimized if we select the vectors u_i that correspond to the largest eigenvalues of the covariance matrix. As a result of PCA transformation, the original data set is represented in fewer dimensions (typically 2-3), and the measurements can be plotted in the new coordinate system. This plot shows the relation between different observations or experiments. Groupings of data points in this plot suggest some common properties and can be used for classification.

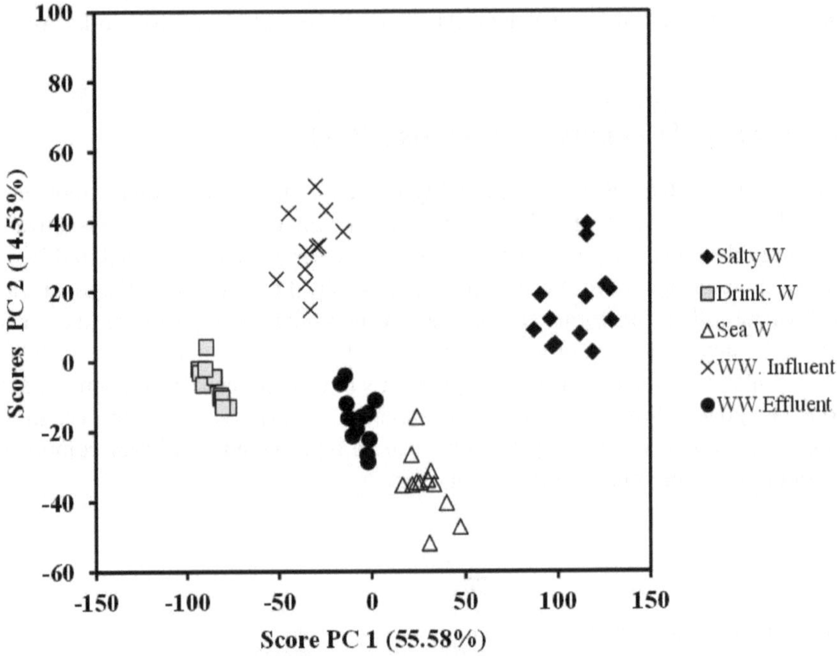

Fig. 1. A PCA score plot of different water samples using data from a VET

The PCA results are expressed as observation parameters (scores) and weights of variables (loads). To be able to interpret the results, these are expressed graphically in two dimensions (PC1 and PC2), but depending on the application or the result, sometimes a three-dimensional diagram (PC1, PC2, PC3) is necessary. In the graph of the observations (see Fig. 1 as an example), possible spontaneous groupings of actions, called "clusters", are found so that when points are close together in the PC1-PC2 plane, it means they have common characteristics. The PC1 component is always associated with the highest percentage of variance explained, and then the variations on the PC1 axis have much more significance than those on the PC2 axis. The graph of the weights (loads) provides information on the relationship of the sensors' behavior, meaning that if they are close on the graph, there is a correlation between them and they offer similar information. The position of the two points in relation to the center sensors offers significant information. If they form a small angle in relation to the center, it means that there is correlation and they offer similar information. If an angle of approximately 180° is formed, this means that both variables have a negative correlation; i.e., if the sensor signal increases, the signal of another declines. Finally, if the experimental points of both sensors are at an angle of approximately 90°, then the sensors provide orthogonal information; i.e., they are independent of each other and there is no correlation between them. Fig. 2 shows an example of all the cases. Strictly speaking, although the PCA data processing algorithm is a simple dimension reduction technique, it is used more in practice as a data classification method.

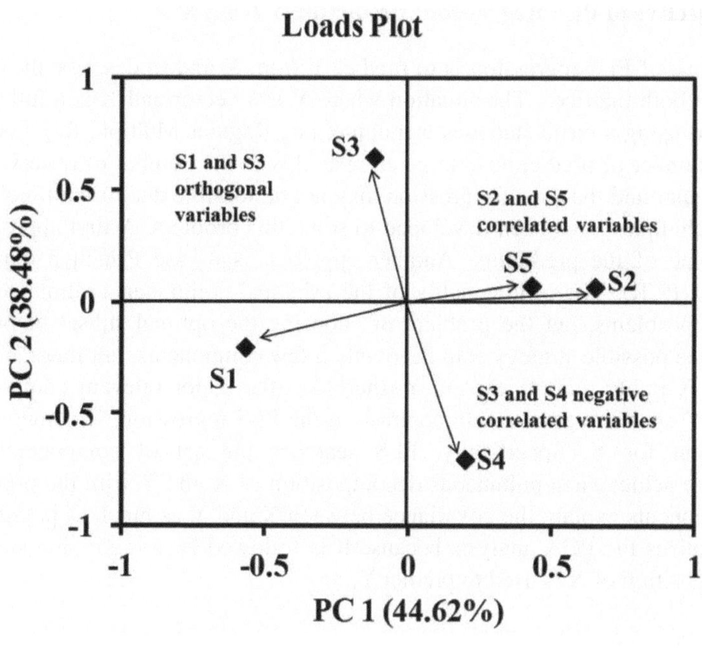

Fig. 2. PCA loads plot for a PET based on five different electrodes

2.2 *Partial Least Square Analysis*

Partial Least Square (PLS) regression is a technique that generalizes and combines features from PCA analyses and multiple regressions. The main objective of PLS is to predict or analyze a set of dependent variables from a set of independent variables or observables. Prediction is achieved by extracting a set of orthogonal factors, called latent variables, from the observed variables that represent the best predictive power on the dependent variables. PLS regression is particularly useful when you need to predict a set of dependent variables from a large set of independent variables (predictors). This algorithm was developed by Herman Wold in 1966 and, originally, the technique was applied mainly to social sciences. Its popularity has grown and today it is widely applied in the fields of chemometrics [39] and sensory evaluation [40].

The main difference to the Principal Component Regression (PCR) method is that it ensures that the main components are informative for predicting matrix Y. Accordingly during the calibration step, the PLS algorithm uses the information contained in the matrix of the property to be determined (Y) by obtaining latent variables which act as coefficients. Before performing factorization, arrays X and Y are centered or auto scaled as in PCA. Each array is decomposed simultaneously on a number of factors. The decompositions of both arrays are not independent, but are simultaneously performed by setting an internal relation between the scores of blocks X and Y.

2.2.1 Objective of PLS Regression: Predicting Y from X

The objective of PLS regression is to predict Y from X and to describe the common structure of both matrices. The situation where Y is a vector and X is a full rank can be achieved using a basic statistics technique: i.e., Regular Multiple Regression. Yet when the number of predictors is large compared with the number of observations, X can be singular and the use of regression may not be feasible due to multi-colinearity. Various techniques have been developed to solve this problem. A first approach is to remove some of the predictors. Another approach is to use Principal Component Regression (PCR). The orthogonality of the principal components eliminates multi-colinearity problems, but the problem of choosing the optimal subset of predictors remains. One possible strategy is to keep only a few components, but these are chosen to express X rather than to explain Y, therefore, the major relevant components to explain X Y are not warranted. In contrast, in the PLS regression, X components are also relevant for Y. Specifically, PLS searches the set of components (latent variables) to achieve a simultaneous decomposition of X and Y with the proviso that these components explain the covariance between X and Y as much as possible. This step generalizes the PCA analysis because it is followed by a regression step where the decomposition of X is used to predict Y.

2.2.2 Model Validation

After choosing the optimal number of variables and building the prediction model, the next step involves assessing its robustness. In some works, this assessment is carried out using cross-validation techniques. However, cross-validation uses the same data used to build the model, and a rigorous data validation should be done entirely outside the model. Thus, model validation is normally performed by a new set of samples, called a validation set, in which the characteristics (parameter which is to be predicted) are known. The validation set is applied to the built model and the resulting values are shown in a graph of observed data versus predicted data. The analysis of this graph can help draw important conclusions about the robustness of the model.

A perfect model is that in which the points predicted in the graph versus those observed appear to be aligned on the line of slope 1 and intercept 0; i.e., the predicted value of each sample is equal to its measured or observed value (Fig. 3). For real models, what it is usually observed is a cloud of points distributed along the line of the best fit or regression (continuous line). Model quality can be assessed visually by observing:

- The closeness of the points to the trend line. If the points are far from the trend line, this means that the prediction is not good.
- The degree of similarity that exists between the trend line and line 1:1. If the data are in line with the trend line, but the line is separate from line 1:1, again (either the slope or the intercept) prediction is not good. Moreover a numerical evaluation of the model can also be based on the following parameters:
- R^2 or coefficient of determination: it is the square of the regression coefficient of the trend line. This coefficient indicates the closeness of the points on the graph to the trend line.
- Slope (A) and intercept (B) of the trend line: they numerically indicate the degree of similarity between the trend line and line 1:1.
- Root mean square error of prediction tells us the average difference between the predicted and the observed values; see (4).

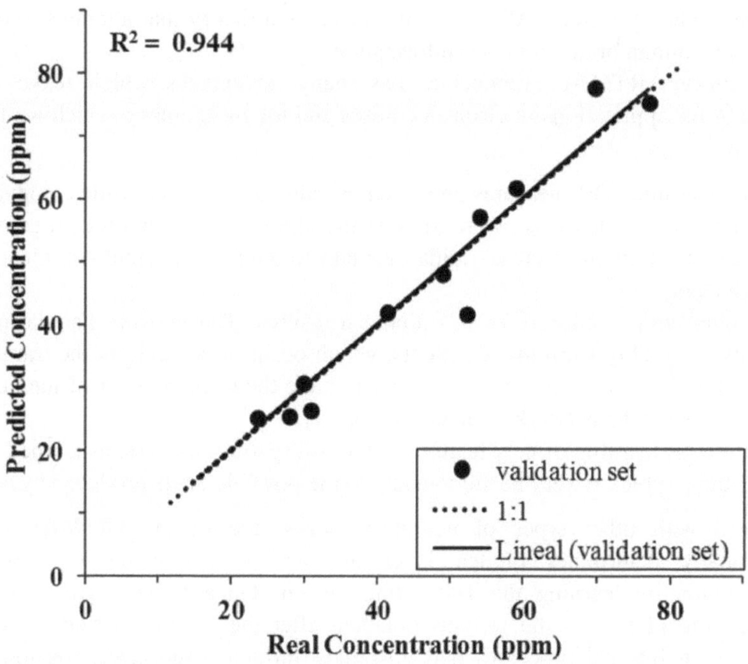

Fig. 3. Predicted versus real concentration plot

$$RMSEP = \frac{\sum_i^M (y_{OBSi} - y_{PREDi})^2}{M} \quad (4)$$

Where *YOBSi* is a parameter value, which is observed (known) for sample *i*, *PREDi* is a parameter value, which is predicted for sample *i* and *M* is the number of samples.

2.3 Neural Networks Analysis

A widely used data classification method in electronic tongues is the so-called artificial neural networks. Neural networks require a previous training stage; i.e., they must learn the relationship between inputs and outputs, and must connect them through coefficients or weights. In particular, an artificial neuron comprises a series of *n* inputs which act on a weighted basis according to a certain weight (W_i) on a transfer function. The most common type of artificial neuron is the Perceptron, where the core of the neuron acts as a summation of inputs (*I*), which are weighted according to their weight (*W*), plus a dc component or threshold (θ) as in (5).

$$x = \sum_{i=1}^{n} W_i I_i + \theta \quad (5)$$

ARTMAP-type networks are a class of neural networks that implement supervised learning and the subsequent classification of input vectors in a multidimensional set of output categories. The operation of these networks is based on the Adaptive Resonance Theory (ART). ART was introduced as a theory that attempts to emulate the way the human brain processes information.

The fuzzy ARTMAP architecture has many advantages which make it very interesting for applications in electronic noses and tongues, among which we find the following:

- Quick learning: The measures presented in training entail very little computational burden, which allows us to programme the algorithm in low-cost programmable devices, to implement cross-validations and to attempt different combinations of parameters.
- Learning with a reduced set of training measures: The network has the peculiar ability to quickly learn unique events, which occur very rarely in the training set. Therefore in this set, it is not necessary to have the same number of measures for each class for the network to function properly.
- Continuous learning of new features without forgetting the lessons learned before, something which is very useful to adapt to the possible drifts produced by sensors.

Compared with other types of neural networks, the Fuzzy ARTMAP network automatically determines hidden layer neurons. It maximizes the power of generalization by learning the 100% training set. Once trained, you can extract classification rules from the weights obtained after the training period, which may shed light on internal processes and how they influence the categorization of the results.

3 Operating Principles of VETs

The basis of pulse voltammetry lies in the difference in the decay rate of the charging and Faradaic currents following a potential step. The charging current, associated with non-Faradaic process, decays exponentially, whereas the Faradaic current normally decays as a function of $1/t^{1/2}$.

The most important parameters for pulse voltammetry are:

- **Pulse Amplitude:** The step pulse height. This may be constant or not depending on the pulse array design.
- **Pulse Width:** The potential pulse duration.
- **Sample period:** The time between two consecutive measurements of the current.

The pulse voltammetry technique usually works with three electrodes. The electrochemical system comprises a sensor or working electrode, upon which the electric potential required is applied, an auxiliary electrode, required to apply the potential step and, finally, a reference electrode, which is needed to control the electric potential applied to the working electrode.

It is sometimes convenient to work with the two-electrode technique (the Reference Electrode is not in use). Under these conditions, the exact potential of each electrode is not known, yet the results are useful for studying non-Faradaic processes.

3.1 *Double Layer Capacitance and Charging Current in Electrochemical Measurements*

When a metal electrode is immersed in an electrolytic solution (for example, in KCl 0.25 M), a series of electrostatic and electrochemical processes leads to the formation of the so-called electrical double layer. It is considered that the electric double layer is

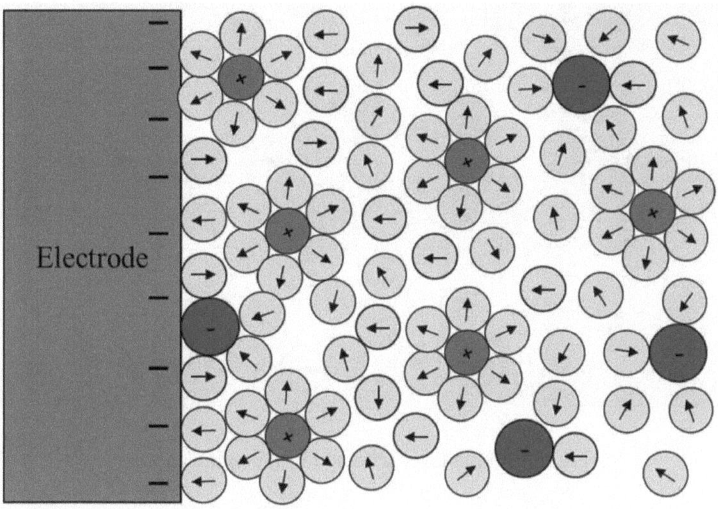

Fig. 4. Double layer scheme for a neutral metal or a metallic cathode

formed by two layers of ions adsorbed on the metal surface: a rigid layer formed by ions which are adsorbed on the metal and the diffuse layer formed by ions near the surface by electrostatic, but thermally disordered interactions (Fig. 4).

The rigid layer (also called the Helmholtz Layer) is sometimes formed by an internal plane, built with anions that are chemically adsorbed on the metal surface, and the outer plane, in which a large amount of solvated cations come into direct contact with the metal surface, which is rich in negative charge excess [41,42]. This double layer structure is stable, but can be altered if an electrical potential is applied to the electrodes because the Fermi level and the charge density of the electrodes change.

3.2 Faradaic and Non-faradaic Processes

In order to generate a potential difference between two initially balanced electrodes, it is necessary to produce the displacement of an amount of electric charge from one electrode (the working electrode) to another (the auxiliary electrode). Let us assume that the transfer of electric charge has taken place, the working electrode is positively charged and the auxiliary electrode is negative. Obviously in this new situation, the structure of the electrical double layer described in the previous section is not stable. Indeed, the positive ions adsorbed onto the working electrode undergo a repulsion of the positive electric field and they quickly move away from the electrode (Fig. 5). Immediately, positive ions are replaced with anions, which take their place on the electrode.

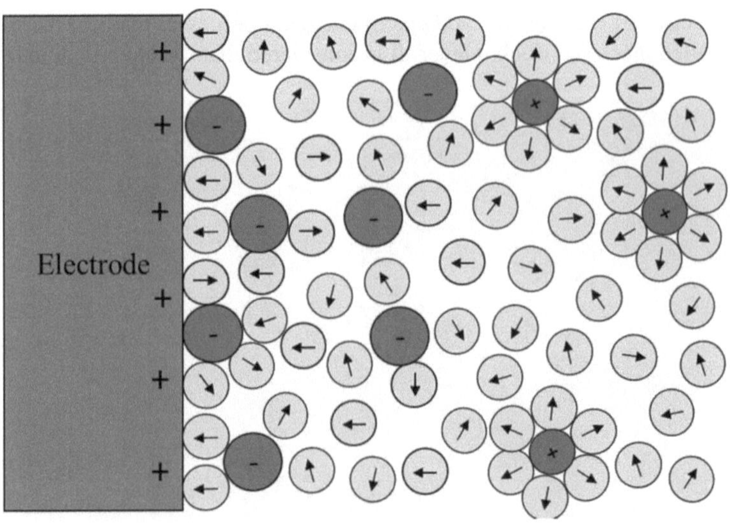

Fig. 5. Double layer scheme for a metallic anode

Simultaneously, the cation concentration increases in the vicinity of the auxiliary electrode due to the increased negative charge on the metal surface. Therefore, the result of applying a potential step produces a current transient between the working and auxiliary electrodes, which brings about changes in the electric double layer structure and tends to move rapidly to zero if the applied potential is unable to produce a redox process.

This type of process is called non-Faradaic, a mechanism type where the electric current passing through the system is associated only with the electric charge and discharge of the double layer. Fig. 6 shows the variation of current intensity as a function of time when a potential step of 50 mV is applied between two electrodes of equal surface immersed in a 0.25 M of KCl solution. The shape of the curve suggests that the variation of current versus time obeys a kinetic model of exponential decay.

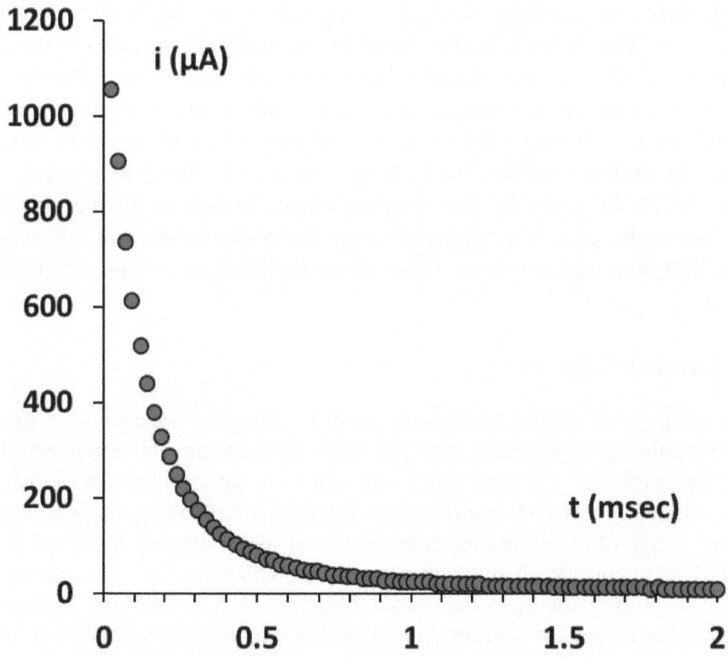

Fig. 6. Electric current when a pulse of 50 mV is applied to a KCl 0.25 M solution using a pair of platinum electrodes of 0.75 cm^2

Now assume that we add to the electrolytic KCl solution any compound capable of being oxidized or reduced. If we now apply a suitable electrical potential step, the total current passing through the system results from the combination of non-Faradaic processes with the current needed to produce the redox reaction. This type of charge transport mechanism is called a Faradaic process.

The relationship between current and time when a Faradaic current flows through a system was established by Cottrell [43]. The Cottrell model proposes that, for a Faradaic process, the current depends on the inverse square root of time in accordance with (6).

$$I = \frac{nFAC_0\sqrt{D}}{\sqrt{\pi t}} \quad (6)$$

Where n is the number of electrons exchanged during the redox process, F is the Faraday constant, A is the electrode surface, C_0 is the initial redox compound concentration and D is its diffusion coefficient. This equation has been tested on many occasions; its validity has been shown and helps obtain good results for the behavior of redox systems. However, the Cottrell equation presents a major problem: as time goes to zero, the current tends to infinity.

These problems, and others, have occasionally led to having to revise the model [44]. We have shown in a recent paper that the Cottrell equation is not satisfied in the initial stages of applying a potential step [45]. This behavior has been justified by suggesting that the working electrode does not reach the potential step value instantaneously. This is because the charge of the electrode potential first causes the restructuring of the electric double layer because this is the fastest process. Afterwards, if the electrode potential suffices to produce an electrochemical reaction, the Faradaic process begins. Moreover, it is important to note that it is necessary to coordinate the transfer of electrical charge during the Faradaic process; i.e., the electrical charge delivered by the working electrode has to be recovered by the auxiliary electrode. This is a requirement for the conservation and electroneutrality principles. Furthermore, this factor slows down the Faradaic versus the non-Faradaic process.

3.2.1 Equivalent Circuits

Variation in the intensity of the current passing through a solution as a function of time after applying a potential step provides information about the physical and chemical properties of the system; it can offer us information about the type of dissolved ions and their concentration, the presence of redox agents, etc. Equivalent circuits are arrays of electronic elements obtained by combining resistive, capacitive or inductive elements. With this conveniently organized array of components, it is possible to reproduce the electrochemical behavior of real chemical systems. Thus, the system is better understood and its properties can be modeled [46,47]. They can also be used to analyze and understand the processes of conduction and charge transport through the solution, and can sometimes greatly facilitate making predictions of the system's electrochemical behavior.

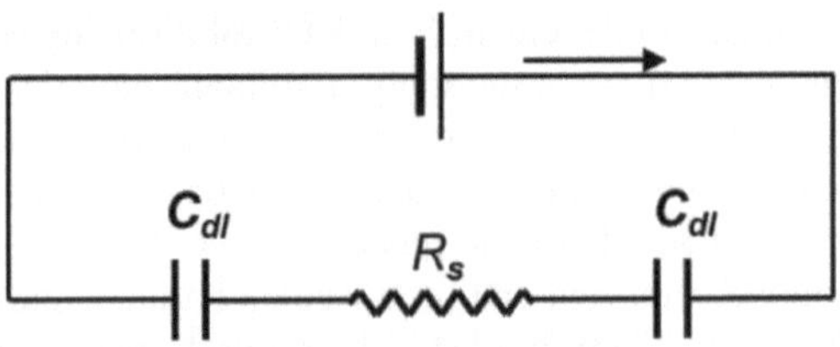

Fig. 7. Simplified equivalent circuit for an electrochemical cell (useful model for the interpretation of non-Faradaic processes working with a potential DC step)

Fig. 7 shows the generally accepted equivalent circuit to model non-Faradic processes in a system comprising two noble metal electrodes immersed in a solution [48].

It is easy to demonstrate that the intensity of the current passing through the circuit depends on the applied potential step, the resistance of the medium (R_S) and the double layer capacity (C_{dl}) of the electrodes forming the system. With DC current laws and Kirchhoff rules (meshes and nodes on current DC laws, and Kirchhoff rules [for meshes and nodes]), (7) describes the solution for the equivalent circuit.

$$I = \frac{V_{AD}}{R_S} e^{\frac{-2t}{R_S C_{dl}}} \quad (7)$$

Where, V_{AD} is the potential difference (in volts) of the potential step applied between the working electrode and the auxiliary electrode, R_S is the resistance (in Ohms) of the solution, and C_{dl} is the value of the capacity (in Farads) of the electrical double layer. In this case, we assume two electrodes of equal area, and t is the time in seconds.

However, when this model is applied to adjust the experimental data of current intensity versus time, the proposed model presents serious deficiencies. The problem is related to the total current intensity measured, regarded as the result of the combination of two components, one anodic and one cathodic current, just as Kholrausch proposed early on in the 20th century. Thus the equivalent circuit representing the non-Faradaic behavior of a solution is slightly more complex than that commonly used. The simplest representation that complies with the independent migration law of ions is shown in Fig. 8.

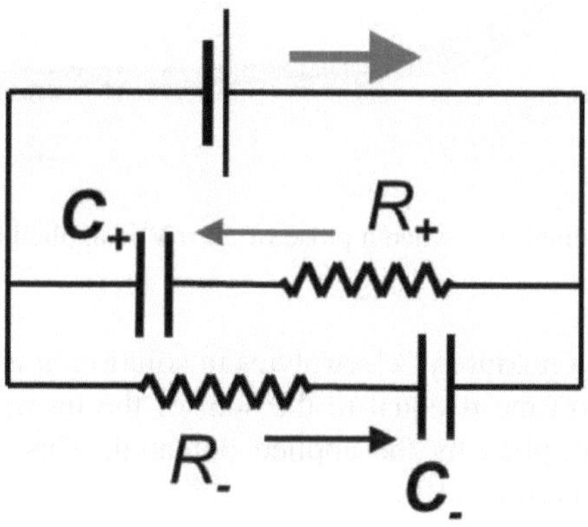

Fig. 8. Equivalent circuit for non-Faradaic processes of an electrochemical cell with separate ion migration

It is not difficult to verify that in this equivalent circuit, the current satisfies (8), where V_{AC} is the potential step applied between the working electrode and the auxiliary electrode, R_+ and R_- are the resistance of the medium to the ionic migration, and C_+ and C_- are the values of the electrical double layer capacity for cations and anions, respectively. The fit of the experimental data using this model is usually excellent (see Fig. 9). From (8), it is suggested that the initial intensity of the pulse for a non-Faradic process can also be expressed as in (9).

$$i = \frac{V_{AC}}{R_+} e^{-t/R_+C_+} + \frac{V_{AC}}{R_-} e^{-t/R_-C_-} \tag{8}$$

$$i_0 = \lim_{t \to 0} \left(\frac{V_{AC}}{R_+} e^{-t/R_+C_+} + \frac{V_{AC}}{R_-} e^{-t/R_-C_-} \right) = V_{AC} \cdot \left(\frac{1}{R_+} + \frac{1}{R_-} \right) \tag{9}$$

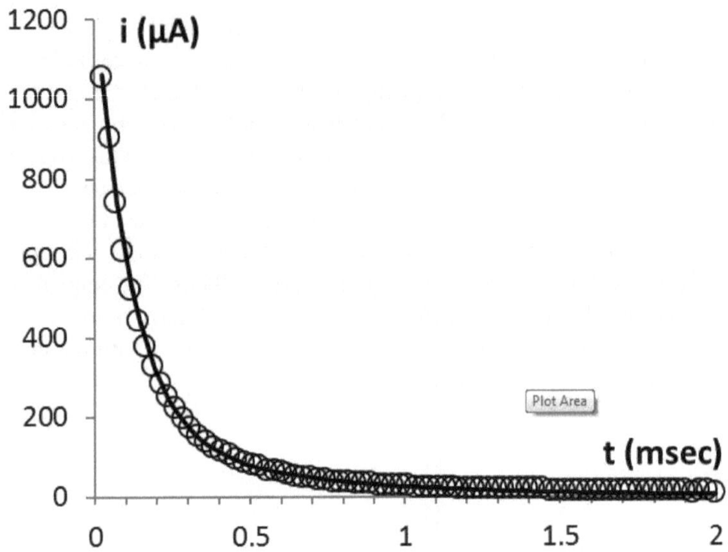

Fig. 9. Fit of the experimental data when a pulse of 50 mV is applied to 0.25 M of KCl, using (8)

In a general case of a mixture of electrolytes in solution, it is easily shown that the total current at the zero time is equal to the sum of the inverse of the resistance of cations and anions multiplied by the applied potential. This result agrees with the model proposed by Kholrausch (10).

$$i_0 = \lim_{t \to 0} = V_{AC} \cdot \left(\sum_{i=1}^{n} \frac{1}{R_{i+}} + \sum_{j=1}^{m} \frac{1}{R_{j-}} \right) \tag{10}$$

These results suggest that the initial intensity of a pulse contains information about the concentration of dissolved ions and must, therefore, be correlated with the conductivity of the sample. These equations are valid when electrochemical experiments are carried out using two electrodes. To apply these relationships, it is necessary to know the value of the applied potential step and to assume that this is applied instantly in the system. Let us take an example that applies the pulse array shown in Fig. 10 between two platinum electrodes of equal area to an aqueous KCl solution. Fig. 11 shows the intensity recorded working with two electrodes for the array of pulses applied. The graphical representation of the peak intensity of all the pulses applied against the value of the potential step is a straight line. From the slope of this line, the resistance value of the solution can be deduced, as shown in Fig. 12.

The molar conductivity value obtained from the resistance data is similar to the tabulated conductivity of this salt. However, the response intensity/time obtained by applying the pulse sequence shown in Fig. 10 provides much more information. For example, Fig. 13 shows the current measured at the end of each pulse as a function of applied potential.

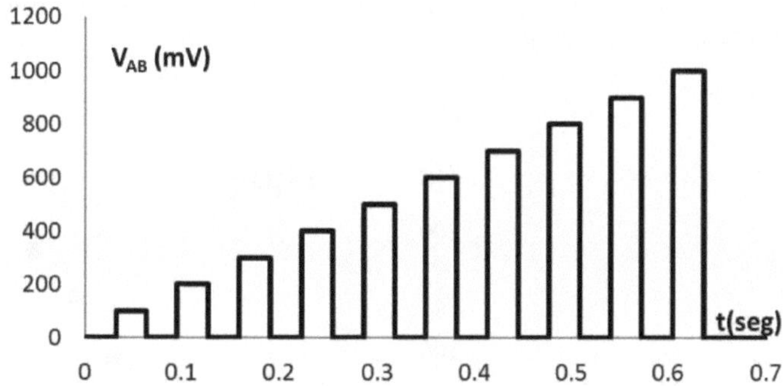

Fig. 10. Example of a pulse array of large amplitude

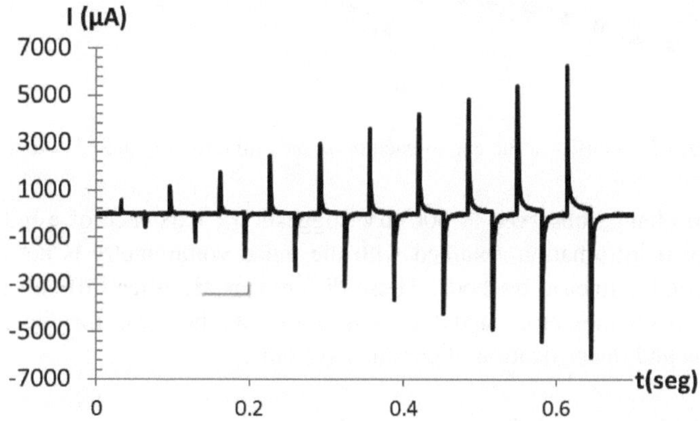

Fig. 11. Current intensity obtained with the pulse array in Figure 10, applied to a KCl 0.25M solution

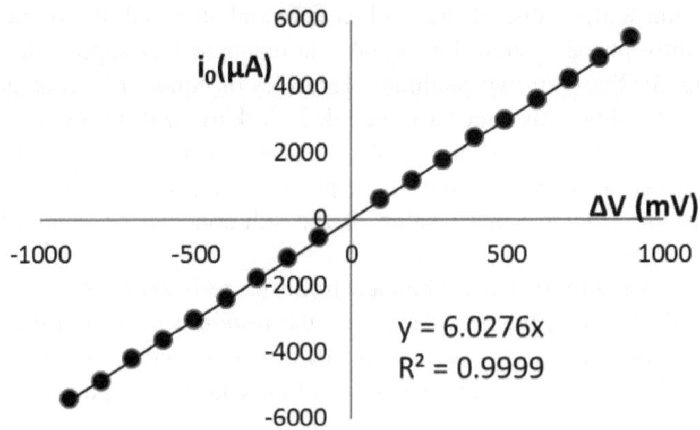

Fig. 12. Plot of the peak current vs. the potential step. KCl 025 M. The resistance value obtained is 165.9 Ω.

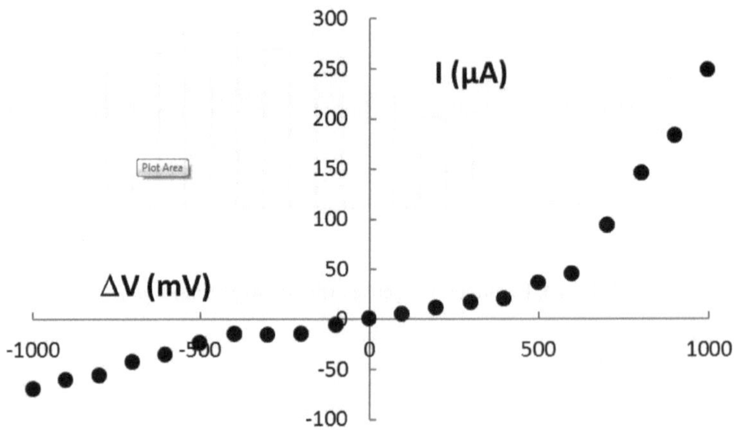

Fig. 13. Intensity at the end of each pulse as a function of applied potential

The slope change observed at 600 mV suggests the existence of a redox process. This additional information obtained with the pulse voltammetry is not detected by traditional conductimetric methods. These differences are often difficult to interpret, but thanks to chemometric analysis tools, they can be used for the recognition, classification and quantification of complex systems.

3.2.2 Equivalent Circuit for a Faradaic Process

The simplest useful equivalent circuit to interpret the electrochemical behaviour of a solution when subjected to a sufficiently high potential to produce a Faradaic process is shown in Fig. 14.

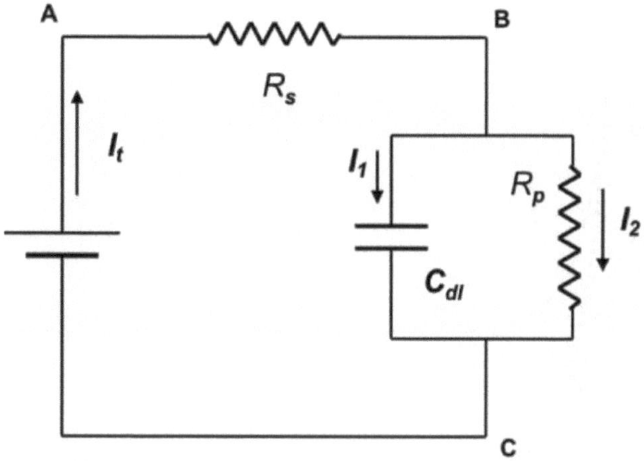

Fig. 14. A Randles simplified circuit for an electrochemical cell in which Faradaic and non-Faradaic processes occur

This is a similar circuit to that proposed by Randles, but the Warburg Impedance introduced by AC studies is eliminated. In this equivalent circuit, R_S is the resistance of the solution, C_{dl} is the electric double layer capacity and R_p is the resistance to electron transfer. The mathematical solution for this circuit helps obtain non-Faradaic current I_1 from (11). The non-Faradic current can be obtained from (12).

$$I_1 = \frac{V_{AC}}{R_s} e^{-t\left(1+R_s/_{Rp}\right)/R_sC_{dl}} \tag{11}$$

$$I_2 = \frac{V_{AC}}{R_s + R_p}\left(1 - e^{-t\left(1+R_s/_{Rp}\right)/R_sC_{dl}}\right) \tag{12}$$

This equation indicates that, at very short times, Faradaic current intensity is zero and that gradually, as time passes, it increases to a limit equal to $V_{AC}/(R_S+R_p)$. The total current passing through the cell is the sum of currents I_1 and I_2 which can be found via (13).

$$I = \frac{V_{AC}}{R_S + R_p}\left(1 - e^{-t(1+R_s/_{R_p})/R_sC_{dl}}\right) + \frac{V_{AC}}{R_S} \cdot e^{-t(1+R_s/_{R_p})/R_sC_{dl}} \tag{13}$$

Figure 15 shows the various components of the current when a Faradaic process is caused by a potential step.

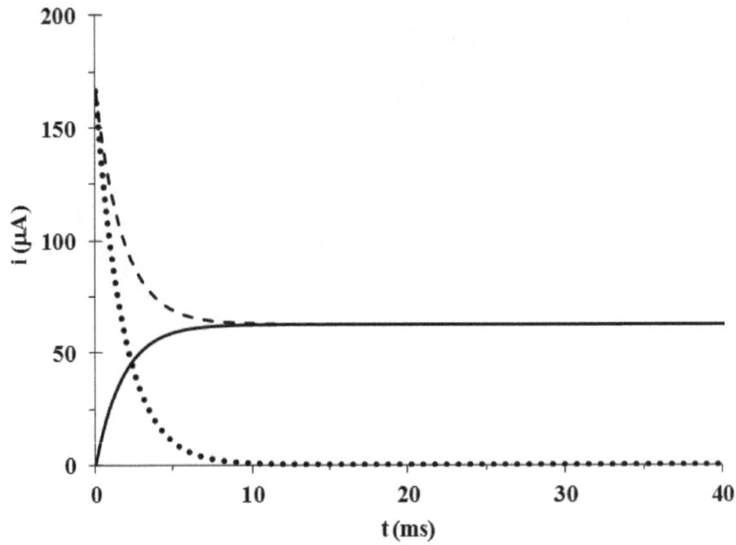

Fig. 15. Faradaic, non-Faradaic and total current for a Randles equivalent circuit with C_{dl} = 30 µF, R_s and R_p 3 and 5 KOhm, respectively. *Continuous line:* Faradaic current (I_2); *dotted line:* non-Faradaic current (I_1); *dashed line:* total current (I_t).

Unfortunately, the Randles model is not accurate as it does not take into account the variation of the redox compound concentration versus electrolysis time. An equation that takes into account the Randles model, and which also meets the Cottrell equation for long electrolysis times, is the following. This equation can adjust the relaxation curves of intensity versus time satisfactorily.

$$I = \frac{nFAC_0\sqrt{D}}{\sqrt{\pi t}} \cdot \left(1 - e^{-kt}\right) + \frac{V_{AC}}{R_s} e^{-t\left(1+R_s/Rp\right)/R_sC_{dl}} \tag{14}$$

In this equation, K is an adjustment constant that must be optimized. Accordingly, the non-Faradaic current is the main component at the start of the potential step, while the Faradaic current is dominant over long times. So if we choose the sampling time of the pulse properly, we can select the information obtained to gain a better understanding of the composition of the system.

Two factors may affect the shape of the curve of the current versus time; (i) temperature and (ii) viscosity of the medium. This is because these two parameters affect the ionic or molecular mobility in solution. As temperature variations can significantly affect the electric signal, this should be controlled and compensated.

3.3 *Noble and Non-noble Metals: Mechanisms and Electrode Processes*

A Faradaic process is a type of electrochemical reaction that occurs with the passage of an electric charge through the metal interface solution. Normally, this process occurs when a dissolved species reaches the electrodes and there is a suitable electrical potential. A Faradiac transfer usually results from a series of linked physical, chemical and electrochemical processes that enables the interfacial redox reaction [49].

The set of steps through which molecules pass to be subjected to an electrochemical process at electrodes is called *electrochemical reaction mechanism.* To determine the mechanism reaction of an electrochemical process, it is necessary to know all the steps involved for the electronic transfer and their speeds. This helps not only establish the overall reaction rate, but finds the relationships between the proposed mechanism and the measured current.

Generally, the first stage of an electrochemical process involves the transport of molecules from the bulk solution to the electrode's vicinity. This transport may be accomplished by any of the usual mass transport mechanisms, e.g., convection, migration or diffusion, or by any combination of these processes.

Typically, the second stage is a molecule process adsorption that seeks the most favorable position and orientation on the electrode surface to produce the transfer of electrons. Sometimes this process is preceded by a chemical reaction that favors product adsorption.

Once the molecule on the electrode surface is adsorbed, the electron transfer occurs, which is usually a quick step. Following the exchange of electrons, the next step often involves the separation of the molecule from the metal surface. Occasionally, desorption is followed by a chemical reaction. This can happen when the new species is of low chemical stability and forms new products. The final stage involves the removal of the product from the interface into the bulk solution.

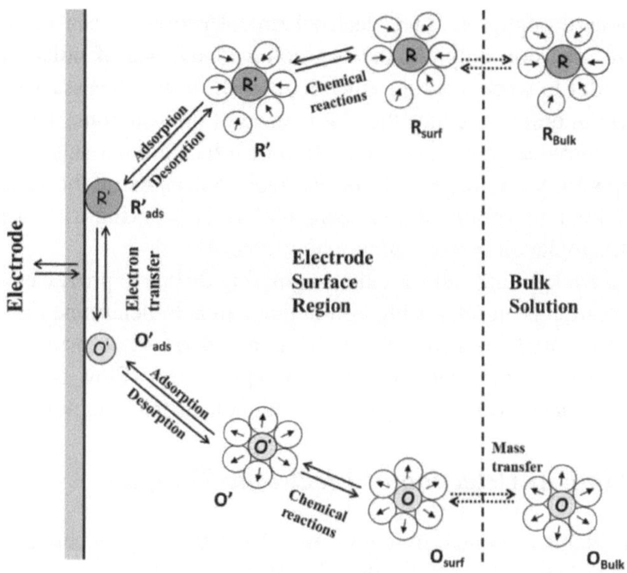

Fig. 16. A common Faradaic process mechanism

3.3.1 The Rate-Controlling Step

In an electrochemical process, the lowest stage can often be decisive in the net reaction rate. For instance, we may state that a process is controlled by diffusion when this step is the slowest one, so it imposes a limit to the current passing through the system. From Fig. 16, it is obvious that the slowest step can be either adsorption or may be associated with a chemical reaction. Hence one can speak of control by adsorption or by reaction kinetics. This classification is very useful because the complete mathematical modeling is complex and this allows us to make simplifications that are typified in accordance with the model that best fits the electrochemical mechanism governing the kinetics of the process.

3.3.2 Mass Transport

As previously mentioned, for the electron transfer to occur, species must move from the bulk solution towards the electrodes by different mechanisms: the first mechanism is the so-called ion migration, which is caused by the electric field gradients in the solution, which cause ions to undergo displacement, whose direction depends on the sign of both the ion and electric charge on the electrodes. The second mechanism is the so-called transport by diffusion, which takes place because of the concentration gradients created in the electrode's vicinity if the Faradaic transfer process commences. Finally, the third mechanism is known as convection, which is a mass transport process caused by either the thermal gradients or the mechanical agitation caused intentionally or accidentally in the system [50].

In the absence of mechanical stirring or vibration, and given the specific case where the current flowing through the system is very low, transport by convection may be considered negligible, and electrochemical processes can involve migration and control by diffusion only. At the beginning of the use of polarography, it was found that in order to work in a reproducible manner, it was necessary to introduce an inert electrolyte in order to reduce the ohmic drop of the solutions. The way to reduce the effect is to introduce a high concentration of a background salt. In this way, it is possible to separate the influence of non-Faradaic processes on the Faradaic electron transfer. The technique applies similar principles when working with organic solvents and uses, for example, quaternary ammonium salts [51].

The use of a background salt for either regulating diffusive processes or controlling the pH of the system is an interesting option that can help determine the concentration of a redox active product accurately. In fact, based on these principles, there is a number of published works which combines typical FIA (Flow Injection Analysis) techniques with detectors commonly used to build electronic tongues [52-54].

3.4 The Limit of Detection of Electronic Tongues

One of the problems presented by conventional voltammetric techniques is that their limit of detection is usually poor for the control of many parameter types. These limits of detection can be improved in a range of concentrations of about 10^{-5} M when the electronic tongue is constructed with a rotating electrode or by a hydraulic system which increases the flow of material on the electrode surface. In fact, it is well-known that in the current associated with a diffusive Faradaic process, the current increases with the square root of the electrode's angular velocity according to the Levich equation [55]. A simpler and cheaper alternative to the rotating electrode is based on the use of channel electrodes, where the sensors of the electronic tongue can be confined in a chamber through which the sample flows by a pump at a controlled rate [56].

However for Faradaic processes, the current increases as the flow velocity of the liquid or the angular rotation rate of the electrode increases, while variations in the intensity of the charge and discharge of the electric double layer remains practically constant for non-Faradaic processes. This difference especially contributes to distinguish one electrochemical process type from another. One consequence of this fact is that the combination of rotating disk electrodes, or channel electrode, with static electrodes can help distinguishing between non-Faradaic and Faradaic processes and can, therefore, assist in the detection and quantification of target dissolved substances.

3.4.1 Chemical Reaction Prior to the Electron Transfer

The step which normally follows the transport of matter to the electrode surface is the adsorption on the electrode. However on many occasions, prior to adsorption on the electrode chemical reactions, which favor the adsorption of the formed products, occurs. Such processes are called CE because a chemical (C) reaction occurs before the electron transfer (E). Reduction of formaldehyde in an aqueous solution can be considered an example of this CE process. In water, formaldehyde exists as a non-reducible hydrated form. Dehydration can occur, which leads to the generation of the planar form ($H_2C{=}O$), and this species adsorbs on the electrode and is reduced. Other examples of this case are the reduction of some weak acids or aldoses. In such process types, the values of the equilibrium constants of the preceding reaction can bring about changes in the redox potential or current. This mechanism is also quite common in the electrochemistry of metal complexes [57,58].

3.4.2 Adsorption

For many years, researchers have been interested in studying the electrochemical behavior of various organic substances capable of being adsorbed on the surface of noble metals, such as Rh, Ir, Pd, Pt and Au. It is known that these noble metals can suffer passivation forming oxide layers under certain pH conditions and when subjected to a suitable electrical potential [59]. Examples of such interest include the use of platinum electrodes, which have been widely used to study the adsorption, orientation and type of binding with the surface of a wide range of aromatic and quinonoid compounds. Furthermore as an interesting facet, the existence of electrocatalytic processes, when the current passes through these electrodes, has been verified [60,61]. A further result of this study shows that the adsorption rate is relatively high and, for example, that the adsorption equilibrium status of some products on a flat polycrystalline platinum electrode can be reached in less than 1 min. In a more recent study, the adsorption of alkenes, alkynes, aromatics, heterocyclic aromatics, alcohols, phenols, sulphides, amines, amides, nitriles, aldehydes, carboxylic compounds, CO, esters, among others, has been proved to cause significant changes in the oxidation potential of the metal and in current intensity [62].

The same paper proposed that chemisorption constants for a particular noble metal decrease as so: Aromatic mercaptans >Aromatic-N > Aromatic Ring > Alkynes > Alkenes > R_2S > Arylcyanides > N-Aliphatic > Alcohols > Alkylcyanides > Carboxylic acids > Esters > Alkanes. This study proposed that, for a given functional group, the increasing molecular weight of the adsorbate slowed down the metal corrosion rate. It also was observed that, for a particular functional group, adsorption strength was dependent on the metal used according to this order:

$$Pt > Pd \sim Rh \sim Ir > Au$$

Significant adsorption of ionic compounds on noble metal electrodes was also found. Indeed, some published papers have studied the adsorption of sulphates, carbonates, phosphates, halides, etc., and anions on different metal surfaces [63,64]. A typical species such as ammonia has also been studied using different metal electrodes [65]. These works have demonstrated that the oxidation potential of ammonium ion is dependent on the metal used. Moreover, it has been shown that the equilibrium adsorption constant follows this order:

$$Ru > Rh > Pd > Ir > Pt >> Au, Ag, Cu$$

It is interesting to note that this order is not the same as those established in studies conducted with organic compounds. Moreover, gold, silver and copper do not appear to significantly adsorb these species, and absorption seems to occur only when a metal oxide layer is formed. The product obtained by ammonia oxidation is nitrogen, which is obtained on electrodes with no oxide layer [66]. However, ammonium is transformed into different nitrogen oxides, nitrites and nitrates on the formed metal oxide layer.

3.4.3 Reactivity of Non-noble Metals

As the results of the studies conducted with some noble metals suggest, it is sometimes convenient to oxidise the working electrode. This proves most interesting when utilising non-noble electrodes because a non-noble electrode is easy to oxidise, and spontaneous reactions quite often occur between the cation generated by oxidation and some of the species present in solution, which can be either ionic or neutral. As a result of these chemical reactions, it is generally found that the corrosion potential of the metal electrode is substantially modified. In fact cyclic voltammetry performed with non-noble electrodes displays new peaks and waves, which occasionally leads to a progressive displacement of the redox potential when the metal is found to be in the presence of different concentrations and different chemical species.

Non-noble metal electrodes have been used occasionally for the detection of chemical species in solution. For example, the ability of copper and nickel electrodes to detect amino acids and peptides [67], or the capacity of copper to detect and quantify certain herbicides such as glyphosate [68], has been experimentally demonstrated. One explanation proposed to justify these electrodes' ability to detect specific compounds is that the species generated by the oxidation of the metal, Cu^+, Cu^{2+}, Ni^{3+}, their oxides or hydroxides, react with the dissolved species, thus modifying the metal corrosion rate. Glyphosate has been shown to form highly stable complexes with copper (II) [69], while the electrodes modified with nickel hydroxide have been successfully used in the amperometric detection of carbohydrates in solution [70].

Recent studies have revealed the feasibility and reproducibility of the copper electrode to determine the Electrochemical Oxygen Demand (EOD) in natural and wastewaters from different sources [71,72]. Some studies have shown that EOD correlates well with the Chemical Oxygen Demand (COD) and it has been suggested that electrochemical measurements can be a good, cheap alternative to test COD. Using this electrode, high reproducibility has been justified based on the following reaction sequences:

$$2\,Cu^0 + H_2O + 2\,OH^- \xrightarrow{-2e} Cu_2O + 2\,H_2O$$

$$Cu_2O + H_2O + 2\,OH^- \xrightarrow{-2e} 2\,CuO + 2\,H_2O$$

$$CuO_{(s)} + OH^-_{ads} \xrightarrow{-1e} CuOOH^{\cdot}_{(s)}$$

$$CuOOH^{\cdot}_{(s)} + Organic_{(red)} \rightarrow CuO_{(s)} + Organic_{(ox)} + H_2O$$

This particular process is electrocatalytic and can continue to transform the organic compound into carbon dioxide. The applications of copper in the reduction zone are also of much interest. For instance, it has been observed that copper catalyzes the electrochemical reduction of nitrate to nitrite, which can be converted into ammonia if the electrolysis process is allowed to proceed [73]. These studies well justify copper's capacity to identify and quantify these species in solution. The electrochemical reduction efficiency is so high that some attempts have been made to exploit the ability of copper electrodes by forming various bronze or brass alloys to remove nitrates and nitrites from water by electrolytic processes, be it with limited success as only a 36% conversion of nitrate into nitrogen gas was observed with some bronzes, and a 43% efficiency was reported for the same transformation while working with aluminum electrodes [74].

4 Wastewater Monitoring Using Electronic Tongues

As we have shown in previous sections, there are a number of studies showing applications of different metal electrodes types for monitoring complex processes in natural or wastewaters samples. According to the World Health Organization or to the particular legislation of each state, water quality control requires monitoring a minimum set of parameters and the performance of a comprehensive set of controls, which must be carried out periodically. It is clear that a VET is not able to replace the classical methodologies and instrumentations used for systematic controls. However it is a powerful tool because, for routine monitoring, its use can reduce the economic costs of implementing measures, and it can cut the time required to carry out necessary controls. In the following sections, we provide details of how an electronic tongue based on simple metal electrodes can be used to monitor certain parameters in wastewater samples, such as pH, conductivity, COD and BOD, phosphates, total nitrogen, nitrates, nitrites, sulphates and sulphides.

4.1 Electric Conductivity of the Solution

From the above considerations, it can be established that a set of pulses, like that shown in Fig. 10, can be used to obtain the conductivity of a solution quite accurately. Pulse length should be approximately one millisecond. Intensity must be measured in the first instants after applying the pulse, and the electrode must be a noble metal; e.g., a platinum electrode (see Fig. 17).

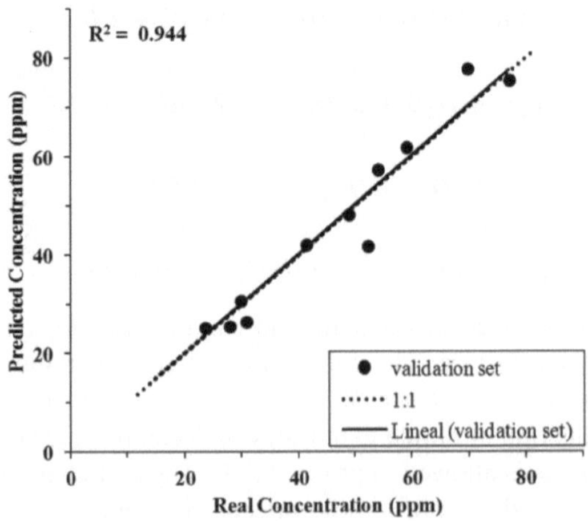

Fig. 17. Predicted conductivity versus real conductivity using a VET for several waste water samples

The set of pulses must be applied using two electrodes. In a previous step, the obtained conductivity should be normalized using standard KCl solutions. This step is necessary because the obtained values are dependent on: the surfaces of the used electrodes, the distance between them and their geometric distribution inside the electrochemical cell.

4.2 Chemical Oxygen Demand

As discussed previously, the electrochemical oxygen demand (EOD) determination has been reported to be calculated satisfactorily by a relatively simple coulombimetric process using copper electrodes in a basic medium. Moreover, it is well-known that several organic compounds can modify the corrosion potential of noble metals by a chemisorption process. These reactions result in electrocatalytic processes when a reasonably positive potential is applied. Therefore, the combined use of noble electrodes with other metal electrodes, such as copper or nickel, can provide information about the total content of the oxidizable compounds present in water. Given the fact that the intensity measured using the tongue is proportional to the organic content, the complete electrolysis of the products dissolved in water can be omitted. Fig. 18 shows the correlation obtained between the real COD and the predicted COD calculated form experimental data using an electronic tongue based on four noble electrodes and on four other non-noble electrodes.

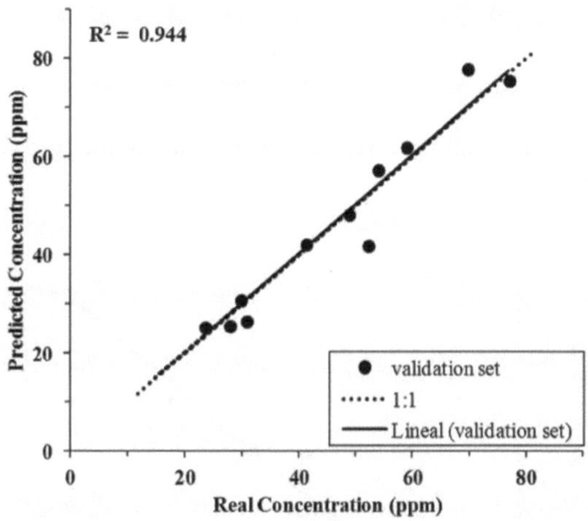

Fig. 18. Predicted COD versus real COD determined by a VET

In this case, samples were collected from urban wastewater over a 1-year sampling period. The real COD was obtained following the classical potassium dichromate method. Moreover, this study proves that the correlation between COD and current intensity values can be obtained in a short period of time without the need for comprehensive electrolysis.

4.3 Total Nitrogen, Nitrate, Nitrite and Ammoniacal Nitrogen

Sources of nitrogen in water can take various forms: ammonia, which may occur in the ionized (NH_4^+) or the non-ionized (NH_3) form, whose relative concentrations are pH-dependent; and in oxidized forms normally associated with nitrates and nitrites. Finally, organic nitrogen can be found in a dissolved form or associated with suspended particles. The electrochemical processes associated with nitrate and nitrite reduction and ammonium and nitrite oxidation can be detected by different distinct potentials values using noble electrodes [65,66,73,74]. Furthermore in the presence of ammonium ions, non-noble electrodes show peculiar oxidation-reduction processes based on electrocatalytic or complex formation phenomena.

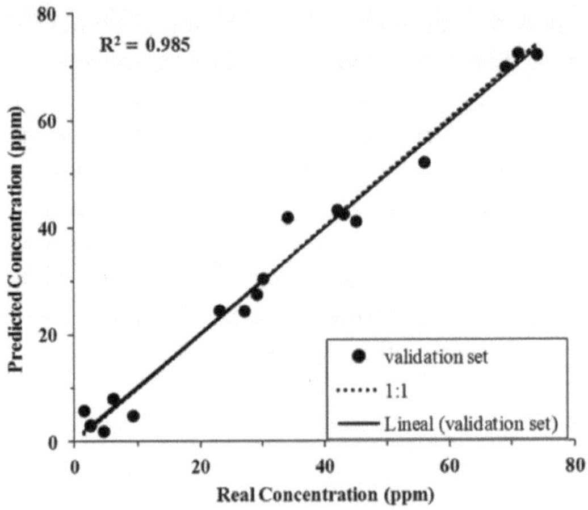

Fig. 19. Predicted ammoniacal nitrogen versus real ammoniacal nitrogen determined by a VET in wastewater samples

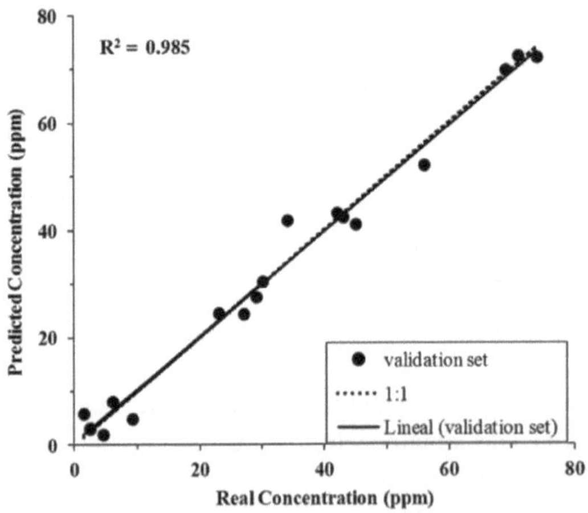

Fig. 20. Predicted total nitrogen vs. Real total nitrogen determined by a VET in domestic laundering samples

This rich electrochemical behavior facilitates the detection of different species and their independent quantification, as shown in Fig. 19, which displays a good correlation found between predicted ammoniacal nitrogen determined by a VET using Pt, Rh, Ir, Au, Cu, Co, Ag and Ni electrodes in wastewater samples and Real ammoniacal nitrogen. Moreover, Fig. 20 shows the good correlation obtained between the predicted and experimental values for total nitrogen in several wastewater samples from domestic launderings using Pt, Rh, Ir and Au electrodes in an electronic tongue format. This correlation seems to indicate that Total Nitrogen is a parameter that can be quickly determined using electronic tongue technology.

This result is not surprising considering the demonstration that the detection of amino acids and peptides is possible when copper electrodes are used [67,68]. The authors of this paper have shown that under neutral of slightly basic pH conditions, and in the presence of amino acids or peptides, an anodic dissolution process of the copper electrode surface occurs due to the highly coordinating ability of these compounds with Cu(II). Similar results were obtained with Pt and Au electrodes [75].

4.4 Phosphate

The importance of this anion lies in its ability to act as a nutrient and to favor, when circumstances are right, the appearance of eutrophication processes in aquatic environments. The detection and quantification of this species using VETs are possible because of not only physisorption phenomena, but also chemical reactions in order to form slight soluble species on the metallic electrode surface. Fig. 21 displays the correlation found between the experimental data obtained by classical techniques and the calculated predicted values using a voltammetric pulse electronic tongue and Pt, Rh, Ir, Au, Cu, Co, Ag and Ni electrodes.

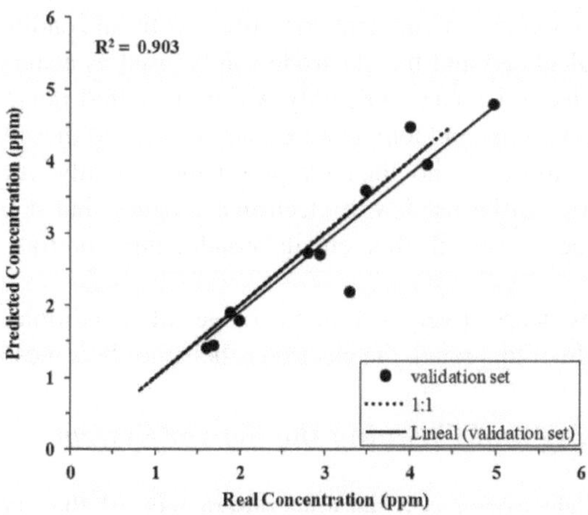

Fig. 21. Predicted total phosphate vs. Real total phosphate determined by a VET

4.5 Sulphates and Sulphides

Numerous organic and inorganic sulphur-containing compounds are adsorbed on metallic surfaces, which may be oxidized or not. Gold and platinum surfaces have been used to detect this type of compounds, including thioalcohols, thioeters, thiophenes, thiocarbamates, organic thiosulphates and several inorganic compounds.

5 Reference and Pseudo-reference Electrodes

As discussed above, most physicochemical parameters of interest for water quality control entail performing experiments using a three-electrode technique. This implies the introduction of a reference or pseudo-reference electrode. Reference electrodes are non-polarizable systems whose electric potential should be both practically constant over time and independent of the medium into which it is introduced. Reference electrodes should also have a low coefficient of variation of potential as regards temperature. Pseudo-reference electrodes are systems whose basic difference to the ideal reference electrode lies in that its potential depends on the medium in which it is immersed. They can be used when the general medium characteristics do not change appreciably. The most commonly used reference electrodes are of the metal/metal insoluble salt/soluble salt type of a certain anion. Examples of reference electrodes that are, or have been, widely used are Calomel electrode and Ag/AgCl/KCl electrode. The utility of other electrodes, such as the standard hydrogen electrode, is limited given its complexity and maintenance. Fundamentally, the main problems in employing reference electrodes are the pollution of the filling solution, and the contamination or clogging of the porous membrane of frit glass.

A pseudo-reference electrode is an electrode whose potential can change depending on the working environment conditions. If these conditions are known, the potential can be calculated and the electrode can be used as a suitable reference. The majority of such electrodes can work only within a limited range of conditions; for example, within certain pH and temperature ranges. Beyond these limits, the behavior of this type of electrodes is no longer reproducible. Finally, reference or pseudo-reference electrodes can be used with electronic tongues, but it is preferable to use references electrodes, even if this entails conducting controls and their regular maintenance.

Most electrodes work over a limited range of conditions, such as pH or temperature, outside of this range the electrodes behavior becomes unpredictable.

5.1 Maintaining and Cleaning the Sensor System

Contamination of electrodes is a natural consequence of the chemical or physical adsorption processes of the different components present in wastewater. For example, anionic or cationic surfactants tend to be fixed onto the electrode surface. On this formed layer, other organic compounds, like fats, may also fix. Furthermore after performing the voltammetric scans sequence, oxides can be generated, which can react with the anions or cations present in the water, or may produce catalytic reactions of interest [76]. It should also be noted that the reactivity of non-noble metals is much greater than that of noble metals, and that the oxide surface naturally grows as a result of oxygen attack in non-noble metals, although no potentials are applied. Subsequent substitution reactions of metal oxide or hydroxide by carbonate, phosphate, chloride salts, etc., are also feasible. Moreover, the ability of halide anions to contaminate gold surfaces or the ability of sulphur-containing derivatives to contaminate noble electrodes is also known [77].

Electrodes can be cleaned by different procedures; i.e., mechanically, by laser, ultrasound or microwaves; using chemical or electrochemical treatments. For example, electrochemical cleaning has been widely used in liquid chromatography to quickly regenerate electrode surfaces. This has also proved to be a reproducible method for the regeneration of metallic electrode surfaces when following certain electrochemical methods.

The literature shows two different ways to polish metal surfaces in electrodes. The first implies cleaning the sensing metallic surfaces in the electronic tongue by using electro-polishing procedures [27]; the second involves the use of mechanical polishing. A detailed procedure of the latter was proposed by Olson et al [26]. The authors constructed an electronic tongue formed by four noble metals wires (i.e., Rh, Ir, Pt and Au) housed in a cylindrical stainless steel container, which was coupled with a T-shaped motor which supported sandpaper. The engine started automatically after each voltammetric sweep, and performed a periodical, abrasive mechanical cleaning of the electronic tongue. Olson et al also proposed that this mechanism also cleans non-noble electrodes, such as metallic copper.

5.2 Voltammetric Waveform Design

In a series of recent works, we have demonstrated that the waveform design applied to a complex system can improve the amount and quality of the information that can be extracted from an electronic tongue [78]. There are a number of proposed waveforms that have been applied in VETs [45]. Yet despite the differences that each waveform presents, they can be classified into two main groups: Large Amplitude Pulse Voltammetry (LAPV) and Small Amplitude Pulse Voltammetry (SAPV). In addition, the second critical parameter to extract information from the current measured by an electronic tongue is the time of the applied pulse duration. We have discussed above that if the pulse duration is short, the measured current is mainly related with non-Faradaic processes, and is also mainly related with the conductivity of the solution. However for longer pulses, Faradaic processes of redox active species present in the solution are detected.

Fig. 22 displays a typical design of an LAPV waveform. It starts from an initial potential of 0 V, and after a cycle of potential pulses, it finally returns to the starting point. From the initial value, the potential jumps to +1600 mV, and then to -1600 mV. Then these jumps are repeated by increasing ±320 mV from the starting potential. This and other similar waveforms have been previously reported to be applied to the study of compounds and have shown reversible and irreversible electrochemical processes. LAPV waveforms have also been applied by using different sweep frequencies (MLAPV); for instance, to study different food samples [79]. The use of multisweep frequencies allows the selection of Faradaic or non-Faradaic sectors, and the coexistence zones of both.

Fig. 23 shows a typical SAPV waveform. In this case, the designed waveform was produced by starting from an initial potential of 0 V and, by increasing every 100 milliseconds, +270 mV to the initial applied potential to reach a final maximum value of +1620 mV vs. the reference electrode. Decreasing steps of -270 mV were applied to reach the potential of -1620 mV. Finally, steps of +270 mV were applied until the initial potential value was attained.

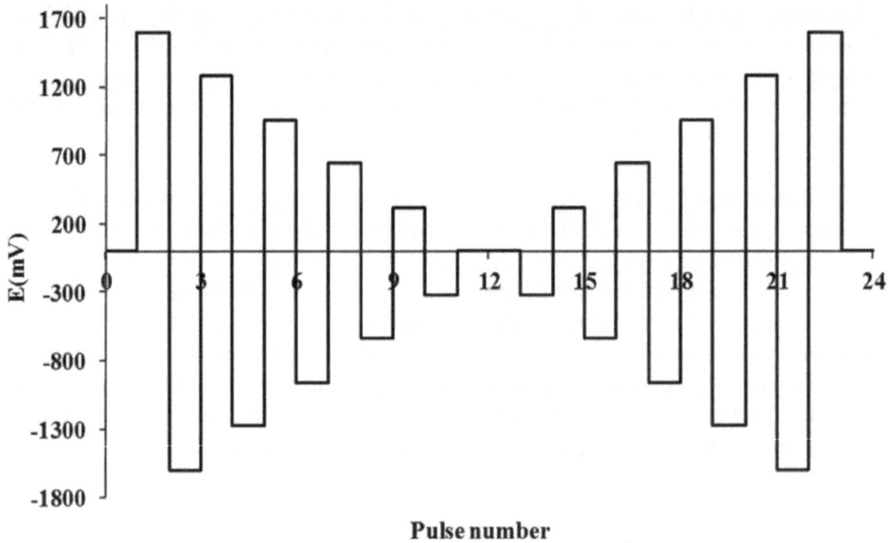

Fig. 22. Scheme of applied electrical pulses in accordance with an LAPV model

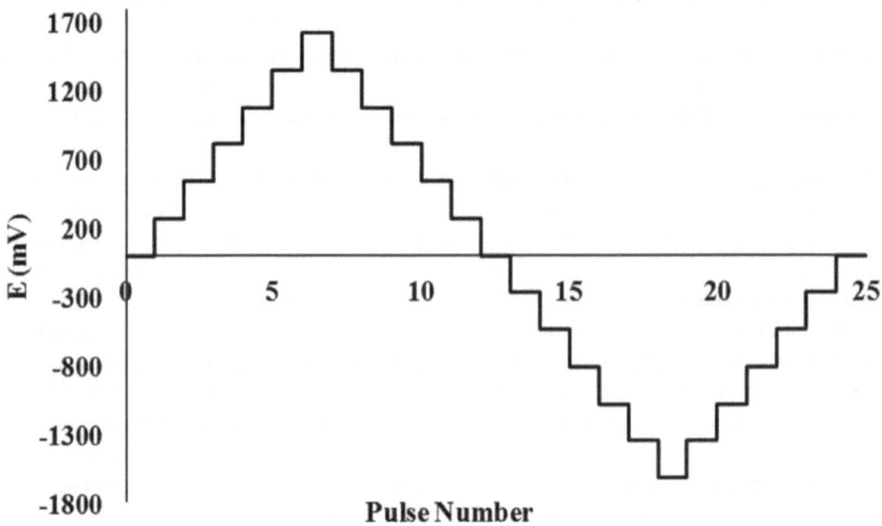

Fig. 23. Scheme of applied electrical pulses in accordance with an SAPV model

In order to design the denominated smart waveforms, it is necessary to obtain electrochemical information of the compounds integrating the problem to be studied. This information can be easily acquired by adopting classical techniques, such as cyclic voltammetry, from which it is possible to obtain the potential values at which a certain compound oxidizes or reduces. Usually the different peaks of the certain redox active compounds registered by cyclic voltammetry may change depending on the metallic electrode used. Therefore, a study of the electrochemical response of the different compounds to be detected in water should be performed. Then the smart waveform can be constructed after bearing in mind the potential where the redox processes occur.

6 Conclusion

VETs are a powerful tool to control and predict levels of important parameters in aqueous samples. In this chapter, we have attempted to outline the basis for the design and performance of VETs bearing metallic electrodes for their general application in water samples and for monitoring wastewater parameters, such as those usually controlled in wastewater treatment plants. Traditionally, treated wastewater quality is defined by the measurement of certain parameters in influent and effluent wastewaters. However, most of these parameters are determined by traditional laboratory analytical techniques, which are off-line, and some, such as BOD, require between 5 and 20 days to obtain the measurement, thus rendering these tests inadequate for automated monitoring and control purposes.

Recently, rapid advances have been made in the field of innovative equipment for online monitoring. However, not all parameters can be measured online and, in general; current analyzers require considerable economic resources, whose use and maintenance prove tedious and time-consuming. As an alternative, the use of electronic tongues is emerging as a rapid, low-cost, simple and suitable alternative. As this chapter explains, the electronic tongues strategy relies on the use of semi-specific sensors that produce a signal pattern when subjected to a sample, which can be related to either a specific compound or a quality aspect, achieved by subjecting the signal pattern to multivariate data analyses. In fact, the utilization of simple electronic tongues to monitor and control the biological processes done in wastewater can help increase their pollutant removal capacity and can achieve a consistent, stable operation which meets effluent requirements at a minimum cost.

References

[1] Alrababah, M.A., Bani-Hani, M.G., Alhamad, M.N., Bataineh, M.M.: Boosting seedling survival and growth under semi-arid Mediterranean conditions: Selecting appropriate species under rainfed and wastewater irrigation. Journal of Arid Environments 72, 1606–1612 (2008)

[2] Al-Lahham, O., El Assib, N.M., Fayyad, M.: Impact of treated wastewater irrigation on quality attributes and contamination of tomato fruit. Agricultural Water Management 61, 51–62 (2003)

[3] Aucejo, A., Ferrer, J., Gabaldon, C., Marzal, P., Seco, A.: Diagnosis of boron, fluorine, lead, nickel and zinc in citrus plantations in Villarreal, Spain. Water, Air, and Soil Pollution 94, 349–360 (1997)

[4] Boursier, H., Beline, F., Paul, E.: Piggery Wastewater characterization for biological nitrogen removal process design. Bioresource Technology 96, 351–358 (2005)

[5] Orhon, D., Okutmand, D., Insel, G.: Characterization and biodegradation of settleable organic matter for domestic wastewater. Water S.A. 28(3), 299–306 (2002)
[6] Orhon, D., Okutmand, D.: Respirometric assessment of residual organic matter for domestic sewage. Enzime and Microbial Technology 32(5), 560–566 (2003)
[7] Roeleveld, P.J., van Loosdrecht, M.C.M.: Experiences with guidelines for wastewater characterization in the Netherlands. Water Science & Technology 45(6), 77–87 (2002)
[8] Zhuiykov, S.: Solid-state sensors monitoring parameters of water quality for the next generation of wireless sensor networks. Sensors and Actuators B: Chemical 161, 1–20 (2012)
[9] Bartlett, J., Gardner, P.: A history of electronic noses. Sensors and Actuators B 18, 211–220 (1994)
[10] Dewettinck, T., Van Hege, K., Verstraete, W.: The electronic nose as a rapid sensor for volatile compounds in treated domestic wastewater. Water Research 35(10), 2475–2483 (2001)
[11] Ozturk, Z.Z., Tasaltin, C., Engin, G.O., Gurek, A.G., Atilla, D., Ahsen, V., Ince, M.: Evaluation of a fast wastewater odourcharacterisation procedure using a chemical sensor array. Environmental Monitoring and Assessment 151(1-4), 369–375 (2009)
[12] Legin, A., Rudnitskaya, A., Vlasov, Y.: Electronics tongues: sensors Systems and applications. Sensors Update 10, 143–188 (2002)
[13] Vlasov, Y., Legin, A., Rudnitskaya, A., Di Natale, C., D'Amico, A.: Non-specific sensor arrays "Electronic Tongue" for chemical analysis of liquids. Pure and Applied Chemistry 77(11), 1965–1983 (2005)
[14] Rodriguez-Mendez, M.L., Arrieta, A., Parra, V., Vegas, A., Villanueva, S., Gutierrez-Osuna, R., de Saja, J.A.: Fusion of three sensory modalities for the multimodal characterization of red wines. IEEE Sensors Journal 4, 348–354 (2004)
[15] Hayashi, K., Yamanaka, M., Toko, K., Yamafyji, K.: Multichannel taste sensor using lipid membranes. Sensors and Actuators B: Chemical 2, 205–213 (1990)
[16] Martínez-Máñez, R., Soto, J., Garcia-Breijo, E., Gil, L., Ibáñez, J., Llobet, E.: An electronic tongue design for the qualitative analysis of natural waters. Sensors and Actuators B 104, 302–307 (2005)
[17] Garcia-Breijo, E., Atkinson, J., Gil-Sanchez, L., Masot, R., Ibañez, J., Garrigues, J., Glanc, M., Laguarda-Miro, N., Olguin, C.: A comparison study of pattern recognition algorithms implemented on a microcontroller for use in an electronic tongue for monitoring drinking waters. Sensors and Actuators A: Physical 172, 570–582 (2011)
[18] Di Natale, C., Macagnano, A., Davide, F., D'Amico, A., Legin, A., Vlasov, Y., Rudnitskaya, A., Selezenev, B.: Multicomponent analysis on polluted waters by means of an electronic tongue. Sensors and Actuators B 44, 423–428 (1997)
[19] Moreno, L., Merlos, A., Abramova, N., Jimenez, C., Bratov, A.: Multi-sensor array used as an electronic tongue for mineral water analysis. Sensors and Actuators B 116, 130–134 (2006)
[20] Bratov, A., Abramova, N., Ypatov, A.: Recent trends in potentiometric sensor arrays. A review. Analytica Chimica Acta 678, 149–159 (2010)
[21] Mimendia, A., Gutiérrez, J.M., Leija, L., Hernández, P.R., Favari, L., Muñoz, R., del Valle, M.: A review of the use of the PET in the monitoring of environmental systems. Environmental Modelling & Software 25, 1023–1030 (2010)
[22] Bard, A.J., Faulkner, L.R.: Electrochemical Methods: Fundamentals and Applications. WileySec (2000)
[23] Winquist, F., Wide, P., Lundström, I.: An electronic tongue based on voltammetry. Analytical Chimica Acta 357, 21–23 (1997)

[24] Krantz-Rülcker, C., Stenberg, M., Winquist, F., Lundström, I.: Electronic tongue for environmental monitoring based on sensor arrays and pattern recognition: a review. Analytica Chimica Acta 426, 217–226 (2001)
[25] Olsson, J., Ivarsson, P., Winquist, F.: Determination of Detergents in Washing Machine Wastewater with a VET. Talanta 76, 91–95 (2008)
[26] Olsson, J., Winquist, F., Lundtröm, I.: A self-polishing electronic tongue. Sensors and Actuators B: Chemical 118, 461–465 (2008)
[27] Holmin, S., Krantz-Rülcker, C., Winquist, F.: Multivariate optimization of electrochemically pretreated electrodes used in a VET. Analytical Chimica Acta 519, 39–46 (2004)
[28] Gutés, A., Cespedes, F., del Valle, M., Louthander, D., Krantz-Rulcker, C., Winquist, F.: A flow injection VET applied to paper mill industrial waters. Sensors and Actuators B 115, 390–395 (2006)
[29] Campos, I., Masot, R., Alcañiz, M., Gil, L., Soto, J., Vivancos, J.L., Garcia, E., Labrador, R.H., Barat, J., Martinez-Mañez, R.: Accurate concentration determination of anions nitrate, nitrite and chloride in minced meat using a VET. Sensors and Actuators B: Chemical 149, 71–78 (2010)
[30] Campos, I., Alcañiz, M., Aguado, D., Barat, R., Ferrer, J., Gil, L., Marrakchi, M., Martínez-Mañez, R., Soto, J., Vivancos, J.: A VET as tool for water quality monitoring in wastewater treatment plants. Water Research 46, 2605–2614 (2012)
[31] Palit, M., Tudub, B., Bhattacharyya, N., Dutta, A., Dutta, P.K., Jana, A., Bandyopadhyay, R., Chatterjee, A.: Comparison of multivariate preprocessing techniques as applied to electronic tongue based pattern classification for black tea. Analytica Chimica Acta 675, 8–15 (2010)
[32] Palit, M., Tudu, B., Dutta, P.K., Dutta, A., Jana, A., Roy, J.K., Bhattacharyya, N., Bandyopadhyay, R., Chatterjee, A.: Classification of Black Tea Taste and Correlation with Tea Taster's Mark Using VET. IEEE Transactions on Instrumentation and Measurement 59, 2230–2239 (2010)
[33] Pomerantsev, A.L., Rodionova, O.Y.: Chemometrics and Intelligent Laboratory Systems. Chemometrics and Intelligent Laboratory Systems 103, 19–24 (2010)
[34] Mardia, K., Kent, J., Bibby, J.: Multivariate Analysis. Academic Press, s.n., London (1989)
[35] Jackson, A.J.: User's Guide to Principal Components. Wiley, New York (1991)
[36] Miller, P., Swanson, R., Heckler, F.: Contribution plots: the missing link in multivariate quality control. In: 37th Annual Fall Conference ASQC. s.n., Rochester (1993)
[37] Westerhuis, J., Gurden, S., Smilde, A.: Generalized contribution plots in multivariate statistical process monitoring. Chemometrics and Intelligent Laboratory Systems 51, 95–114 (2000)
[38] Bishop, C.M.: Neural Networks for Pattern Recognition. Oxford University Press Inc., New York (1995)
[39] Wold, S.: Personal memories of the early PLS development. Chemometrics and Intelligent Laboratory Systems 58, 83–84 (2001)
[40] Martens, H., Naes, T.: Multivariate calibration. Wiley, London (1989)
[41] Gramahe, D.C.: The Electrical Double Layer and the theory of electrocapillarity. Chemical Reviews 41(3), 441–501 (1947)
[42] Devanathan, M.A.V., Tilak, B.V.S.R.A.: The Structure of the Electrical Double Layer at the metal-solution interface. Chemical Rewievs 65(6), 635–684 (1965)
[43] Cottrell, F.G.: Der ReststrombeigalvanischerPolarisation, betrachtetalsein diffusions problem. Z. Phys. Chem. 42, 385 (1903)

[44] Myland, J.C., Oldham, K.B.: Cottrell's equation revisited: an intuitive, but unreliable novel approach to the tracking of electrochemical diffusion. Electrochemistry Communications 6, 344–350 (2004)
[45] Campos, I., Alcañiz, M., Masot, R.L., Soto, J., Martínez-Máñez, R., Vivancos, J.L., Gil, L.: A method of pulse array design for VETs. Sensors and Actuators B: Chemical 161, 556–563 (2012)
[46] Barsoukov, E., Hoboken, J.R.M.: Impedance Spectroscopy. Theory, Experiment and Applications. Wiley, New Jersey (2005)
[47] Lasia, A.: Electrochemical Impedance Spectroscopy and Its Applications, Modern Aspects of Electrochemistry, vol. 32, pp. 143–248. Kluwer Academic/Plenum Publishers, New York (1999)
[48] Bard, A.J., Faulkner, L.R.: Electrochemical Methods: Fundamentals and Applications. Wiley, s.l. (1980)
[49] Rieger, P.H.: Electrochemistry. Prentice-Hall International Inc., s.l. (1986)
[50] Reddy, J., Bockris, O.M., Amulya, K.N.: Modern Electrochemistry. Plenum Press, New York (1998)
[51] Nicholson, R.S.: Polarographic Theory, Instrumentation, and Methodology. Analytical Chemistry 42(5), 130R–138R (1970)
[52] Moreno Codinachs, L., Baldi, A., Merlos, A., Abramova, N., Ipatov, A., Jiménez-Jorquera, C., Bratov, A.: Integrated Multisensor for FIA-Based Electronic Tongue Applications. IEEE Sensors Journal 8, 608–615 (2008)
[53] Buratti, S., Benedetti, S., Scampicchio, M., Pangerodb, E.C.: Characterization and classification of Italian Barbera wines by using an electronic nose and an amperometric electronic tongue. Analytica Chimica Acta 525, 133–139 (2004)
[54] Winquist, F., Rydberg, E., Holmin, S., Krantz-Rülcker, C., Lundström, I.: Flow injection analysis applied to a VET. Analytica Chimica Acta 471, 159–172 (2002)
[55] Levich, V.G., Cliffs, E.: Physicochemical Hydrodynamics. Prentice-Hall, Inc., N.J. (1962)
[56] Schersona, D.A., Tolmachev, Y.V., Wang, Z., Wang, J., Palencsar, A.: Extensions of the Koutecky–Levich equation to channel electrodes. Electrochem. Solid-State Lett. 11(2), F1–F4 (2008)
[57] Survila, A., Kanapeckaite, S., Pileckiene, J., Budiene, J.: Kinetics of Hydrogen Evolution on Copper Electrode Involving Organic Acids as Proton Donors. International Journal of Electrochemistry 2011, 1–9 (2011)
[58] Benito, A., Lloris, J.M., Martínez-Máñez, R., Soto, J.: Switching and tuning processes in the interaction of proton with ferrocenyl amines. Polyhedron 17(4), 491–495 (1998)
[59] Pourbaix, M.J.N., Van Muylder, J., de Zoubov, N.: Electrochemical Properties of the Platinum Metals. Platinum Metals Rev. 3, 100–106 (1959)
[60] Soriaga, M.P., Binamira-Soriaga, E., Hubbard, A.T., Benzinguer, J.B., Peter Pang, K.W.: Surface Coordination Chemistry of Platinum Studied by Thin-Layer Electrodes. Adsorption, Orientation, and Mode of Binding of Aromatic and Quinonoid Compounds. Inorganic Chemistry 24, 65–73 (1985)
[61] Soriaga, M.P., White, J.H., Song, D., Chia, V.K.F., Arrhenius, P.O., Hubbard, A.T.: Surface Coordination Chemistry of Platinum Studied by Thin-Layer Electrodes. Surface Chemical Reactivity of Aromatic and Quinonoid Compounds Adsorbed in Specific Orientational States. Inorganic Chemistry 24, 73–79 (1985)
[62] Michelhaugh, S.L., Bhardwaj, C., Cali, G.J., Bravo, B.G., Bothwell, M.E., Berry, G.M., Soriaga, M.P.: The Influence of Chemisorbed Organic Monolayers on Electrode Surface Oxidation. Corrosion 47(5), 322–328 (1991)

[63] Arihara, K., Kitamura, F., Ohsaka, T., Tokuda, K.: Characterization of the adsorption state of carbonate ions at the Au(III) electrode surface using in situ IRAS. Journal of Electroanalytical Chemistry 510, 128–135 (2001)
[64] Moraes, I.R., Nart, F.C.: Vibrational study of adsorbed phosphate ions on rhodium single crystal electrodes. Journal of Electroanalytical Chemistry 563, 41–47 (2004)
[65] de Vooys, A.C.A., Koper, M.T.M., van Santen, R.A., van Veen, J.A.R.: The role of adsorbates in the electrochemical oxidation of ammonia on noble and transition metal electrodes. Journal of Electroanalytical Chemistry 506, 127–137 (2001)
[66] Gerischer, H., Mauerer, A.: Untersuchungen Zuranodischen Oxidation von Ammoniakan Platin-Elektroden. Journal of Electroanalytical Chemistry and Interfacial Electrochemistry 25(3), 421–433 (1970)
[67] Luo, P., Zhang, F., Baldwiin, R.P.: Constant potential amperometric detection of underivatizad amino acids and peptides at a copper electrode. Analytical Chemistry 63(17), 1702–1707 (1991)
[68] Coutinho, C.F.B., Silva, M.O., Calegaro, M.L., Machado, S.A.S., Mazo, L.H.: Investigation of Copper disolution in the presence of glyphosate using hydrodinamicvotammetry and chronoamperometry. Solid State Ionics 178, 161–164 (2007)
[69] Daniele, P.G., De Stefano, C., Prenesti, E., Sammartano, S.: Copper (II) complexes of N-(phosphonomethyl)glycine in aqueous solution: a thermodynamic and spectrophotometric study. Talanta 45, 425–431 (1997)
[70] Vidotti, M., Cerri, C.D., Carvalhal, R.F., Dias, J.C., Mendes, R.K., Córdoba de Torres, S.I., Kubota, L.T.: Nickel hydroxide electrodes as amperometric detectors for carbohydrates in flow injection analysis and liquid chromatography. Journal of Electroanalytical Chemistry 636, 18–23 (2009)
[71] Silva, C.R., Conceição, C.D.C., Bonifácio, V.G., Filho, O.F., Teixeira, M.F.S.: Determination of the chemical oxygen demand (COD) using a copper electrode: a clean alternative method. Journal Solid State Electrochemistry 13, 665–669 (2009)
[72] Lee, K.-H., Ishikawa, T., McNiven, S.J., Nomura, Y., Hiratsuka, A., Sasaki, S., Arikawa, Y., Karube, I.: Evaluation of chemical oxygen demand (COD) based on coulometric determination of electrochemical oxygen demand (EOD) using a surface oxidized copper electrode. Analytica Chimica Acta 398, 161–171 (1999)
[73] Bouzek, K., Paidar, M., Sadilkova, A., Bergmann, H.: Electrochemical reduction of Nitrate in weakly alkaline solutions. Journal of Applied Electrochemistry 31, 1185–1193 (2001)
[74] Kyriacou, C., Polatides, G.: Electrochemical reduction of nitrate ion on various cathodes – reaction kinetics on bronza cathode. Journal of Applied Electrochemistry 35, 421–427 (2005)
[75] Johnson, D.C., La Course, W.R.: Liquid Chromatography with Pulsed Electrochemical Detection at Gold and Platinum Electrodes. Analytical Chemistry 62(10), 591 A–597 A (1990)
[76] Adams, R.N.: Electrochemistry at Solid Electrodes. Marcel Dekker, New York (1969)
[77] Lamy-Pitara, E., Barbier, J.: The electrocatalytic reactions of oxidation and evolution of hydrogen on iridium electrodes modified by sulphur adsorption. Electroanal. Chem. 416, 47–51 (1996)
[78] Tian, S.Y., Deng, S.P., Chen, Z.X.: Multifrequency large amplitude pulse voltammetry: A novel electrochemical method for electronic tongue. Sensors and Actuators B: Chemical 123, 1049–1056 (2007)
[79] Ivarsson, P., Holmin, S., Höjer, N.E., Krantz-Rülcker, C., Winquist, F.: Discrimination of tea by means of a VET and different applied waveforms. Sensors and Actuators B: Chemical 76, 449–454 (2001)

Automatic Water and Wastewater Quality Monitoring Systems

V. Cerdà[1,*], J. Avivar[1], L. Ferrer[1], and L.O. Leal[2]

[1] Department of Chemistry, University of the Balearic Islands,
Cra. Valldemossa Km 7.5, 07122, Palma de Mallorca, Spain
[2] Renewable Energy and Environmental Protection Department,
Advanced Materials Research Center (CIMAV) S.C.,
Miguel de Cervantes 120, Chihuahua, Chih. 31109, Mexico
victor.cerda@uib.es

Abstract. Water and wastewater quality monitoring programs have to be established in order to fulfill current legal requirements as well as environmental and social concerns. This entails a large number of samples to be analyzed in a short period of time. Flow analysis techniques allow the automation of the analytical methodologies providing great advantages, i.e. improved reproducibility, low reagents consumption and waste generation, as well as high analysis frequency. Sequential Injection Analysis (SIA) is an excellent tool for developing automatic analyzers with multiparametric capabilities. In this chapter, four multiparametric SIA systems which are able to determine key parameters in water and wastewater such as BOD, COD, TOC, TSS, global N, total P, nitrate, nitrite, ammonium, orthophosphate, anionic detergents, sulfate, pH and conductivity are presented and discussed in detail.

Keywords: Sequential Injection Analysis, automation, water and wastewater analysis, multiparametric monitoring.

1 Introduction

Water pollution is an environmental issue, since it affects plants and organisms living in it, damaging not only individual species and populations, but also natural biological communities. Concerns for environmental conservation and improvement of the environmental health have increased during the last decades, and this has led to an increase in environmental monitoring programs. Drinking water is subject to strict quality controls in order to guarantee human health. Thus, the European Council Directive 98/83/EC [1], the World Health Organization (WHO) [2] and the United States Environmental Protection Agency (USEPA) [3] stipulate reference limit values for the most relevant physicochemical parameters in water for human consumption. Regarding wastewater quality monitoring, the European Council Directive 91/271/EEC [4] stipulates that all EU Members States have to appropriately equip

* Corresponding author.

S.C. Mukhopadhyay & A. Mason (Eds.): *Real-Time Water Quality Monitoring*, SSMI 4, pp. 105–123.
DOI: 10.1007/978-3-642-37006-9_5 © Springer-Verlag Berlin Heidelberg 2013

wastewater treatment plants in all cities with a certain number of equivalent-inhabitants, in order to monitor the required parameters. Usually wastewater quality is characterized both by global pollution parameters, such as biological oxygen demand (BOD), chemical oxygen demand (COD), total organic carbon (TOC) or total suspended solids (TSS), and by nitrogen and phosphorous compounds. The characterization of wastewater at the inlet and outlet of the treatment plants is an effective way to control the process efficiency and to guarantee the final quality of treated water. Also in some industrial processes, the quality control of the water involved is of great relevance to guarantee the good performance and the safety of the equipment. All these entailed an increasing demand for control analyses and the large number of samples with which analysts can be confronted imposes various challenges to analytical chemists such as the need to develop new methods exhibiting as much selectivity, sensitivity, sample and reagent economy, throughput, cost-effectiveness, simplicity and environmental friendliness as possible.

Since the beginnings of automation of analytical methods, various different flow techniques have been developed and used for analytical and monitoring applications. They have gained importance for clinical, industrial and environmental purposes as they allow highly reproducible fast determinations. Automation and miniaturization of fluidic-based analysis are essential to make them fast and efficient for routine and research tasks [5]. The development of automatic methods of analysis arose in the 1950s, when clinical tests started to be increasingly used for diagnostic purposes in medicine that led to a spectacular increase in the demand for laboratory tests which, for obvious economic reasons, could not be met simply by hiring more laboratory staff. The solution to this problem was provided by segmented flow analysis (SFA) [6], which afforded not only substantially increased throughput, but also substantial savings in samples and reagents. SFA laid the foundations for modern flow techniques. Although originally flow techniques were performed for off-site measurements, their outstanding feature is the capability for both in-situ and real-time monitoring [7,8] of chemical parameters in waters, with some such flow analyzers being included in the United States Environmental Protection Agency (USEPA) directions [9,10].

Earliest automatic methods used devices suited to particular applications. This restricted their scope to very specific uses such as the control of manufacturing processes or to those cases where the number of samples to be analyzed was large enough to justify the initial effort and investment required. Ever since, SFA was developed by Skeggs in 1957 [6], flow techniques have been in continuous evolution towards new developments such as those of Flow Injection Analysis (FIA) [11] by J. Ruzicka and E.H. Hansen in 1975; Sequential Injection Analysis (SIA) [12], by J. Ruzicka and G.D. Marshall in 1990; Multicommutated Flow Analysis (MCFIA) [13], by B.F. Reis et al. in 1994; Multisyringe Flow Injection Analysis (MSFIA) [14], by V. Cerdà et al. in 1999, Multi-Pumping Flow Systems (MPFS) [15], by R. Lapa et al. in 2002 and Lab-On-Valve (LOV) [16] by J. Ruzicka in 2000, and have aroused special interest in relation to many other automatic methodologies of analysis. Moreover, substantial advantages can be derived from their combination.

FIA is undoubtedly the most widely accepted flow technique. Its widespread success can be ascribed to its ease of implementation and, especially, to its computer-less working capability. By contrast, all subsequent flow techniques require computer control. This initially hindered further development of the latter owing to unavailability of suitable commercial software and general lack of experience in coupling personal computers to instruments. However most of the advantages of current flow techniques are in part consequence of the incorporation of computers; in fact, computers flexibility allows the implementation of the same analytical method (hardware), with little or no alteration, on different types of samples simply by software modification.

Nevertheless, no other flow technique can match SIA in multiparametric determination capabilities. Unlike FIA, SIA can be turned into a true multiparametric analysis system simply by using a switching valve with an appropriate number of channels to hold the different analytical reagents, delivery of which can be precisely programmed via the associated computer. In this respect, SIA is much closer than FIA to the original SFA systems that afforded the determination of up to twenty parameters per sample -except that SIA operates in a much simpler and, especially, more economical manner.

All in all, flow techniques have gone through two generations. On the one side, flow systems were initially operated exclusively by hand (e.g. in SFA and FIA). On the other side, computers facilitated the development of the so-called second generation of flow techniques such as SIA, MCFIA, MSFIA or LOV, based on multicommutation operation. Multicommutated flow techniques have shown great potential in comparison with previous flow techniques in minimizing reagents consumption and waste production, providing more environmentally friendly methodologies, since liquids are only propelled to the system when required and returned to their reservoirs when not. All of them have common components such as impulsion pumps, which act as liquid drivers and a series of plastic tubes or manifold intended to carry liquid streams to the detector.

Wastewater sample pretreatment plays a major role in their analysis. Suspended matter in wastewater samples may include different kind of solids: floating, sedimentable and colloidal solids. Suspended solids effects are related to their size and origin, but from a physical point of view, their interference with the light penetration (turbidity) is the most notable effect. So, when a spectrophotometric detection is performed, it is mandatory to carry out a sample pretreatment in order to eliminate the suspended matter. Most widely used procedures are centrifugation, filtration and flocculation [17]. Usually, the choice depends on the type of analyte to determine, e.g., if the analyte is easily oxidized, flocculation is preferred. However, if the pretreatment is intended to be carried out on line, filtration is the best option. Also, there are devices which allow the liquid to flow through them while retaining the suspended matter. These devices have to be placed on the input line of the sample. For example, a sampling system (particle size exclusion sampling – PSES) was developed by Wurl et al. in 2000 to prevent an obstruction of the manifold by big sediment particles ($d > 0.8$ mm) when analyzing river water samples [18]. This PSES device consists of sixteen induction tubes, with an internal diameter (i.d) of 0.3 mm smaller than the i.d. of the manifold, in order to serve as a filter. If an induction coil is obstructed the further induction tubes are available to sample the water. There are other devices which consist of a piece of PVC with hundreds of perforations leading to a central duct connected to the system. These devices require a frequent cleaning, in order to avoid the blocking of the conduits of entrance and to ensure a constant sample flow rate. In the case of volatile analytes, it is possible to use a diffusion gas cell. In such methodologies, the donor stream could be raw wastewater since the target analyte passes through the membrane to the acceptor stream without requiring any kind of pretreatment.

Several methods based on SIA have been developed for the determination of key environmental parameters with different detection systems [19]. In this chapter, several automatic monitoring systems based on SIA for water and wastewater quality determination are presented and discussed in detail.

2 Sequential Injection Analysis (SIA)

Sequential injection analysis was developed by J. Ruzicka and G. Marshall as an alternative to FIA. SIA has over time proved that its scope departs markedly from that of the former technique.

Fig. 1 shows a schematic depiction of a typical SIA system. A two-way piston pump is connected to the central port of a multiposition valve via a holding coil (HC) which is made to address the peripheral ports of the unit, for sequential aspiration of the sample and reagents, via the central communication channel (CC) in the selection valve.

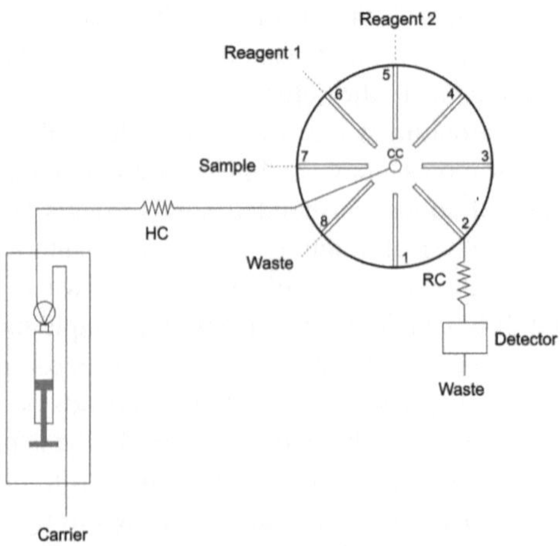

Fig. 1. Schematic depiction of a typical SIA system. HC: Holding Coil; RC: Reaction Coil; CC: Central Conduit.

One of the side ports of the valve is connected to the detector. Side ports can also be used for other purposes such as discharging waste or connection to other devices (e.g. a microwave oven, photo-oxidation system, another multi-position valve or mixing chamber). In a SIA system using a single reagent, the central port of the switching valve is connected to the sample channel and the pump is set to aspirate a preset volume of sample at a fairly low flow-rate in order to avoid the formation of bubbles. Then, the central port is connected to the appropriate side port in order to aspirate a preset volume of reagent. Next, the valve is actuated in order to connect its central port with the channel leading to the detector and an appropriate volume of carrier is dispensed to drive the sample and reagent to the detector. All this occurs under a laminar flow; as a result, the detector profile is no longer rectangular, but rather exhibits the typical asymmetric shape of FIA peaks. As can be seen in Fig. 2, dispersion in a SIA system leads to the sample and reagent plugs overlapping and forming the reaction product to be detected. A typical SIA manifold includes two types of coil, namely: a holding coil (HC) inserted in the channel connecting the

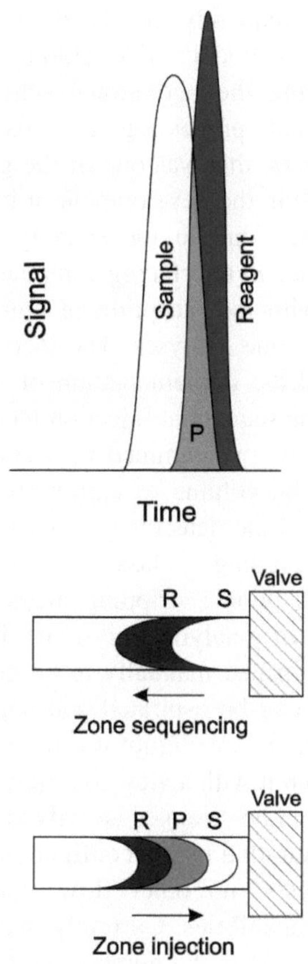

Fig. 2. Overlapping sample and reagent plugs in a SIA system. R: Reagent; P: Product; S: Sample.

piston pump to the central port of the switching valve that is used to prevent the sample and reagents from reaching the piston pump -cleaning of which otherwise would be labor-intensive and time-consuming- and a reaction coil (RC) in the channel leading to the detector that is intended to ensure adequate overlap between sample and reagent plugs in order to allow a quantitative amount of reaction product to form.

One of the essential features of SIA is its computerized control. The power of computers has grown so dramatically that nowadays they allow an automatic method to be set for adaptation to quite different needs; such a high flexibility simplifies the development of effective solutions at substantially reduced costs. The computer controls the connection of the central port to its side ports, starts and stops the pump in order to aspirate or dispense liquids, selects their volume and adjusts the flow-rate. Thus, residence times need no longer to be controlled via the length of the manifold tubes and flow-rates of the peristaltic pump; rather, by the computer being controlled and adjusted in each step of the process at will in a highly reproducible manner, making SIA a highly flexible analytical tool.

A major advantage of SIA over FIA is its ease of implementation, whereas in FIA using a different method very frequently entails altering the configuration of the manifold, switching to another method in SIA seldom requires more than using a different computer file containing the operational settings to be used with each procedure. Data acquisition and processing are also possible thanks to its computerized control. Nevertheless, this was one of the greatest initial hindrances to SIA development, which resulted in the development of barely a few tens of methods during its first year of existence due to the scarcity of commercially available software and the lack of experience in interfacing computers to analytical instruments. Only during the past decade, with the inception of commercial software, SIA has gained ground in the field of routine analyses. The incorporation of computers into SIA systems has also facilitated the implementation of stopped-flow methods. It is shown that sample handling in the sequential injection mode, which employs forward, reversed and stopped flow, can be programmed to accommodate a wide variety of assays. It suffices to calculate the volume of carrier to be delivered and stops the system when a peak is obtained at the detector in order to readily implement various analytical methodologies including classical kinetic, spectrophotometric, polarographic, voltammetric and anodic stripping methods. In addition, computers have increased the flexibility of analytical systems by allowing a number of operations mimicking those performed manually to be programmed by easy on-line implementation. Thus, samples can be aspirated and supplied with a reagent; their mixture driven to a photoreactor; and an aliquot withdrawn, sent to a solid-phase pre-concentration unit and eluted from it with a view to sequentially detecting the analytes by using several detectors that can be arranged serially at the same port or radially at different ports -or even in a serial/radial mixed configuration.

Moreover, as stated previously, no other flow technique can match SIA in multiparametric determination capabilities. Currently available switching valves can have more than twenty side ports. Also, the number can be increased by connecting a side port in a valve to the central port of several others. Such a high degree of expandability is exclusive of SIA.

Another advantage derived from the way a SIA system operates is the dramatically reduced consumption of sample and reagents. In a FIA system, sample and reagent consumptions are virtually independent of the analysis frequency as the peristaltic pump continuously propels the sample and reagents at a constant flow-rate throughout. In SIA, however, the piston pump only works during the time strictly needed to aspirate the amount of sample and reagents needed for a given determination. As an example, a SIA monitor for determining ammonium ion in wastewater needs ten times less reagents than does a comparable FIA monitor [20]; this is of high economic and practical significance, especially with equipment that is intended to operate unattended over long periods (e.g. automatic analyzers for field measurements).

Also, the fact of using piston pumps makes SIA a more robust technique. Since, its predecessors (FIA and SFA) use peristaltic pumps as liquid drivers and these operate with tubing made of materials which are relatively easily damaged by some fluids (viz. acids, bases and, especially, solvents); by contrast, piston pumps use glass or polytetrafluoroethylene (PTFE) tubing, which are highly inert and ensure a long life time. Furthermore, in SIA, the propulsion system holds mostly the carrier solution, that is to say, the sample, reagents and solvents seldom reach it.

One difficulty of SIA operation arises from the way plugs are stacked; this hinders mixing of the sample and reagents (especially with more than two, which require using a sandwich technique that consists on inserting sample plugs between two reagents zones, improving the mixture of them). When many reagents are required a feasible solution could be using a mixing chamber [21] in one of the side ports to homogenize the different sample/reagent mixtures with the aid of a magnetic stirrer.

SIA is versatile, capable of accommodating a wide variety of assays without the need for system reconfiguration, and compatible with a wide range of detectors. Nowadays, the only disadvantages of SIA in front of FIA are the lower injection frequency due to the sequential working mode and the lower mixing capacity between solutions also due to the sequential operational mode. Nonetheless, the benefits of minimizing reagents consumption and so waste generation, and the multiparametric capabilities of SIA make it a powerful tool for developing automated multiparametric environmental analyzers for field measurements.

3 SIA System for Simultaneous Determination of Nitrate and Nitrite

An automated sequential injection system for monitoring nitrate and nitrite in water samples was developed using the Griess reagent [22]. A sandwich arrangement allowed the simultaneous determination of the two parameters. Thus, nitrite was determined in one end of the sample zone by diazotization-coupling reactions and spectrophotometric detection of the azo dye formed. In the other end, nitrate was similarly determined after its on-line reduction to nitrite using a copperized cadmium column.

The system included a liquid driver, a selection valve and a spectrophotometer. The liquid driver consisted of an autoburette equipped with a 5-mL syringe and connected to a personal computer via an RS-232C serial interface. The software package AutoAnalysis (Sciware Systems, Spain) [23] was used for instrumental control, data acquisition and processing. A six-port selection valve, computer-controlled through a PC-9225 board, was used as liquid distributor. The manifold was constructed with PTFE tubing. The activation of the reduction column was carried out in three steps: first, the cadmium granules were washed with HCl and subsequently coated with copper by dipping them into a 1% $CuSO_4$ solution for ca. 1 min. Finally, the copperized cadmium granules were washed with deionized water, and packed into a methacrylate tube (40 mm length and 1.5 mm i.d.), plugged with glass wool. The reduction column was inserted between RC1 and RC2. Fig. 3 shows a schematic depiction of the SIA system for nitrate and nitrite determination.

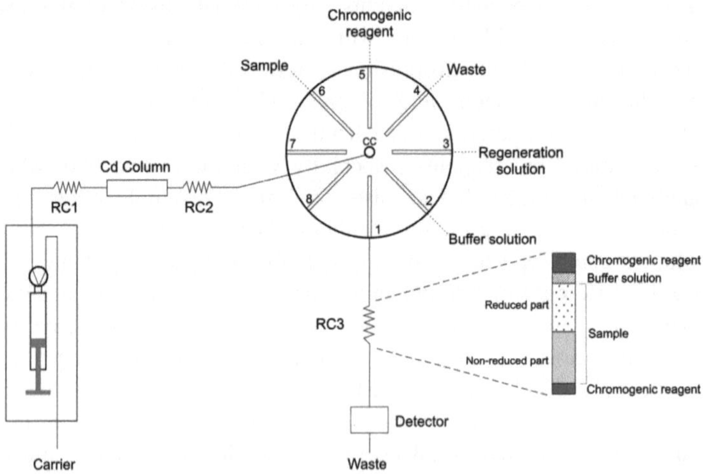

Fig. 3. Schematic depiction of the SIA system for nitrate and nitrite determination. RC: Reaction Coil.

Once the instrumental blank was set, the measuring cycle was started. The basic steps of the cycle comprised the aspiration of chromogenic reagent (Griess reagent, 50 µl) through port 5 of the selection valve, and its storage into the reaction coil RC3. Then, the sequential aspiration of buffer solution (port 2, NH_4Cl + $Na_2B_4O_7$.10 H_2O + EDTA, 40µl) and sample (port 6, 1800 µl) was carried out. Next via RC2, part of the sample passed through the reduction column where nitrate was converted into nitrite. Afterwards, the flow is reversed and the whole volume aspirated in the previous step (1840 µl) is dispensed to RC3. The buffer and sample zones were thus stacked between the valve and the chromogenic reagent introduced in the reactor previously. Next step is the aspiration of chromogenic reagent (port 5, 80 µl). Finally, the injection of the content of RC3 towards the detector takes place and absorbance measurements are registered at 540 and corrected at 420 nm.

Thereby, the sandwich arrangement, with the sample placed between two Griess reagent zones, led to the formation of the azo dye in both sub-zones of the injected sample and hence a readout with two peaks per injection was obtained, the first corresponding to nitrite alone (non-reductive conditions) and the second to nitrite plus nitrate (reductive conditions). The accuracy with which both peaks could be resolved depended on the parameters governing the mutual dispersion of reagent and sample, namely: sample and reagent volumes, zone sequencing, flow rates and length of the tubing.

Since the reduction column seemed to be the most problematic part of the entire system, its long-term stability was examined in detail. The lifetime of the reductor depended on the kind of samples analyzed. When the system was applied to the analysis of natural waters there was no need to replace the column for several months. In contrast, severe problems arose when applying the method to the analysis of wastewaters, probably due to contamination of the column. To solve this drawback a reactivating solution (port 3, 2 mL) passing through the column provided new copper to regenerate the external coating of the cadmium granules and EDTA to avoid the precipitation of $Cd(OH)_2$ or other hydroxides on the reduction column [24]. The column was finally washed with water, leaving the system ready for further analysis. Two different slopes of calibration curves for nitrite and nitrate, both under reductive conditions, were obtained. This fact indicates incomplete reduction of nitrate to nitrite, probably due to short contact times of the sample with the cadmium granules. The reduction efficiency was ca. 50%.

The main interference was that of phosphate, which caused an adverse effect even at low levels, showing a maximum tolerated concentration of 1 mg L^{-1}. The magnitude of the interference was found to be proportional to the concentration of phosphate in the sample and to the time of contact between this ion and the column. However, the column activity was restored with the regeneration process described above.

Most relevant figures of merit are shown in Table 1.

Table 1. Most relevant figures of merit of some SIA monitoring systems for the determination of key environmental parameters in water samples

SIA	Parameter	LOD	Linear working range	Repeatability (RSD, n=10)	Ref.
2 param.	Nitrate	0.006 mg N L^{-1}	0.028-1.400 mg N L^{-1}	1.3 %	[22]
	Nitrite**	0.001 mg N L^{-1}	0.007-0.560 mg N L^{-1}	0.8 %	
3 param.	Nitrate	0.021 mg N L^{-1}	0.14-1.82 mg N L^{-1}	2 %	[25]
	Nitrite**	0.002 mg N L^{-1}	0.01-0.42 mg N L^{-1}	0.5 %	
	Sulfate	3 mg SO_4^{2-} L^{-1}	75-300 mg SO_4^{2-} L^{-1}	1.3 %	
8 param.*	Hydrazine	0.01 mg N_2H_4 L^{-1}	0.03-2 mg N_2H_4 L^{-1}	1.2 %	[33]
	Ammonium	0.13 mg NH_4^+ L^{-1}	0.4-8 mg NH_4^+ L^{-1}	1.6 %	
	Iron	0.04 mg Fe L^{-1}	0.13-1 mg Fe L^{-1}	1.7 %	
	Silicate	0.03 mg SiO_2 L^{-1}	0.1-1.5 mg SiO_2 L^{-1}	2.5 %	
	Phosphate	0.05 mg PO_4^{3-} L^{-1}	0.2-1.5 mg PO_4^{3-} L^{-1}	2.2 %	

*Plus: pH, specific and acid conductivity. ** Non-reductive conditions.

4 Multiparametric Monitoring System for Sulfate, Nitrite and Nitrate Determination

A multiparametric flow analysis method for the automatic determination of nitrite, nitrate and sulfate ions in wastewater and drinking water samples based on SIA was developed [25]. Again, SIA allowed the integration of different techniques in a single manifold for monitoring of multiple environmental parameters. However, there are scarce applications to determine nitrite, nitrate and sulfate in drinking water and wastewater within the same system; mainly, due to the difficulty to keep the manifold free of precipitate after the sulfate determination by the turbidimetric method.

This analyzer exploited spectrophotometric detection after a previous chemical transformation of the analytes of interest. For the determination of nitrite, the colorimetric modified Griess reagent (sulphanilamide, N-(1-naphthyl) ethylene diamine and HCl) was used, which consists in the formation of colored azo reddish purple complex. Nitrate ion was reduced to nitrite in the presence of cadmium (Cd). Commercial cadmium grains coated with copper sulfate ($CuSO_4$) were used to fill a methacrylate column (55 mm length and 4 mm i.d.). The concentration of nitrate was calculated by the difference of TN determination and nitrite measurement. The turbidimetric method was carried out to determine sulfate. The sulfate ion precipitates in acetic acid with barium chloride ($BaCl_2$), forming barium sulfate crystals ($BaSO_4$) of uniform size. To prevent the coating of $BaSO_4$ on tubing walls a colloidal emulsifier such as arabic gum (gum acacia) was added during the sulfate analysis.

The system included a piston pump type, a selection valve and a spectrophotometer, all of them connected to a personal computer (Fig. 4). The piston pump type consisted on an autoburette equipped with a 10-mL syringe which was

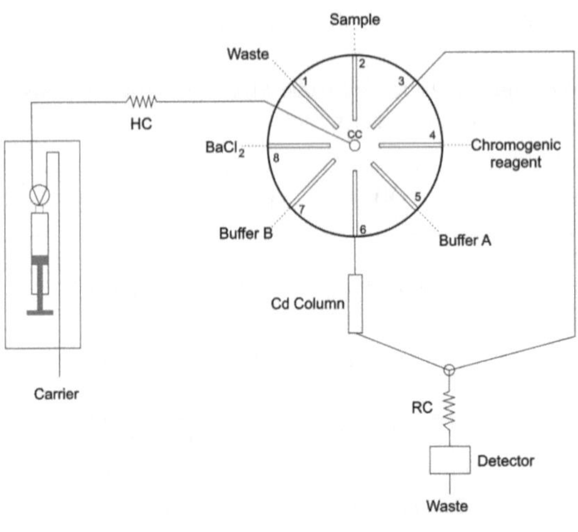

Fig. 4. Schematic depiction of the three-parametric SIA analyzer. HC: Holding Coil; RC: Reaction Coil; CC: Central Conduit.

used as liquid driver. Spectrophotometric measurements were made at 540 nm. An eight-port selection valve was used as a liquid distributor. The reduction column was inserted in the detector line between the eight-port selection valve and the reaction coil, using a three-way connector. Instrument control is performed using the software package AutoAnalysis (Sciware Systems).

Management of the sample and reagents through the selection valve was as follows: port 2 was used for sample aspiration, port 4 for the chromogenic solution, buffer solution A (NH_4Cl, $Na_2B_4O_7{\cdot}10\ H_2O$ and EDTA) and buffer solution B ($MgCl_2{\cdot}6H_2O$, $CH_3COONa{\cdot}3H_2O$, KNO_3, CH_3COOH and 5% (m/v) arabic gum dissolved in water) flowed through port 5 and port 7, respectively, and the $BaCl_2$ solution was aspirated through port 8. Port 1 was employed as waste and for loading cleaning solution. The reduction column was connected to port 6. Port 3 was connected to the reaction coil and the detector.

The measuring cycle was carried out using a sandwich technique. For NO_2^- determination, the sample (2 mL) and the chromogenic solution (0.5 mL) were loaded and sent to the reaction coil and then to the detector through port 3. For NO_3^- determination, sample (2 mL), chromogenic reagent (0.5 mL) and buffer solution A (0.08 mL) were loaded from their containers and then impelled through the reduction column located in port 6; after the reduction of NO_3^- to NO_2^-, they were passed through the reaction coil and the detector. For SO_4^{2-} determination, sample (1 mL), $BaCl_2$ solution (0.5 mL) and buffer solution B (0.2 mL) were loaded through their respective ports and then sent to the reaction coil and after that to the detector. After each SO_4^{2-} determination, a washing process with the cleaning solution (EDTA in NaOH) was performed to remove the precipitate that potentially could have been added to the tubing, leaving the system ready for further assays. The column activation and regeneration were accomplished as in the previous described system for nitrate and nitrite determination proposed by Cerdà et al. [22].

Thus, the accomplished repeatability (as relative standard deviation, RSD) of 1.9% for nitrate, was lower than those acquired in different works which vary in a range of 2.3–3.7% [26-28]. However, the repetitive running of the regeneration solution not only decreased the sampling rate but also shortened the lifetime of the column due to the breakdown of the cadmium granules into smaller particles.

As described previously, the most important interference for the analytical determination of nitrate was the effect of PO_4^{3-} on the reduction column. In this approach, a concentration up to 300 mg PO_4^{3-} L^{-1} was studied without signal variations in the determination of 1.12 mg N L^{-1} of nitrate. This could be due to the higher flow rate used, causing a short contact time between the ion and the column, despite the high concentration of the phosphate ion. The determination of sulfate could be affected by the presence of cations such as Ca^{+2}, Mg^{+2} and K^+, which are commonly found in water samples. The study of interferences in the analytical determination of 200 mg L^{-1} of sulfate showed that concentrations of 500 mg L^{-1} Ca^{+2} and 40 mg L^{-1} K^+ had an interfering effect in the precipitation of $BaSO_4$, whereas 800 mg L^{-1} Mg^{+2} did not interfere.

Table 1 summarizes the major figures of merit presented by each parameter of this system. The detection limits established in the standard methods (0.1 mg N L^{-1},

0.18 mg N L^{-1}, and 1 mg L^{-1} for nitrate, nitrite and sulfate, respectively) [17] were almost completely surpassed by the ones achieved in this work, with the exception of the sulfate detection limit.

The multiparametric method was applied to tap water, groundwater and wastewater (from inlet and outlet of a treatment plant). Recovery tests were performed and results showed recoveries of 95-109%, 102-109% and 95-103% for sulfate, nitrite and nitrate, respectively. This proved a good accuracy in the working range for monitoring of these three ions by the proposed system, making it suitable for drinking water and wastewater samples.

Although regeneration process of the column and washing steps were mandatory in order to avoid the decreasing efficiency of nitrate reduction to nitrite and to prevent the adhesion of $BaSO_4$ to the tubes walls, respectively, these minor drawbacks did not affect the system performance. Moreover, the cleaning cycles were carried out in a fully automated approach. Besides, the developed system did overcome the typical interference of phosphate ion in the determination of sulfate reported at bibliography.

5 A Wastewater Quality Monitoring System

A simple portable wastewater quality monitoring system based on spectrophotometric detection was developed by Thomas et al. [20] for the measurement of BOD, COD, TOC, TSS, global N and total P. Other parameters such as nitrate, nitrite, ammonium, orthophosphate or anionic detergents can be also directly measured with this system.

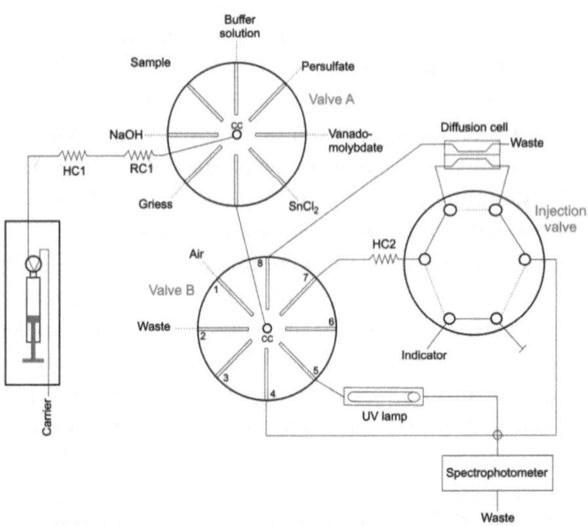

Fig. 5. Schematic depiction of the SIA wastewater quality monitoring system. RC: Reaction Coil; HC: Holding Coil; CC: Central Conduit; S: Syringe.

The proposed multiparametric monitoring analyzer was a SIA system coupled with two simple spectrophotometric detectors especially designed for the system. The SIA system consisted of a 5-mL automatic burette, a two valve multiposition module, an injection valve, a diffusion cell and a photo-oxidation reactor. Parameters were measured either by directly exploiting the UV-spectrum of the sample or by a chemical reaction with a reagent. All the operations and data acquisition were PC-controlled. Fig. 5 shows the scheme of the SIA multiparametric monitoring system.

For the analysis of BOD, COD, TOC, TSS, nitrate and detergents, a direct exploitation of the UV spectrum was used. In practice, the acquisition of spectra was carried out between 200 and 350 nm with the UV detector. The general procedure included a saturation test, a test of restitution quadratic error (between measured and restituted spectra) for the autovalidation of the procedure, and calculation of the parameters. Results for the different parameters were displayed simultaneously 30 seconds after the beginning of the spectrum acquisition.

For TN and organic phosphorus determination, UV photo-digestion was required. The digestion with UV radiation was chosen as the most convenient option for the automatic monitoring system in order to avoid potential problems caused by high temperatures and gas bubble formation usually associated with other methods such as microwave digestion. The photo-oxidation reactor consisted of a 15-W mercury lamp (λ = 254 nm) placed inside an aluminum cylinder, a PTFE tube (150 length and 1.5 mm i.d.) coiled around it and a cooling fan, which was used to keep its temperature low. The TN determination procedure can be briefly described as follows: first, the sample is sandwiched between two reagents and buffer zones, as follows: buffer (100 μL sodium tetraborate, pH 9) - oxidant reagent (30 μL potassium persulfate) - sample (50 μL) - oxidant (30 μL potassium persulfate) - buffer (100 μL sodium tetraborate, pH 9) then the mixture is propelled to the UV reactor, after 3-5 min, the mineralized mixture is pushed through the UV detector where organic and inorganic nitrogen compounds are oxidized into nitrate which is afterwards measured using the previous described UV multiwavelength procedure.

In order to determine either inorganic orthophosphates or total phosphorus (also as orthophosphates after mineralization) a reagent has to be added. The vanadomolybdate method was chosen for its good reproducibility and reagent stability to determine orthophosphates. The absorbance was read at 430 nm. For total phosphorous determination the SIA manifold was basically the same as used for orthophosphate determination, but with UV digestion. Potassium persulfate was used as oxidant and the flow was stopped for a certain time (150 s) to oxidize organic phosphorus compounds into orthophosphate. Finally, the mixture was transported to the detector for orthophosphate determination by means of the vanadomolybdo-phosphate method.

Nitrite determination was carried out using a standard colorimetric method based on a modified Griess reaction. The Griess reagent used was a mixture of sulfanilamide (SAM), N-(1-Naphthyl)ethylenediamine dihydrochloride (NED) and HCl. Absorbance was read at 530 nm [29,30]. The difference between the TN and the sum of the concentrations of nitrates and nitrites before photooxidation is the Kjeldahl nitrogen (TKN). The organic nitrogen can also be calculated.

Finally, ammonium was analyzed by a colorimetric method after conversion to ammonia with 0.01 mol L^{-1} NaOH and its diffusion through a hydrophobic membrane into the acceptor shell containing an acid-base indicator solution (bromothymol blue). The change in the absorbance, at 640 nm, of the acid-base indicator solution used as acceptor stream was measured and referred to the ammonium content in the sample.

In order to validate the proposed monitoring analyzer, results of the different parameters were compared with those of reference methods. The comparison was carried out for each parameter by calculating the correlation coefficient and by testing the validity of the following hypotheses: value of slope and intercept of the regression line (respectively 1 and 0), and validity of the linearity. For global parameter estimation (TSS, BOD, COD, and TOC) and for nitrate determination, the proposed system was tested with 85 treated wastewaters samples from several treatment plants. The correlation coefficient values (r^2), respectively 0.87, 0.73, 0.91 and 0.92, showed that the correlation coefficient were quite good for all the global parameters but for BOD, probably because the measurement principle of the reference method is very different. Moreover, this parameter was the only one with a negative test of validity (slope different from 1). For the other parameters (ammonium, nitrite, organic nitrogen, orthophosphate and organic phosphorus), SIA procedures were optimized and tested with a series of standard solutions and of real samples [29-32]. In conclusion, for almost all the parameters tested, the correlation coefficient values were close to or greater than 0.9 and the validity test was always positive, showing that the proposed monitoring system could be an excellent alternative to current reference methods for wastewater quality monitoring. Furthermore, being a much simpler, economic, handy multiparametric analyzer.

6 SIA System for Monitoring the Water Quality in the Energy Cogeneration System of a Municipal Waste Incinerator

Leading-edge urban solid waste ashing plants use burning heat energy to obtain electrical power. Water fed to their boilers for conversion into steam should be highly pure in order to minimize corrosion, scaling and similar phenomena, which can lead to malfunctioning and a reduced useful life. This highlights the need of a proper management and control of the water supply. A multiparametric monitor based on sequential injection analysis for the sequential determination of up to eight parameters, namely: pH, specific and acid conductivity, hydrazine, ammonium, phosphate, silicate and total iron, was developed by Bauzá et al. [33].

Acid conductivity was determined by passing the sample through a cation-exchange resin in order to retain ammonium ion and release protons. This parameter was deemed the most accurate indicator of dissolved solids in boiler water. Chemical parameters were determined spectrophotometrically: hydrazine by reaction with p-dimethylaminobenzaldehyde, ammonium by the modified Berthelot reaction, iron with o-phenanthroline, and phosphate and silica by formation of a molybdoheteropoly blue dye in the presence of ascorbic acid as reductant.

The system consisted of a multisyringe burette equipped with four 10-mL syringes, a selector module consisting of two 10-way selection valves, an autosampler, a conductimeter equipped with a temperature probe, a homemade conductivity cell made of poly(methyl)methacrylate (PMMA) and gold-coated bronze tubular and wall-jet electrodes, a column (53 mm length and 4 mm i.d.) used to measure acid conductivity also made from PMMA, a pH-meter equipped with a cell which closely mimicked a batch system involving immersion of the electrode in solution, a glass–Ag/AgCl microelectrode (3mm diameter), a Hewlett–Packard HP 8452A diode array spectrophotometer equipped with a 18μl flow-cell (10 mm light path) and a personal computer to govern the system and acquire data by using three communication channels (viz. two RS232 serial ports for the pH-meter and the multisyringe, and an HP-IB parallel port for the spectrophotometer). Instrumental control and data acquisition was carried out by software AutoAnalysis (Sciware Systems). The manifold was constructed with PTFE tubing. Fig. 6 shows the scheme of the SIA monitor.

The procedure for conductivity and acid conductivity involved using valve B and syringe 4 to drive the sample (7 mL) toward the measuring cell. Conductivity measurements were made under continuous-flow conditions since they avoided baseline drifts. Acid conductivity was determined similarly except that the sample (5 mL) was passed through the cation-exchange resin. The resin was found to be able to process up to 60 samples without the need for regeneration. However, the column was regenerated (7 mL of 1mol L^{-1} HCl and 100 mL of distilled water) after each working day in order to ensure optimum conditions operation of the system. Hydrazine was determined by reaction with p-dimethylaminobenzaldehyde (DAB), using valve A and syringe 2 in the multisyringe to aspirate the reagent (0.8 mL) and the sample (0.2 mL). A wait of 30 s allowed a yellow quinone to form, then the colored compound was propelled to the detector for absorbance measurement at 455 nm (with correction at 650 nm). Iron was determined by reaction with o-phenanthroline. Valve B and syringe 3 were used to aspirate hydroxylamine (0.1 mL), sample (0.3 mL) and o-phenanthroline (0.05 mL). The mixture was driven to the spectrophotometer for measurement of the absorbance difference (A_{510}–A_{650}).

Ammonium determination was based on the modified Berthelot reaction. In the first step, ammonium reacted with hypochlorite to form chloramine, which then reacted with phenol to obtain indophenol blue. This reaction was quite slow and required the presence of a catalyst (sodium nitroprusside) in addition to heating and stopping the flow for 1 min in a water bath at 45 °C in a knotted reaction coil (KRC). By the use of valve A and syringe 2, the hypochlorite (0.2 mL), the sample (0.35 mL) and the salicylate (0.35 mL) were aspirated. Absorbance measurements of the reaction product were made at 640nm (spectral oscillations caused by changes in refractive index were corrected by subtracting the absorbance at 800 nm).

Silicate and phosphate determinations were based on kinetic differences in the formation rates of their molybdoheteropoly blue derivatives. The procedure started by using valve A and syringe 1 to aspirate the molybdate for silicate determination (0.35 mL) and sample (1 mL) via a thermostated coil (1 min at 45 °C). Meanwhile, the determination of phosphate by using syringe 2 to aspirate antimonyl tartrate–oxalic

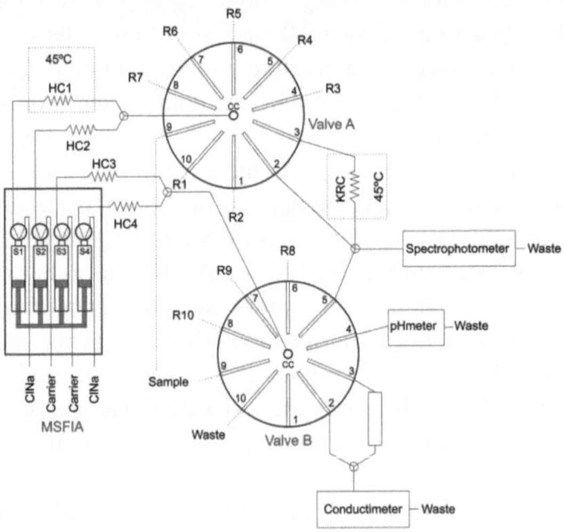

Fig. 6. Schematic depiction of the SIA system for monitoring water in the energy cogeneration system of a municipal waste incinerator. R1: Salicylate; R2: Hypochlorite; R3: Tartrate; R4: Oxalic Acid; R5: Ascorbic Acid; R6: Molybdate; R7: DAB; R8: Hydrochloric Acid; R9: Hydroxylamine; R10: o-phenantroline; HC: Holding Coil; KRC: Knotted Reaction Coil; S: Syringe.

acid buffer (0.25 mL), ascorbic acid (0.30 mL), ammonium molybdate (0.20 mL) and sample (0.75 mL) was performed. Finally, a reaction mixture volume of 4 mL was propelled to the detector for measurement of the absorbance difference (A_{650}–A_{774}). This procedure was followed to determine silicate, which involved aspirating ascorbic acid (0.7 mL), and oxalic acid (0.35 mL), adding to the mixture of the thermostated coil and driving to the detector for spectrophotometric measurement (A_{810}–A_{550}).

The pH was simultaneously determined with silicate and phosphate, using valve B and syringe 4, the sample (3.0 mL) was propelled to the pH electrode for measurement ensuring that it would not be diluted on its way to the measuring electrode and that the cell holding the sample would be properly rinsed with it.

Potential interferences to determine hydrazine were examined (i.e. iron, silicate, phosphate and ammonium). Silicate and phosphate concentrations up to 10 mg L^{-1} resulted in no signal change. Silicate levels in the studied sample type are usually in the μg L^{-1} region and phosphate is usually present in samples from the reboiler, at concentrations close to 1 mg L^{-1}. Iron and ammonium exhibited no interfering effect at concentrations 100 times higher than that of hydrazine. Any interference from iron would have been rather unexpected; in fact, this metal is rarely encountered among these analyses unless the system is corroded. Potential interferences for iron determination were studied at concentrations up to 50 mg L^{-1} for ammonium, and 10 mg L^{-1} for silicate, phosphate and hydrazine, none was found to have an appreciable effect on the analytical signal. In the case of ammonium determination, silicate exhibited no interfering effect at concentrations 50 times higher than that of ammonium. Phosphate and iron at concentrations up to 20 times higher than that of the analyte were found not to interfere, and so was hydrazine at equal concentrations to ammonium—which is usually the case in water from energy cogeneration plants. In the simultaneous determination of phosphate and silicate, the addition of tartaric acid was found to prevent the formation of molybdosilicate in the time span of the experiment—and hence the interference of silicate with the determination of phosphate. For the determination of silicate, the molybdophosphate was ligand-exchanged with oxalate in order to obtain phosphate and molybdo-oxalate. Because molybdosilicate is an inert species not amenable to ligand exchange with oxalate, it remained in solution for further reduction and detection. Thus, reactions required careful optimization of the working conditions in order to avoid mutual interferences.

Main figures of merit are shown in Table 1. The determinations integrated in the described monitor departed slightly from individual and separate SIA determinations due to the compromises adopted in order to be able to perform several determinations at once. Overall, the sensitivity was adequate for determining the five chemical parameters in water from a water–steam cycle at a cogeneration waste incinerator plant. Also, the precision was acceptable. The proposed monitor affords a triplicate determination of eight target parameters within 30 min.

7 Conclusions

Water and wastewater characterization is of great environmental and public health interest, thus the need of water and wastewater quality monitoring systems is highlighted. Automation provides a fast response to a large volume of samples. Particularly, SIA has proved to be an outstanding tool for developing automatic monitoring analyzers. Moreover, SIA expansibility and flexibility provide to this flow technique multiparametric capabilities unaffordable by other automation techniques. Also, SIA outfits each system with accurate control of reagents and sample volumes, of their flow rate, as well as of injection throughputs, resulting in low consumption of sample and reagents, a minimal waste generation, optimum sensitivity, good reproducibility and repeatability. Furthermore, sample handling in the sequential injection mode, which employs forward, reversed and stopped flow, can be programmed to accommodate a wide variety of assays allowing a number of operations mimicking those performed manually to be programmed by easy on-line implementation.

The four analyzers described above share some benefits such as simplicity, portability, savings, and effectiveness. It is worth pointing out that one of the proposed analyzers is able to determine up to 11 parameters within the same system by the use of low cost devices and without any system reconfiguration. Another great advantage of these automatic analyzers is the higher analysis frequency in comparison with manual routine analysis, e.g. the eight-parametric SIA system affords a triplicate determination of eight target parameters within 30 min. Therefore, these monitoring systems could be used to monitor key parameters in certain fields where the continuous assessment of these is required, e.g. wastewater treatment plants, drinking water treatment plant or industrial processes.

References

[1] Council Directive 98/83/EC on the quality of water intended for human consumption, pp. 32–54 (1998)
[2] World Health Organization (WHO) Guidelines for Drinking-Water Quality, Incorporating 1st and 2nd Addenda, 3rd edn. Recommendations, vol. 1. (2004)
[3] United States Environmental Protection Agency (USEPA) 40 CFR ch. 1(7-1-09 Edition) (2009)
[4] Council Directive 91/271/EEC concerning urban waste water treatment, pp. 40–52 (1991)
[5] Trojanowicz, M.: Advances in flow analysis. Wyley-VHC, Weinheim, Germany (2008)
[6] Skeggs, L.: An automatic method for colorimetric analysis. American Journal of Clinical Pathology 28, 311–322 (1957)
[7] Trojanowicz, M., Benson, R.L., Worsfold, P.J.: Recent developments in water quality monitoring by flow injection analysis. Trends Analytical Chemistry 10, 11–17 (1991)
[8] Alexander, P.W., Di Benedetto, L.T., Dimitrakopoulos, T., Hibbert, D.B., Ngila, J.C., Sequeira, M., Shiels, D.: Field-portable flow-injection analyzers for monitoring of air and water pollution. Talanta 43, 915–925 (1996)
[9] USEPA, ESS Method 140.4: Chloride - Automated Flow Injection Analysis (1993)
[10] USEPA, ESS Method 220.3: Ammonia Nitrogen and Nitrate + Nitrite Nitrogen, Automated Flow Injection Analysis Method (1991)
[11] Ruzicka, J., Hansen, E.H.: Flow injection analyses. I. New concept of fast continuous flow analysis. Analytica Chimica Acta 78, 145–157 (1975)
[12] Ruzicka, J., Marshall, G.D.: Sequential injection: a new concept for chemical sensors, process analysis and laboratory assays. Analytica Chimica Acta 237, 329–343 (1990)
[13] Reis, B.F., Giné, M.F., Zagatto, E.A.G., Lima, J.L.F.C., Lapa, R.A.: Multicommutation in flow analysis. Part 1. Binary sampling: concepts, instrumentation and spectrophotometric determination of iron in plant digests. Analytica Chimica Acta 293, 129–138 (1994)
[14] Cerdà, V., Estela, J.M., Forteza, R., Cladera, A., Becerra, E., Altimira, P., Sitjar, P.: Flow techniques in water analysis. Talanta 50, 695–705 (1999)
[15] Lapa, R.A.S., Lima, J.L.F.C., Reis, B.F., Santos, J.L.M., Zagatto, E.A.G.: Multi-pumping in flow analysis: concepts, instrumentation, potentialities. Analytica Chimica Acta 466, 125–132 (2002)
[16] Ruzicka, J.: Lab-on-valve: universal microflow analyzer based on sequential and bead injection. Analyst 125, 1053–1060 (2000)
[17] Standard Methods for the Examination of Water & Wastewater, 22nd edn. American Public Health Association (APHA), American Water Works Association (AWWA) & Water Environment Federation, WEF (2012)
[18] Wurl, O., Elsholz, O., Baasner, J.: Monitoring of total Hg in the river Elbe: FIA-device for on-line digestion. Fresenius Journal of Analytical Chemistry 366, 191–195 (2000)
[19] Cerdà, V., Cerdà, A., Cladera, A., Oms, M.T., Mas, F., Gómez, E., Bauzá, F., Miró, M., Forteza, R., Estela, J.M.: Monitoring of environmental parameters by sequential injection analysis. Trends Analytical Chemistry 20, 407–418 (2001)
[20] Thomas, O., Theraulaz, F., Cerdà, V., Constant, D., Quevauviller, P.: Wastewater quality monitoring. Trends Analytical Chemistry 16, 419–424 (1997)
[21] Pinto, P.C.A.G., Saraiva, M.L.M.F.S., Lima, J.L.F.C.: A flow sampling strategy for the analysis of oil samples without pre-treatment in a sequential injection analysis system. Analytica Chimica Acta 555, 377–383 (2006)

[22] Cerdà, A., Oms, M.T., Forteza, R., Cerdà, V.: Sequential injection sandwich technique for the simultaneous determination of nitrate and nitrite. Analytica Chimica Acta 371, 63–71 (1998)
[23] Sciware Systems, S.L., `http://www.sciware-sl.com`
[24] Otsuki, A.: A reactivation solution for a copperized cadmium column in the automatic determination of nitrate in natural waters. Analytica Chimica Acta 99, 375–377 (1978)
[25] Ayala, A., Leal, L.O., Ferrer, L., Cerdà, V.: Multiparametric automated system for sulfate, nitrite and nitrate monitoring in drinking water and wastewater based on sequential injection analysis. Microchemical Journal 100, 55–60 (2012)
[26] Ensafi, A., Rezaei, B., Nouroozi, S.: Simultaneous spectrophotometric determination of nitrite and nitrate by flow injection analysis. Analytical Sciences 20, 1749–1753 (2004)
[27] Mesquita, R.B.R., Ferreira, M.T.S.O.B., Segundo, R.L.A., Teixeira, C.F.C.P., Bordalo, A.A., Rangel, A.O.S.S.: Development of a sequential injection system for the determination of nitrite and nitrate in waters with different salinity: application to estuaries in NW Portugal. Analytical Methods 1, 195–202 (2009)
[28] Haghighi, A., Farrokhi-Kurd, S.: Sequential flow injection analysis of ammonium and nitrate using gas phase molecular absorption spectrometry. Talanta 64, 688–694 (2004)
[29] Oms, M.T., Cerdà, A., Cerdà, V.: Sequential injection analysis of nitrites and nitrates. Analytica Chimica Acta 315, 321–330 (1995)
[30] Cerdà, A., Oms, M.T., Forteza, R., Cerdà, V.: Speciation of nitrogen in wastewater by flow injection. Analyst 121, 13–17 (1996)
[31] Oms, M.T., Cerdà, A., Cerdà, V.: Preconcentration by flow reversal in conductometric sequential injection analysis for ammonium. Electroanalysis 8, 387–390 (1996)
[32] Oms, M.T., Cerdà, A., Cladera, A., Cerdà, V., Forteza, R.: Gas diffusion techniques coupled sequential injection analysis for selective determination of ammonium. Analytica Chimica Acta 318, 251–260 (1996)
[33] Bauzá de Mirabó, F.M., Forteza, R., Cerdà, V.: A multisyringe sequential injection method for monitoring water in the energy cogeneration system of a municipal waste incinerator. Talanta 79, 1011–1020 (2009)

Development and Deployment of a Microfluidic Platform for Water Quality Monitoring

J. Cleary*, D. Maher, and D. Diamond

CLARITY Centre for Sensor Web Technologies,
National Centre for Sensor Research,
Dublin City University, Dublin, Ireland
john.cleary@dcu.ie

Abstract. There is an increasing demand for autonomous sensor devices which can provide reliable data on key water quality parameters at a higher temporal and geographical resolution than is achievable using current approaches to sampling and monitoring. Microfluidic technology, in combination with rapid and on-going developments in the area of wireless communications, has significant potential to address this demand due to a number of advantageous features which allow the development of compact, low-cost and low-powered analytical devices. Here we report on the development of a microfluidic platform for water quality monitoring. This system has been successfully applied to in-situ monitoring of phosphate in environmental and wastewater monitoring applications. We describe a number of the technical and practical issues encountered and addressed during these deployments and summarise the current status of the technology.

Keywords: Microfluidics, autonomous sensing, chemical sensors, water quality.

1 Introduction

There is a growing need for increased monitoring of water quality across a broad range of applications, including municipal and industrial wastewaters as well as drinking water and environmental waters (rivers, lakes, groundwater and marine). This demand is fuelled by a range of drivers, with legislation being the most direct and important in many applications. For example, the European Union's Water Framework Directive (WFD) [1], supported by a range of directives such the Urban Wastewater Directive [2], the Nitrates Directive [3], the Drinking Water Directive [4] and the Groundwater Directive [5], establishes a comprehensive basis for the management of water resources within the EU. The priorities of the WFD are to prevent further deterioration of and to protect and enhance the status of water resources and to promote sustainable water use based on long-term protection of water resources. To this end, the Directive requires that an integrated monitoring programme be established within each river basin district, which will enable

* Corresponding author.

S.C. Mukhopadhyay & A. Mason (Eds.): *Real-Time Water Quality Monitoring*, SSMI 4, pp. 125–148.
DOI: 10.1007/978-3-642-37006-9_6 © Springer-Verlag Berlin Heidelberg 2013

collection of the physical, chemical and biological data necessary to assess the status of surface and groundwater bodies in each such district.

Reliable, high quality, and high resolution information about water quality is essential for water management and for maintaining or improving the quality of water resources. However, existing monitoring practices are unsatisfactory for a variety of reasons. Monitoring of environmental waters is still predominantly based on manual sampling followed by laboratory analysis using standard techniques such as colorimetry, atomic absorption spectroscopy, mass spectrometry and ion chromatography. While this approach yields high quality and reliable data (assuming that the appropriate protocols are followed during collection, transport, storage and analysis of the sample) the cost per sample is significant due to the manpower requirement for sample collection as well as the cost of analysis. This approach is therefore incapable of meeting the demand for monitoring at the much higher temporal frequencies and geographical densities which are envisaged [6].

For wastewater monitoring, a range of online sensors and analyzers are commercially available, based on colorimetric chemical detection, electrochemical detection or optical detection. Generally available as single units or a suite of units aimed at monitoring individual parameters such as phosphate, nitrate or ammonia, these are also subject to a number of limitations, including: high cost of purchase and installation, high cost of operation due to the significant volumes of reagents and/or standard solutions required, and varying reliability, leading to frequent intervention by operators in order to clean fouled surfaces, remove blockages, or run calibration protocols. In this sector, the key advantage of microfluidics-based sensing systems is the lower cost of purchase and operation, as well as reduced infrastructure requirements. For this reason, microfluidic systems are likely to find a significant niche in the wastewater sector, particularly in smaller-scale wastewater treatment plants where the cost of conventional online systems may be prohibitive, and in alternative treatment systems such as constructed wetlands [7] where mains power may be unavailable or restricted to parts of the site. The installation of large numbers of new wastewater treatment systems in developing economies also represents a major opportunity for developers of low-cost monitoring systems for key parameters such as nutrients.

Since the "miniaturized total analysis system" (μTAS) concept was introduced by Manz et al. in 1990 [8], it has been recognized that the properties of microfluidic or "lab-on-a-chip" systems make them eminently suitable for the development of compact, autonomous analytical devices for a range of applications including environmental monitoring. The advantages of microfluidic sensing systems can be summarized as follows. (i) The small sample sizes used, typically in the μL range, minimize reagent consumption and waste generation. (ii) The small size of the microfluidic manifold facilitates the development of compact and portable analytical systems. (iii) Fast analysis times result from performing chemical analysis on the μm scale, where diffusion-based mixing can be an efficient process, allowing high sample throughput and/or frequent measurements. (iv) Low-cost sensing devices can be developed by combining microfluidic systems with simple, low cost detectors. This combination of properties makes microfluidic systems highly attractive as a basis for reagent-based monitoring of chemical species in the aquatic environment.

There are also drawbacks associated with operating at the micro-scale. Using μL-scale sample sizes has implications in terms of adequately representing the complete body of water which is to be measured. This issue is significantly mitigated by the higher measurement frequency which is possible using *in-situ* systems; for example, by capturing short-term variations, multi-daily *in situ* measurements which are approximately accurate will provide more information on the levels of a target analyte in a water body than a highly accurate sample-based measurement which is only carried out at monthly intervals. Due to the small dimensions of the microfluidic channels, they are susceptible to blockage or interference by fine particulate matter. The use of fine filters at the sampling point can eliminate or minimize this issue, although at the cost of limiting the analytical parameter to the dissolved fraction of the target species. Issues such as these, and others including interference caused by bubble formation/trapping within the micro-analytical system, have represented significant barriers to the development and uptake of microfluidics-based *in-situ* sensing systems. Although numerous micro-analytical systems have been developed and their performances described under laboratory conditions, examples of integrated micro-analytical devices which have been successfully deployed for extended periods under real environmental conditions are much scarcer. Therefore, while the development of sensing nodes for the aquatic environment which are sufficiently reliable and low in cost to form the basis of extensive wireless sensor networks has been anticipated, realization of this goal has been more difficult to achieve. The challenges in developing such systems include achieving adequate stability and reliability of the analytical system, cost and power consumption of the integrated device, robustness under harsh environmental conditions, and fouling due to particulate matter and microorganisms. In developing an autonomous environmental monitoring device, deployable lifetimes in the order of months to a year are desirable. Achieving such lifetimes without elevating the fabrication cost of the integrated system to unviable levels can be described as a key challenge for the analytical science community [9].

2 Examples of Deployable Microfluidic Devices

A range of reagent-based analytical techniques can be readily implemented in microfluidic systems. However, despite rapid and on-going developments in the area of microfluidic/lab-on-a-chip technology for chemical analyses over the last 20 years or so, and the significant requirements and opportunities which exist for autonomous environmental monitoring systems [10,11], relatively few field deployable microfluidic devices based on reagent chemistry have come to fruition.

One area which has seen significant progress is the integration of colorimetric methods for nutrients such as phosphate, ammonia and nitrate/nitrite into microfluidic manifolds with light emitting diode (LED)/photodiode-based optical detection systems. Gardolinski *et al.* [12] developed a miniaturised, LED-based chemical analyser for *in-situ* monitoring of nitrate. The limit of detection (LOD) with a 20 mm path length flow cell was 2.8 mg L^{-1} N. The linear range could be adjusted to suit local conditions in the field by changing the flow cell path length. The analyser was tested in shipboard mode for mapping nitrate concentration in the North Sea and in submersible mode to conduct a transect of the Tamar Estuary (England).

Doku *et al.* [13] developed a micro-flow injection analysis (μFIA) technique for orthophosphate based on the molybdenum blue reaction. The μFIA manifold was formed by etching of borosilicate glass, and electro-osmotic flow was used both for the mobilization of reagents and for sample injection. A LOD of 0.1 μg mL^{-1} was achieved, with a rapid analysis time of 60 s and low sample/reagent volume (total system volume of 0.6 μL). Greenway *et al.* [14] produced a similar μFIA system for the determination of nitrite, based on the Greiss reaction to form an azo dye. Following optimization of the electro-osmotic flow characteristics, reaction chemistry, and injection time, a LOD of 0.2 μmol L^{-1} was achieved. A further progression of this work was to incorporate a cadmium reduction column within the device. This allowed for the analysis of nitrate in water, for which a LOD of 0.51 μmol L^{-1} [15] was achieved. Daridon et al. [16] investigated the Berthelot reaction for the determination of ammonia in water utilizing a microfluidic device consisting of a silicon chip between two glass plates. The path length in this device was 400 μm, and the integrated system comprised fiber optics coupled to an LED. High aspect ratio channels (30 μm wide × 220 μm deep) were used for sample/reagent mixing in order to achieve efficient diffusional mixing.

Azzaro et al. [17] described an automatic colorimetric analyzer prototype (MicroMAC FAST MP3) for high frequency measurement of nutrients in seawater. This system utilizes the Berthelot method for ammonia detection, the sulphanilamide/ethylenediamine method for nitrate detection, and the blue phosphomolybdate method for phosphate detection, and LODs of 5, 2.5 and 2.5 μg/L were reported for $N\text{-}NH_4$, $N\text{-}NO_3$ and $P\text{-}PO_4$ respectively. The system was based on loop flow analysis technology developed by Systea, an Italian company, and has been further developed to provide a system more suited to deployment on typical monitoring platforms. This multi-nutrient analyser system has been trialled in collaboration with YSI Hydrodata at two locations in the United Kingdom (Hanningfield Reservoir and River Blackwater Estuary) [18].

A number of other systems for *in-situ* nutrient monitoring have been assessed by the Alliance for Coastal Technologies (ACT) at Chesapeake Biological Laboratory, USA [19-21]. ACT conducts technology evaluations which focus on classes of commercially available instruments in order to provide confirmation that each technology meets the manufacturer's performance specifications or claims under real deployment conditions. The American Ecotech NUT 1000 was originally developed at Monash University and commercialised by Ecotech Pty. Ltd. The system can achieve LODs for reactive phosphate of < 3μg L^{-1} with a response time of 30 seconds. The high sampling rate is allowed by the use of rapid sequenced reagent injection in combination with a multi-reflection flow cell. Reagent injection also minimizes the reagent consumption, allowing over 1000 measurements to be performed with only 20 mL of reagent. In the ACT trials, the NUT 1000 was used for surface mapping on board a research vessel in Monterey Bay, California [19]. Very good correlation with validation samples was achieved, after correction for a measurement offset of 50 μg L^{-1} $P\text{-}PO_4$, which was attributed to differences in refractive index of natural seawater versus the reagent grade water used for preparation of internal standards and machine calibration. The ACT has also reported on successful deployments of the WET Labs

Cycle-P nutrient analyzer [20] and the YSI 9600 Nitrate Monitor [21]. Vuillemin *et al.* [22] described a miniaturized chemical analyzer (CHEMINI) for dissolved iron and total sulphide, based on FIA and colorimetric analysis. Designed for deep-sea applications, the system is submersible to a depth of 6000m, with detection limits of 0.3 μM and 0.1 μM for iron and sulphide respectively. It was used to monitor the chemical environment within mussel beds, over a 6 month deployment period during which 8 measurements per day were performed.

Mowlem and co-workers at the National Oceanography Centre, Southampton, UK, have developed *in-situ* sensors for a range of chemical parameters [23-27] including nitrite, nitrate, ammonia, phosphate and iron. The first of this series of microfluidic chemical analysis systems [23] was used to detect nitrate and nitrite with a limit of detection (LOD) of 0.025 μM for nitrate (0.0016 mg L^{-1} as NO_3^-) and 0.02 μM for nitrite (0.00092 mg L^{-1} as NO_2^-). The device was deployed in an estuarine environment (Southampton Water, UK) to monitor nitrate and nitrite concentrations in waters of varying salinity and was able to track changes in the nitrate-salinity relationship of estuarine waters due to increased river flow after a period of high rainfall. In subsequent work, nanomolar detection limits were achieved for iron [24], ammonium [25], phosphate and nitrate [26]. Like the CHEMINI system, these analysers are designed primarily for oceanographic applications, and the level of engineering required to achieve the high analytical performances required in this environment means that the cost of such systems is likely to be prohibitive in terms of deploying large numbers of devices for routine monitoring applications.

3 Reagent Stability

In many cases, reagent stability represents a major limitation when considering chemical detection methods for environmental monitoring, due to the need for devices to be exposed to environmental conditions for a significant period of time (months or longer) while giving reliable and reproducible results over this time frame. Sequeira et al. [28] examined the stability of the reagents for the Berthelot reaction for ammonia detection in terms of viability for long-term monitoring applications. The three key reagents in the reaction are potassium sodium tartrate, phenol and hypochlorite. While potassium sodium tartrate and phenol are generally regarded as stable if stored properly, hypochlorite solutions are unstable and can decompose to form chlorate ion and oxygen over a period of days or weeks. It was shown that iron and copper, present as contaminants, catalyze the decomposition. Ensuring that the hypochlorite solution was free from these contaminants allowed long-term storage of the hypochlorite solution without significant loss in activity. Salicylate was also used as a replacement for phenol, and was found to yield broadly similar analytical performance in terms of sensitivity and kinetics, while having the major advantages of being highly stable and non-toxic. Diamond and co-workers [29,30] evaluated the yellow vanadomolybdophosphoric acid method as an analytical method for the determination of phosphate in water within a microfluidic device with the aim of producing an automated device with a field-deployable lifetime of one year. In this method ammonium molybdate, $(NH_4)_6Mo_7O_{24}.7H_2O$, is reacted with ammonium metavanadate, NH_4VO_3, under acidic conditions. The combined reagent and sample containing orthophosphate react to form the vanadomolybdophosphoric acid complex, $(NH_4)_3PO_4NH_4VO_3.16MoO_3$. The resulting solution has a distinct yellow colour arising from the strong absorbance of this complex below 400 nm. It was shown that batches of the reagent could be used for over a year without significant loss in performance [29]. This method was therefore selected in preference to the molybdenum blue method, due to the greater stability of the reagents used in the analysis, and resulted in an assay with an LOD of 0.2 mg L^{-1} and a dynamic linear range of 0–50 mg L^{-1}.

4 First Generation Autonomous Microfluidic System for Phosphate Analysis

Following on from investigations on the microfluidic application of the vanadomolybdophosphoric acid method for phosphate detection, Diamond and co-workers have developed an autonomous microfluidic analyzer for phosphate, which has been successfully deployed in wastewater, freshwater and estuarine locations. An early prototype for this system consisted of bottles for storing the reagent, calibration solutions and cleaner, a sample port for collecting the water sample to be analyzed, and an array of solenoid pumps for pumping the required liquids through the microfluidic chip. The microfluidic chip allowed for the mixing of the reagent and sample, and the chip also presented the reacted sample to a photodiode and LED for an absorbance measurement. The analyzed sample is then pumped to the waste storage container. All of the fluid handling and analytical components were controlled by a microcontroller that also performed the data acquisition and stored the data in a flash memory unit. A GSM modem was used to communicate the data via the SMS protocol to a laptop computer. The layout of the new system is shown schematically in Fig. 1.

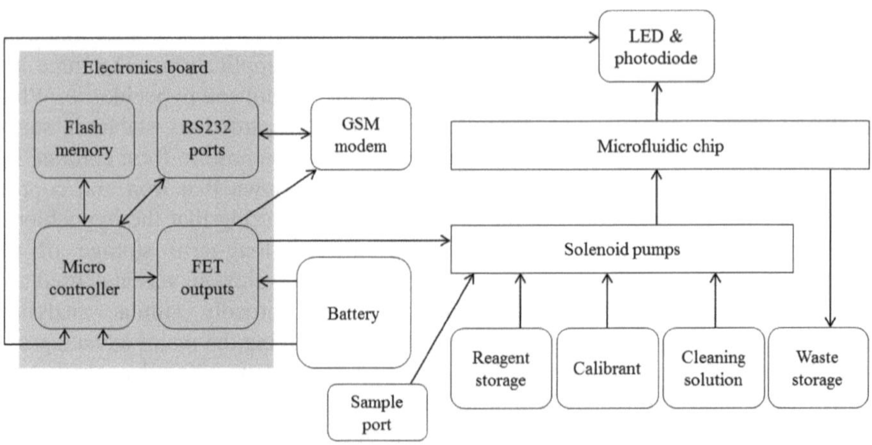

Fig. 1. Schematic detailing the components of the phosphate sensor and their relation to each other

The microfluidic chip used to perform the colorimetric measurement is shown in Fig. 2(a). This allows for the colorimetric reaction to be performed using 20 μL each of sample and reagent. The reagent and sample are mixed in a T-mixer which leads into a serpentine channel where the resulting solution is presented to the emitter and photo-detector for the absorbance measurement. Each side of the T-mixer has three inlet channels leading into them. On one side are the two phosphate standard solutions for a two point calibration. A third inlet channel is for the water sample. On the other side of the T-mixer there is an inlet for the reagent and two inlets for deionized water for cleaning the micro-channels. All channels are of 200μm in width and depth. The chip was manufactured from a 1 mm thick PMMA sheet (ME303011, Goodfellow, Cambridgeshire, England). The microfluidic channels were milled into the sheet using the CAT-3D-M6 micromill (DATRON Electronic, Mühltal, Germany) using a 200μm endmill (00680029, DATRON Electronic, Mühltal, Germany). The 35×20 mm chip was then cut from the sheet. The channels are sealed by bonding a second 35×20 mm PMMA chip layer using a pressure sensitive adhesive (ARcare® 8890, Adhesives Research, Limerick, Ireland). PEEK tubes are inserted into the inlet and outlet ports and are held in place using epoxy adhesive.

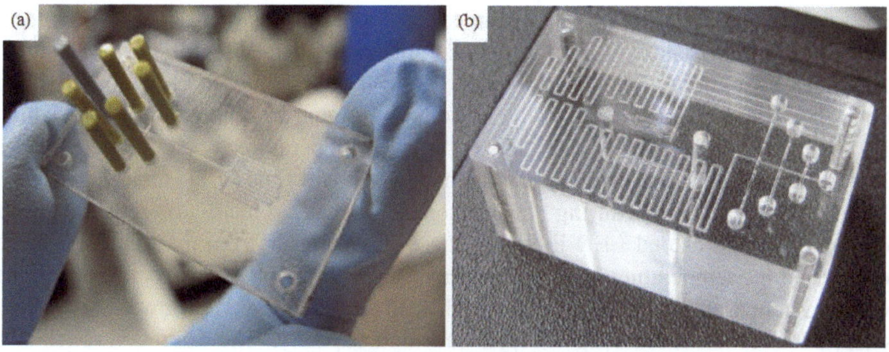

Fig. 2. Two versions of the microfluidic chip used for phosphate detection. (a) The prototype microfluidic chip described in this section, with serpentine channel visible on the right of the chip; the six inlets and single outlet can be seen on the left. (b) A redesigned version of the chip in which a serpentine channel was used for mixing and reaction of sample and reagent, and a cylindrical optical cuvette replaced the serpentine detection area.

The reagent, calibration solutions and cleaners were stored in five square 250 ml HDPE bottles (NALGENE 2018-0250, Nalge Nunc, New York, USA). This capacity of reagent storage allowed for over 11,000 phosphate measurements. Two square 1L HDPE bottles were used for waste storage (NALGENE 2018-1000, Nalge Nunc, New York, USA). The bottle caps (Q-Series, Bio-Chem Valve, New Jersey, USA) used were fitted with Teflon check valves, which allowed air pressure to equalize between the bottle and its surroundings as the liquid is being pumped out of or into the various bottles. Teflon tubing with ¼"-28 UNF connectors (Omni-Lok, Bio-Chem Valve, New Jersey, USA) fitted at the ends, were used to connect the various fluidic components in the system. Six independently controlled miniature solenoid pumps

(120SP, Bio-Chem Valve, New Jersey, USA) were used to control fluid movement in the system. The fluid-contacting parts of these pumps are made from chemically inert PEEK for high chemical resistance. The pumps deliver 20 µL for each stroke and have ¼"-28 UNF ports for tubing interconnects. The system used dedicated pumps for the water sample, two calibration solutions, and the reagent. The remaining two pumps were used to deliver a cleaning solution to the microfluidic chip between assays.

The microcontroller used for the control of the system was the MSP430F449 (Texas Instruments), which was chosen for its low power consumption during operational and sleep mode. The MSPGCC compiler also provided an open source platform for software development (mspgcc.sourceforge.net). This microcontroller has, along with numerous digital inputs and outputs, an 8 channel 12-bit ADC for data acquisition and two UART ports for communicating with peripheral devices. A 2 megabyte SPI flash chip mounted on the PCB with the microcontroller allowed for 16,384 data points to be logged. The solenoid pumps and the power to the GSM modem were controlled via an array of field effect transistors (FETs). A photodiode (S1227-33BR, Hamamatsu Photonics UK Limited, Hertfershire, UK) and a 370nm LED (NSHU550E, Nichia Corporation, Tokushima, Japan) were used for the absorbance measurement. A transimpedance amplifier circuit based around a TLV2772 operational amplifier (Texas Instruments) was used to condition the signal from the photodiode. This circuit was built on a separate PCB from the main controller board, in order to limit the noise on the signal from the photodiode, which is mounted beside the microfluidic chip. A temperature sensor (LM335, ST Microelectronics) is also mounted on the daughter board. The LED is controlled via a BJT and was pulsed during the absorbance measurement to reduce power consumption and extend LED lifetime. The voltage signals from the transimpedance circuit and the temperature sensor were read by the ADC on the microcontroller. The entire system was powered by a 12V 7Ah lead acid battery. The power source was chosen for its low cost, long lifespan and high capacity. The microcontroller also monitored and logged the battery voltage with each phosphate measurement.

A GSM modem was used for communication during system deployments. The main reason for the choice of this communications standard was the extensive coverage of the GSM network in Ireland and other countries in the European Union, which meant that the phosphate sensor could be deployed in most areas while maintaining data communication capability. The modem used was the MC35IT (Siemens, Berlin, Germany) which communicates via RS232, allowing for easy implementation in the system. The system used the GSM modem's SMS functionality to send the data to another GSM modem connected to a laptop computer which acted as a base-station for data collection. During the operation of the sensor, data was collected and then stored in the flash memory. After a predetermined number of measurements the collected data was transmitted via SMS messages to the base-station. To withstand the environment the sensing system was housed in a rugged enclosure, in the form of a polypropylene "Pelicase" (1430, Peli Products, Barcelona, Spain). This case is completely water tight and resistant to the reagent used. It is lightweight and so does not affect portability. The chemical storage bottles and battery were mounted in the lower part of the case, while the solenoid pumps and microfluidic chip and LED and photodiode were mounted on a custom ABS mounting plate manufactured using a 3D printer (Dimension SST768, Stratasys, Minnesota, USA). The controller board and GSM Modem were mounted in the enclosure lid. The sample port was mounted on the side of the enclosure.

Fig. 3 shows the fully assembled system with Pelicase enclosure. The five storage bottles and two waste bottles can be seen at the bottom of the enclosure. The battery sits between the waste bottles. The top plate is flipped up to reveal the six solenoid

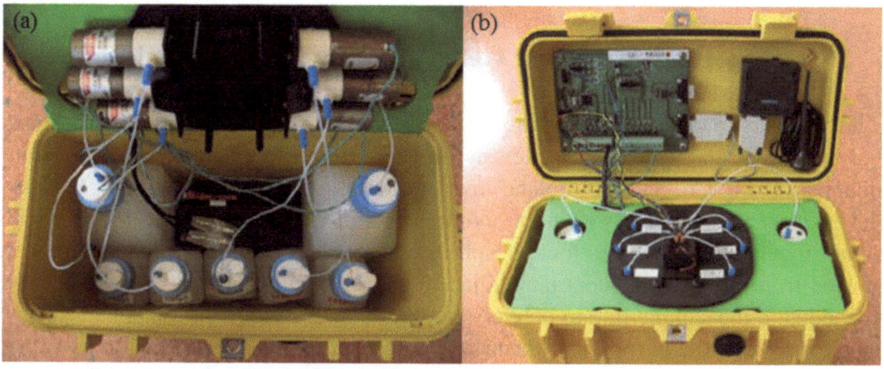

Fig. 3. Fully assembled system. (a) Bottom of enclosure showing bottles and battery. (b) Top plate and lid showing electronics board and GSM modem. Adapted with permission from [31].

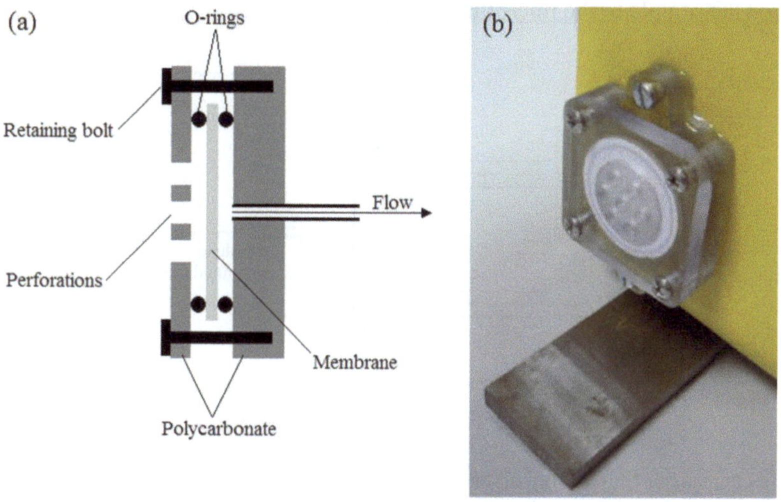

Fig. 4. (a) Schematic of sample port. (b) Sample port mounted on sensor housing. Adapted with permission from [31].

pumps on the underside. These are mounted with the inlets facing down towards the storage bottles. The outlets can be accessed through the top to allow them to be connected to the microfluidic chip. This is housed along with the temperature sensor, LED and photodiode with their accompanying daughter board in the center of the top plate. In the lid of the enclosure the controller board is mounted on the left hand side. The GSM modem along with its antenna is mounted on the right. The sample is drawn into the system through a 0.45 μm pore diameter filter membrane (Supor, PALL Corporation, Michigan, US), held in place between two polycarbonate plates and sealed with O-rings (Fig. 4). The port is attached to the Pelicase and sealed with polyurethane adhesive. The port has a ¼"-28 UNF port for connecting tubing.

5 In-Situ Testing of First Generation Phosphate Analyser

Following laboratory based validation studies [31,32], the first generation phosphate analyser was deployed at a wastewater treatment plant (Osberstown Wastewater Treatment Plant, Co. Kildare, Ireland) over a 6 week period in July and August 2008. The analyser was situated within the monitoring building which houses the WWTP's online monitoring systems for phosphate, ammonia and nitrate. Samples were pumped from the final effluent tank adjacent to the monitoring building, and a valve was added to the sample line to deliver samples to the prototype phosphate analyzer's sample inlet. For the purposes of this trial, the sample inlet was modified to consist of a modified membrane filter holder (shown schematically in Fig. 5) which allowed the sample to contact a membrane filter (Supor membrane filter, 0.45 μm pore size, 25 mm diameter, Sigma-Aldrich) and then exit to waste. The sample was drawn through the membrane and into the analyzer system using a solenoid pump which then delivered the sample to the microfluidic chip.

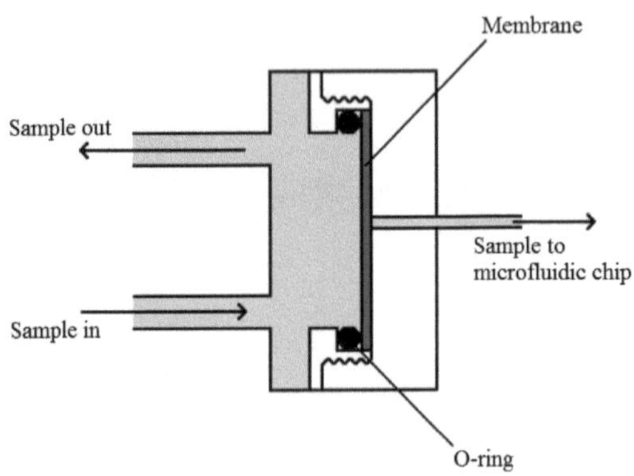

Fig. 5. Schematic cross section of the sampling port. The threaded parts of the membrane holder are screwed together so that the membrane is held in place by the O-ring seal. Samples circulate past the outer side of the membrane and the required volume is drawn into the analyzer by a solenoid pump. Adapted with permission from [45].

In order to compensate for variation in the detector output due to diurnal temperature fluctuations, as well as signal drift over long time periods, a calibration procedure was automatically carried out for each sample measurement. This was performed by carrying out separate reaction and detection runs using blank (0 mg L^{-1} phosphate) and standard (10 mg L^{-1} phosphate) solutions. The absorbance of the standard phosphate solution was calculated using (1).

$$A = log(I_o/I) \tag{1}$$

Where A is absorbance, I_o is the measured light intensity for the blank solution, and I is the measured light intensity for the standard solution. The absorbance for the sample solution was calculated similarly, and was converted to phosphate concentration using the absorbance of the standard solution. Finally, the sample's phosphate concentration value was converted to "phosphorus as phosphate" (P-PO_4) in order to facilitate comparison with the plant's existing online phosphorus monitor (Aztec P100 instrument, Capital Controls, UK). Data from the latter system was downloaded via serial connection to a laptop computer at weekly intervals.

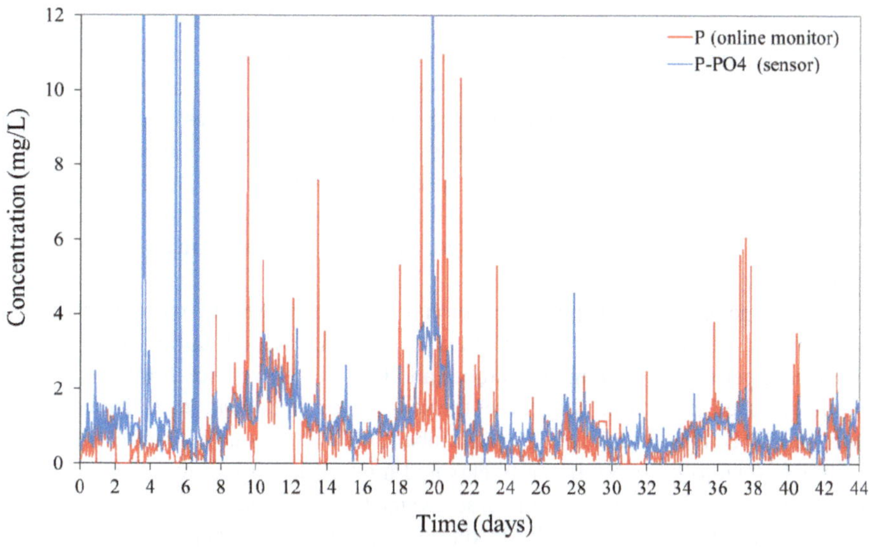

Fig. 6. Output from the prototype phosphate sensor and the plant's online phosphorus monitor during the trial period. Adapted with permission from [45].

The complete dataset from the 6 week trial is presented in Fig. 6. Over the period as a whole there is excellent agreement between the data collected from the prototype sensor and from the online monitoring system. However, a number of discrepancies were also observed. Between days 3 and 7 of the trial there were a number of spurious high readings in the prototype sensor data. These were caused by air bubbles accessing the external sample line and being drawn into the microfluidic chip where they interfered with the passage of light through the optical cuvette. On day 7 of the trial the valve on the sample line was adjusted which resulted in this issue being almost entirely eliminated, with only 2 further such readings observed (days 20 and 28). A more sustained deviation between the sensor and online monitor data was observed over days 18-21 of the trial, and was found to have been caused by the membrane filter becoming clogged with solids. After the filter was replaced, a close agreement in the data was resumed, and the membrane filter was changed at weekly intervals thereafter to prevent a recurrence of this issue. There were also a number of occasions on which the online monitor recorded elevated values which were not detected by the prototype sensor. These discrepancies were caused by solids in the sample affecting the online monitor readings, which has a relatively coarse filter at its inlet point. The online monitor uses heating and acid to digest solids in the sample, followed by analysis using the blue molybdate method for phosphate detection. Any solid material therefore has the potential to increase the measured value (acid-hydrolysable phosphorus). This is not the case with the prototype sensor, as solids are excluded from the system using the 0.45 μm filter membrane at the inlet point, in order to avoid blockage or fouling of the microfluidic channels. Despite the limitations involved in using the online monitoring data as validation for the prototype sensor, this trial data showed that the prototype phosphate sensor had considerable potential to be used for monitoring of phosphate in wastewater streams. The performance of the prototype sensor was shown to be comparable with that of a commercially available phosphorus monitor, while significantly less maintenance was required over the trial period, even after the membrane filter was changed at one week intervals to account for the heavy fouling potential of the sample. Running cost of the sensor was significantly lower than the online monitor, due to the minimal amount of reagent and calibrant solutions required, and the sensor's much lower power consumption.

The first generation phosphate sensor has also been used to monitor phosphate levels in an estuarine environment in Co. Dublin, Ireland during two separate deployment periods in September-October and October-November 2009. Following a laboratory-based calibration procedure, the sensor was placed *in-situ* at Broadmeadow Water in Co. Dublin, Ireland on 25 September 2009. This is an estuarine water body which is known to have significantly elevated nutrient levels due to a combination of inputs from agricultural sources and wastewater treatment plant discharges [33]. The sensor was located at a small islet in one of the estuary channels and held in place using a steel anchoring device. As this was a tidal location, the GSM modem antennae was located outside the box and elevated above the high-water mark to ensure constant network coverage, as the sensor enclosure itself was fully immersed at high tide. The sensor operated with hourly measurement frequency, and data was transmitted by the GSM modem in SMS mode to a laboratory-based laptop computer at 5 hour intervals. For validation purposes, daily manual samples were collected as close as possible to the sensor's sample inlet, and timed to coincide with the sensor's sampling time. Samples were filtered immediately, and analyzed using a Hach-Lange DR890 portable colorimeter and the appropriate reagent pack (amino acid method for high range phosphate).

Fig. 7. The phosphate sensor *in-situ* during a trial at Broadmeadow Water, Co. Dublin, Ireland in September 2009. The sample inlet is visible below the water line.

Fig. 8(a) shows the sensor's output during the trial period. The sensor performed 236 measurements during the period in question. A distinct daily pattern of variation emerged which is ascribed to tidal influences in the estuary. The sensor was located within 100 m of a wastewater treatment plant outfall, and changing tidal levels would significantly affect the degree of dilution which the wastewater discharges were subject to at a given time. Fig. 8(a) shows that there was generally good agreement between the phosphate concentrations measured by the sensor and by the validation method. The correlation coefficient (R^2) of 0.699 indicates that some discrepancies existed between the two datasets, despite the synchronous collection of manual samples from as close to the sensor inlet as possible. Given the rapid rate of change of the concentrations at certain times (Fig 8a), it is not surprising that such differences exist. Other possible sources of discrepancy include the small volume (22 μL) of sample used in the sensor-based assay, which may be less representative, at a particular moment in time, of the general water body than the 250 mL manual sample. Possible sources of error in the manual sampling data include minor inaccuracies in the sample volume used for analysis (25 mL) which, for on-site measurements, is measured using a graduated vial, or in the volume of the reagent solutions (ammonium molybdate solution and amino acid solution, 1 mL in each case). Perhaps more significantly, differing interfering species for the two methods could also give rise to some discrepancies. The molybdenum yellow method used in the phosphate sensor is subject to interference by dissolved organic substances with light absorption in the UV region. This was unlikely to be a significant factor in this study as after filtering to remove suspended solids, all water samples were observed to have good optical clarity without significant colour. Silica (Si), arsenate (AsO_4^{3-}) and ferrous iron (Fe^{2+}) are other potential interferants with the molybdenum yellow method [34]. Silica and arsenate, however, are mainly of concern when the reaction is heated to provide faster reaction rates, which was not the case in this study, and ferrous iron does not interfere at concentrations below 100 mg L^{-1}. The impact of interferants on the sensor data is therefore not likely to have been significant. Potential interferants for the amino acid method include sodium (Na^+), nitrite (NO_2^-) and sulfide (S^{2-}) [35]. The estuarine location used in this trial is a dynamic water body, affected by tidal fluctuations and a significant point source of phosphate and other nutrients (wastewater treatment plant discharge). Altering levels of these interferants over time have the potential to cause discrepancies between the sensor data and manual measurements.

Discrepancies between standard, sampling-based approaches to water quality monitoring and sensor-based approaches clearly represent a barrier to wider uptake of sensor technologies. However, despite these considerations, the data shown in Fig. 8(a) clearly illustrates a key advantage of the deployable monitoring system. The more frequent sensor data showed regular variations in phosphate levels during the trial period which were not captured using daily manual sampling (a relatively high manual sampling rate which is not likely to be practicable in larger-scale monitoring programmes). The sensor-based monitoring is therefore more likely to identify short-term pollution events, as well as providing high-resolution data over long timeframes which can provide valuable information on the long term trends for the analyte of interest.

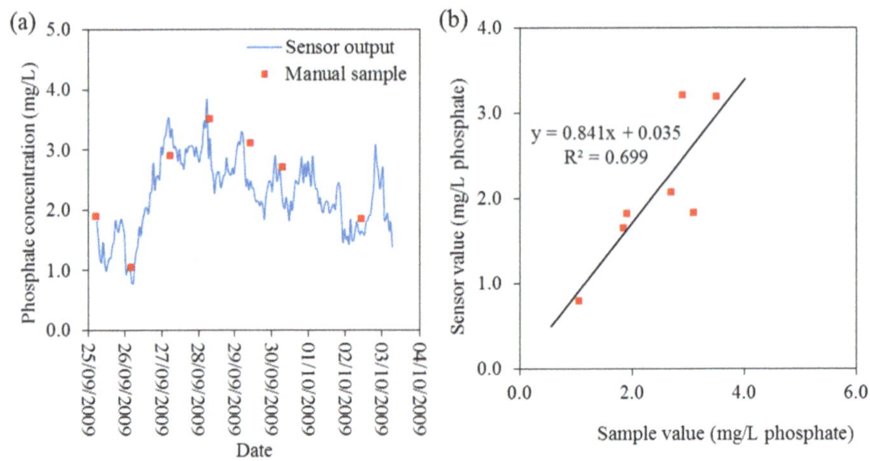

Fig. 8. (a) Phosphate concentrations measured during the first trial period by the sensor and by manual sampling. (b) Sensor output plotted as a function of phosphate concentration as measured by analysis of manually collected samples.

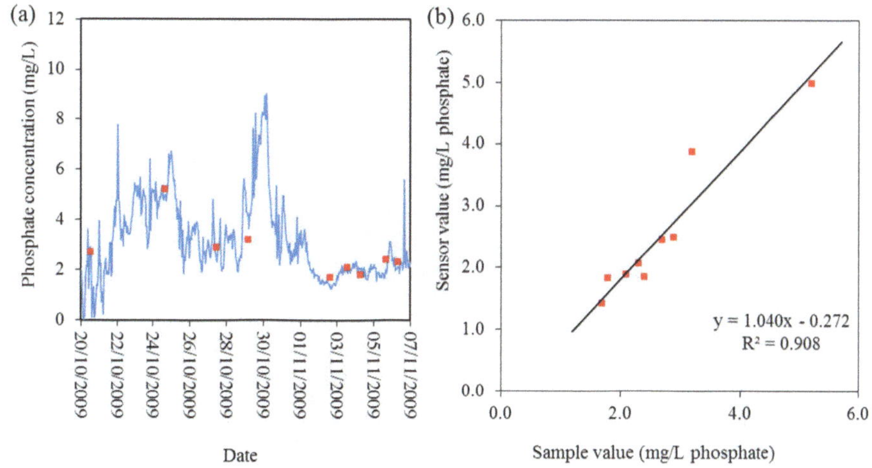

Fig. 9. (a) Phosphate concentrations measured during the second trial period by the sensor and by manual sampling. (b) Sensor output plotted as a function of phosphate concentration as measured by analysis of manually collected samples.

After 8 days of successful operation, the initial trial was terminated due to clogging of the membrane filter by sediment which had collected around the sampling port. To prevent this problem from recurring, some alterations were made to the sampling port design, and the sensor was deployed close to the original site but at a location with lower susceptibility to sediment build-up on the channel bed. Approximately 480 autonomous measurements were carried out during a second trial period. Figure 9(a) shows that the data obtained during the second trial was broadly similar to that obtained in the first trial period, with phosphate levels generally varying between 2 and 6 mg L^{-1}, and significant peaks and troughs observed on a number of occasions. From 07/11/2009, the data became erratic, resulting in the termination of the trial. Subsequent investigations showed that these erroneous data points were caused by a particulate obstruction in the line delivering reagent to the microfluidic chip, an issue which has since been addressed by careful filtering of all reagent and calibrant solutions prior to use in the sensor. Fig. 9(b) shows that the correlation between sensor output and analysis of manual samples was significantly improved relative to the first trial period, with $R^2 = 0.908$.

6 Design and In-Situ Testing of Second Generation Phosphate Analyser

A substantial redesign of the phosphate analyzer was carried out in collaboration with an industry partner (Episensor Ltd., Limerick, Ireland) in order to address some limitations of the system and, in particular, to reduce the component cost of the system to less than €200 per unit while maintaining or improving the analytical performance.

Fig. 10. (a) Second generation phosphate analyzer design; (1) Sample inlet; (2) Control board and detection system; (3) Dual channel peristaltic pumps; (4) Reagent bags; (5) IP68 enclosure. (b). The analyzer *in-situ* at the Broadmeadow Water test site.

The second generation phosphate analyzer is a fully integrated system incorporating fluid handling, microfluidic technology, colorimetric chemical detection, and real time wireless communications in a compact and rugged portable device. The major system components are shown in Figure 10(a). The mode of operation is similar to that of the first generation system. Table 1 shows some of the significant modifications and improvements implemented in the second generation system. The use of a folded sheet metal frame on which the internal components are mounted, collapsible polypropylene bags for reagent, calibrant and waste storage, and low-cost dual channel peristaltic pumps, have resulted in significant improvements in component cost, compactness and manufacturability. The system also features an improved bubble detection and removal protocol to reduce the impact of bubbles entering or forming within the microfluidic system. The system communicates using short range ZigBee® radio to a remote network gateway, which passes the sensor data via GSM or Wifi to a web-database for remote access by the user. This setup allows the user to not only monitor sensor data in real time but also to remotely change the sensor setup parameters, such as sampling rate.

Table 1. Comparison of first and second generation phosphate analyzers

First Generation System	Second Generation System
Internal Volume: 15L	Internal Volume: 2.3L
System Mass: 12 kg	System Mass: 1.7kg
Battery: 12V Lead Acid	Battery: 3.6V Lithium
Battery Life: 2 months	Battery Life: 12 months
Comms: GSM Modem	Comms: ZigBee Radio
Delayed Reporting	Real time 'sensor to database' reporting

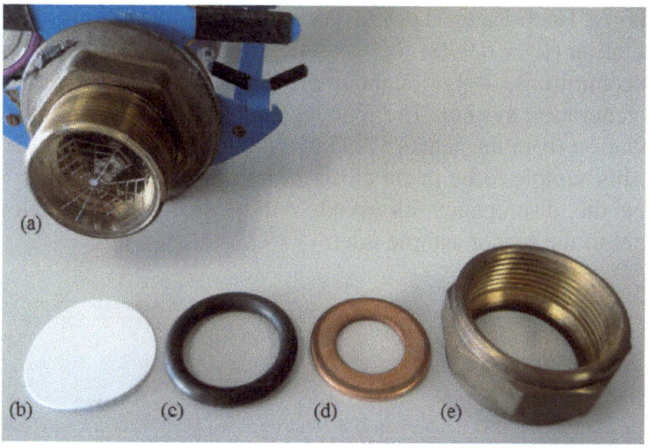

Fig. 11. Components of the sample intake/filter assembly. (a) "spider-web" PMMA filter backing, (b) polyethersulfone membrane filter, (c) O-ring seal, (d) copper washer, (e) brass screw-on fitting.

An additional change was made to the design of the sensor's sample intake module. This involved the use of commercially available copper-containing fitting to hold the membrane filter. In addition to being cost-effective, the copper content was expected to be of benefit in terms of protecting the membrane filter from biofouling. Copper, copper alloys and copper compounds are widely used as biofouling inhibitors for sensors and other devices deployed in aquatic environments [36,37]. The sensor intake was also located at the bottom of the sensor unit. In addition to being a practical placement for deployment purposes, this had the benefit of shielding the membrane filter from ambient light. Light favors growth of micro-organisms which can promote the formation of biofilms on surfaces. The membrane filter used was a hydrohophilic polyethersulfone membrane designed for biological sample preparation, with low protein binding properties [38]. As attachment of macromolecules (organic and inorganic) and microorganisms represent the initial stages in the biofouling process [36,37], it was hoped that this combination of measures would delay and/or limit biofouling of the filter surface. The redesigned sample intake/filter assembly is shown in Fig. 11.

A "spider-web" pattern is formed on the surface of a PMMA disc by micro-milling. The disc is then fitted into a brass connector where it acts as a backing for the polyethersulfone membrane filter. The spider-web pattern allows a higher proportion of the membrane surface to be used for sample intake, while minimizing the dead volume created. The membrane is held in place using an O-ring seal, a copper washer and a brass threaded nut. These copper/brass fittings allow the filter to be easily exchanged / replaced while also offering a degree of protection from biofouling.

Following laboratory validation and calibration, the second-generation analyzer was deployed at Broadmeadow Water Estuary, Co. Dublin, Ireland on 04/09/2010. The sensor output over the 62-hour deployment period is shown in Fig. 12(a). The sensor performed 124 autonomous measurements during the trial, and four manual samples were collected for laboratory analysis. These samples were analyzed in the laboratory using a Hach-Lange DR890 Portable Colorimeter. Fig. 12(b) shows the excellent correlation ($R^2 = 0.9706$) between the sensor output and the manual sample phosphate concentrations, a significant improvement over the correlations achieved with the first generation system. The manual sample collected on 06/09/2010 shows the largest deviation from the sensor value. On that occasion, the tidal level at the site did not allow this sample to be taken close to the sensor sample inlet, which may be responsible for the discrepancy observed. All other manual samples were taken directly adjacent to the sensor sample inlet.

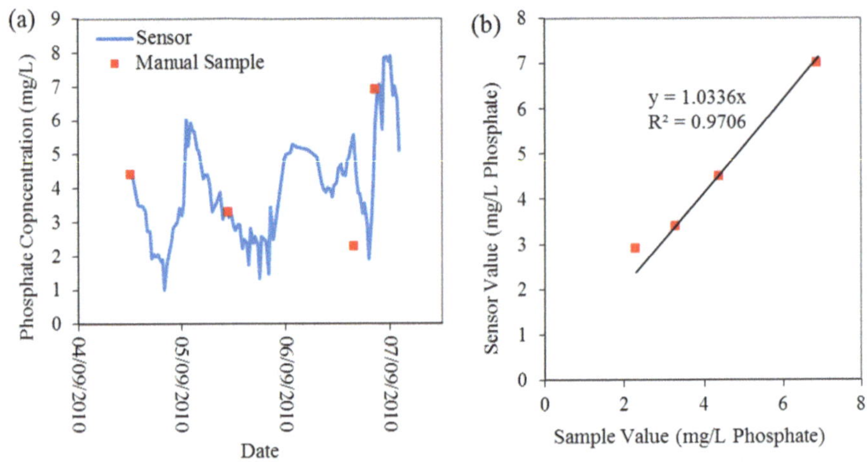

Fig. 12. (a) Phosphate concentrations measured during field trialing of the second generation phosphate analyzer at Broadmeadow Water. (b) Sensor output plotted as a function of phosphate concentration as measured by analysis of manually collected samples.

The second generation phosphate analyzer has also been used in a wastewater monitoring application. This trial was also carried out at Osberstown Wastewater Treatment Plant, Co. Kildare, Ireland. However, in this case, the analyzer was located in the plant's treated effluent discharge tank, as shown in Fig. 13(a). The sensor module was directly immersed in the effluent, while the communications module was mounted above the water level. This allowed wireless communication using Episensor's ZigBee® radio system, to a base-station located nearby. An auto-sampler (Buhler BL 2000, Hach-Lange Ireland Ltd.) was used to collect validation samples at 7-hour intervals, these were stored within the refrigerated compartment of the auto-sampler, collected periodically, and analyzed in the laboratory using the vanadomolybdophosphoric acid method (yellow method) for phosphate detection and a UV-Vis spectrophotometer (UV-1600PC Spectrophotometer, VWR International). For sensor validation purposes, this method was preferable to relying on data from the plant's online phosphorus monitor, as it allowed a more direct comparison between sensor readings and sample values, since (i) the sensor also utilizes the yellow method, and (ii) the collected samples were filtered before analysis, so that the sample values represented the dissolved molybdate reactive phosphate fraction only, again in accordance with the actual sensor measurements.

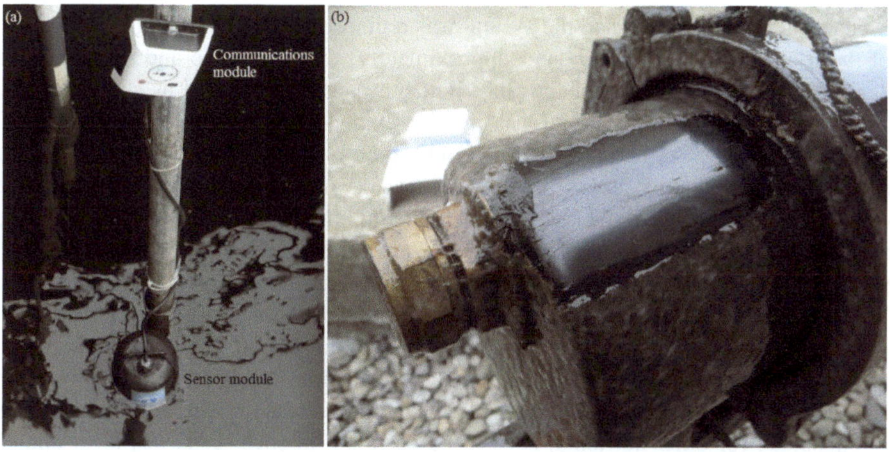

Fig. 13. (a) Second generation phosphate analyser *in-situ* at Osberstown Wastewater Treatment Plant, Co. Kildare, Ireland. (b) Biofouling on the analyser body after approx. 2 weeks immersion in wastewater.

Fouling of the sensor body was a significant feature of this trial. Fig. 13(b) shows the partially cleaned sensor module after removal from the effluent tank after approx. 14 days immersion. A fouling layer approximately 1 mm in thickness was observed. The copper-containing filter housing was noticeably less affected than the sensor body. Fig. 14(a) shows that the membrane filter was also fouled with a thin, adherent layer of material, although there was no discernible effect on the sensor performance.

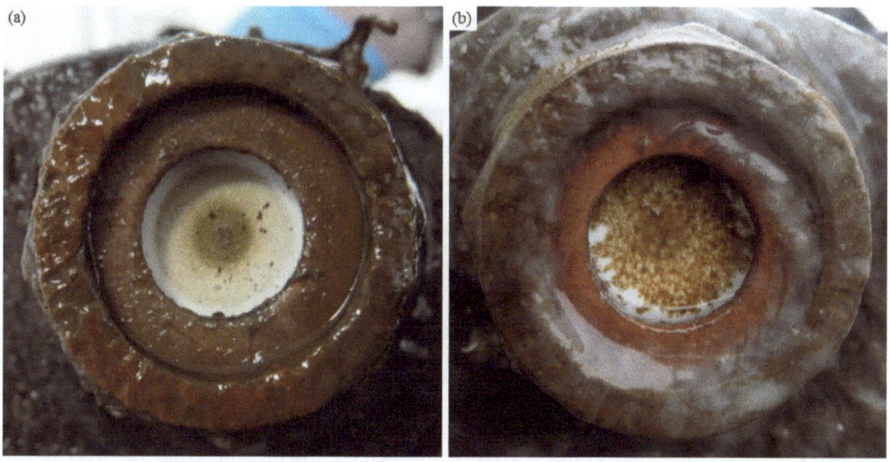

Fig. 14. Fouling of membrane filters and filter housings after approx. (a) 14 days and (b) 10 days immersion in wastewater during a sensor trial at Osberstown Wastewater Treatment Plant

The filter was changed and the system was replaced in the effluent tank. At the end of the trial (after a further 10 days, approximately) the membrane and housing were again examined. At this point the filter holder was heavily fouled as shown in Fig. 14(b). A layer of fouling material was again in evidence on the membrane filter itself. However, in this case the material was found to be only loosely adherent to the filter surface, and was readily removed with a deionized water wash bottle. This qualitative difference between the types of fouling material on the two filters highlights the variability of fouling which can occur over a relatively short period of time, presumably due to changes in sample composition and prevailing weather conditions. While no specific effect of fouling on sensor performance was observed in this trial, the rapid build-up of material suggests this could be an issue in longer-term deployments in wastewater and in other sample matrices. This is especially relevant in applications such as marine waters or oligotrophic freshwaters where very low levels of phosphate (or other nutrients) need to be detected. In such scenarios, local uptake of nutrients by biofilm on sensor or filter surfaces could significantly impact on the measured value.

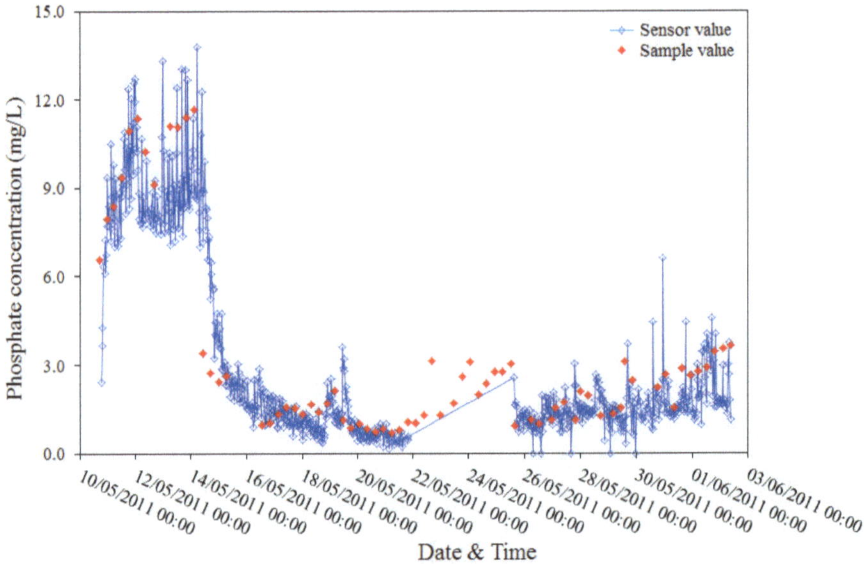

Fig. 15. In-situ sensor data (sensor value) and validation data (sample value) from samples collected using an auto-sampler during a trial of the second generation phosphate sensor at Osberstown Wastewater Treatment Plant

Fig. 15 shows the data collected from the sensor during the trial, as well as the results of samples collected by auto-sampler and analyzed in the laboratory. The sensor and sample values are generally in good agreement. The elevated phosphate levels during the first 5 days of the trial were due to the plant's phosphorus removal system (based on dosing with ferric chloride) being out of operation. After this system was restored to operation on 15/05/2011, the phosphate values quickly returned to normal levels, generally between 0.5 and 3.0 mg/L PO_4^{3-}. Sensor data was not collected during the period 22/05/2011–25/05/2011 due to maintenance works at the plant during this period, resulting in the water level in the effluent tank being below the sensor's sample intake port. As soon as the water level returned to normal and samples were again accessible to the sensor, the readings automatically recommenced and once again are in good agreement with validation sample values for the duration of the trial. This is encouraging as it demonstrates the sensor's robust ability to cope with an interruption of this nature and to clear the air which would have been drawn into the microfluidic system while the sample was inaccessible.

7 Current Status and Further Developments of the System

The data presented here shows that the microfluidic analyzer which has been developed represents a viable, low-cost and reliable method for long-term monitoring of phosphate across a range of applications including wastewater and surface water monitoring. While the current limit of detection of approximately 0.2 mg L^{-1} PO_4^{3-} represents a limitation on the applicability of the system, it can be expected to find significant applications in the monitoring of in-process and discharged effluent from municipal wastewater treatment plants, industry and other point sources of phosphate, as well as in the monitoring of surface waters known to be impacted by phosphate pollution. Moreover, Jordan and co-workers [39,40] have shown the importance of high temporal resolution monitoring for the detection of short-term, high-phosphorus concentration inputs to catchments that typically have low background levels of phosphorus. Detection of this kind of short-lived pollution event can be crucial to understanding the overall nutrient dynamics within a water body, and can only be reliably achieved using *in-situ* analytical devices.

Current applications notwithstanding, on-going efforts are aimed at improving the analytical performance of the phosphate sensor, particularly the limit of detection, so as to broaden the applicability and maximize the marketability of the system. Additional long-term field deployments are also planned, with the objective of validating the long-term performance of the analyzer in different applications, and examining the effect of biofouling during these long-term deployments. Biofouling potential is likely to vary significantly across different applications (e.g. wastewater, marine, freshwater), and between different locations in the same water system, and is also dependent on weather conditions and time of year. In general, however, biofouling is likely to represent a practical limitation to the (maintenance-free) deployable lifetime of this sensor and similar systems. Therefore it is essential that developers of water monitoring technology are familiar with the range of established and emerging [36,37,41-44] approaches to the prevention or limitation of biofouling, and work collaboratively with researchers engaged in this area, which is a key to achieving the long deployable lifetimes which will ultimately facilitate the broader uptake of *in-situ* sensing technologies.

The capabilities of the microfluidic analytical platform are also being expanded to allow detection and monitoring of a range of additional important water quality parameters. This work is being progressed via a number of distinct projects, and the range of parameters being targeted includes nitrate, ammonia, nitrite, pH, chemical oxygen demand (COD), and metals such as cadmium, mercury and lead. Colorimetric techniques for the detection of the various analytes are being assessed or developed, and optimized for application in microfluidic analytical devices, with the aim of developing a suite of single- or multi-parameter sensing systems which can be applied to the monitoring of water quality across a range of applications.

References

[1] Directive 2000/60/EC of the European Parliament and of the Council of 23 October 2000, establishing a framework for Community action in the field of water policy

[2] Council Directive 91/271/EEC of 21 May 1991, concerning urban waste water collection and treatment

[3] Council Directive 91/676/EEC of 12 December 1991 concerning the protection of waters against pollution by nitrates from agricultural sources

[4] Council Directive 98/83/EC of 3 November 1998 on the quality of water intended for human consumption

[5] Directive 2006/118/EC of the European Parliament and of the Council of 12 December 2006 on the protection of groundwater against pollution and deterioration

[6] Moore, T.S., et al.: Marine chemical technology and sensors for marine waters: potentials and limits. Annual Review of Marine Science 1, 91–115 (2009)

[7] Dzakpasu, M., et al.: Nitrogen removal in an integrated constructed wetland treating domestic wastewater. Environ. Sci. Health. A. Tox. Hazard. Subst. Environ. Eng. 46, 742–750 (2011)

[8] Manz, A., et al.: Miniaturised total chemical analysis systems, a novel concept for chemical sensing. Sensors and Actuators B: Chemical B1, 244–248 (1990)

[9] Niessner, R.: Measuring the invisible. Analytical Chemistry 82, 7863–7863 (2010)

[10] Allan, I.J., et al.: A toolbox for biological and chemical monitoring requirements for the European Union's Water Framework Directive. Talanta 69, 302–322 (2006)

[11] Diamond, D., et al.: Integration of analytical measurements and wireless communications - current issues and future strategies. Talanta 75, 606–612 (2008)

[12] Gardolinski, P.C.F.C., et al.: Miniature flow injection analyser for laboratory, shipboard and in situ monitoring of nitrate in estuarine and coastal waters. Talanta 58, 1015–1027 (2002)

[13] Doku, G.N., Haswell, S.J.: Further studies into the development of a micro-FIA (mu FIA) system based on electroosmotic flow for the determination of phosphate as orthophosphate. Anal. Chim. Acta 382, 1–10 (1999)

[14] Greenway, G.M., et al.: Characterisation of a micro-total analytical system for the determination of nitrite with spectrophotometric detection. Anal. Chim. Acta 387, 1–10 (1999)

[15] Petsul, P.H., et al.: The development of an on-chip micro-flow injection analysis of nitrate with a cadmium reductor. Anal. Chim. Acta 428, 155–161 (2001)

[16] Daridon, A., et al.: Chemical Sensing using an Integrated uFluidic System based on Colorimetrics: A Comparative Kinetic Study of the Bertholet Reaction for Ammonia Determination in Microfluidic and Spectrophotometric Systems. Sens. Actuators B 76, 235–243 (2001)

[17] Azzaro, F., Galletta, M.: Automatic colorimetric analyzer prototype for high frequency measurement of nutrients in seawater. Marine Chemistry 99, 191–198 (2006)

[18] Moscetta, P., Thompson, I.: The new in-situ chemical probes and the water quality monitoring platform. Environmental Risk Management Tools for Water Quality Monitoring, Southampton (UK) (March 30, 2009)

[19] Alliance for Coastal Technologies, Performance Demonstration Statement for the American Ecotech NUT 1000 (2008), `http://www.act-us.info/Download/Evaluations/Nutrient//AmerEco/`

[20] Alliance for Coastal Technologies, Performance Demonstration Statement for the WET Labs Cycle-P Nutrient Analyzer (2008), `http://www.act-us.info/Download/Evaluations/Nutrient/WETLabs/`

[21] Alliance for Coastal Technologies, Performance Demonstration Statement for the YSI 9600 Nitrate Monitor (2007), `http://www.act-us.info/Download/Evaluations/Nutrient/YSI/`

[22] Vuillemin, R., et al.: CHEMINI: A new in situ chemical miniaturized analyzer. Deep Sea Research Part I: Oceanographic Research Papers 56, 1391–1399 (2009)

[23] Patey, M.D., et al.: Determination of nitrate and phosphate in seawater at nanomolar concentrations. Trends in Analytical Chemistry 27, 169–182 (2008)

[24] Sieben, V.J., et al.: Microfluidic colorimetric chemical analysis system: application to nitrite detection. Analytical Methods 2, 484–491 (2010)

[25] Bey, A.K., et al.: A high-resolution analyser for the measurement of ammonium in oligotrophic seawater. Ocean Dynamics 61, 1555–1565 (2011)

[26] Floquet, C.F.A., et al.: Nanomolar detection with high sensitivity microfluidic absorption cells manufactured in tinted PMMA for chemical analysis. Talanta 84, 235–239 (2011)

[27] Beaton, A.D., et al.: Lab-on-chip measurement of nitrate and nitrite for in situ analysis of natural waters. Environmental Science & Technology 46, 9548–9556 (2012)

[28] Sequeira, M., Diamond, D.: Progress in the realisation of an autonomous environmental monitoring device for ammonia. Trends Anal. Chem. 21, 816–827 (2002)

[29] Bowden, M., et al.: Analysis of river water samples utilising a prototype industrial sensing system for phosphorus based on micro-system technology. J. Environ. Monit. 4, 767–771 (2002)

[30] Bowden, M., et al.: A prototype industrial sensing system for phosphorus based on micro system technology. Analyst 127, 1–4 (2002)

[31] Slater, C., et al.: Autonomous field-deployable device for the measurement of phosphate in natural water. In: Proc. SPIE 6755, Advanced Environmental, Chemical, and Biological Sensing Technologies V, vol. 67550L, pp. 67550L1–67550L8 (2007)

[32] Slater, C., et al.: Validation of a fully autonomous phosphate analyser based on a microfluidic lab-on-a-chip. Water Sci. Technol. 61, 1811–1818 (2010)

[33] Environmental Protection Agency, Ireland, Water Quality in Ireland, 2001-2003 (2005), `http://www.epa.ie/downloads/pubs/water/waterqua/epa_water_quality_2001-3.pdf`

[34] Neal, C., et al.: Phosphate measurement in natural waters: two examples of analytical problems associated with silica interference using phosphomolybdic acid methodologies. Sci. Total Environ. 251/252, 511–522 (2000)

[35] Hach Company, DR/890 Datalogging Colorimeter Handbook, pp. 523–528 (1999)

[36] Sullivan, T., Regan, F.: Biomimetic design for the development of novel sustainable antifouling methods for ocean monitoring technology. J. Ocean Technol. 6, 41–54 (2011)

[37] Delauney, L., et al.: Biofouling protection for marine environmental sensors. Ocean Sci. 6, 503–511 (2010)

[38] PALL Corporation, Polyethersulfone Membrane (Hydrophilic), http://www.pall.com/main/OEM-Materials-and-Devices/Product.page?id=3949
[39] Jordan, P., et al.: High-resolution phosphorus transfers at the catchment scale: the hidden importance of non-storm transfers. Hydrol. Earth Syst. Sci. 9, 685–691 (2005)
[40] Jordan, P., et al.: Characterising phosphorus transfers in rural catchments using a continuous bank-side analyser. Hydrol. Earth Syst. Sci. 11, 372–381 (2007)
[41] Chapman, J., et al.: Phthalate doped PVC membranes for the inhibition of fouling. J. Membr. Sci. 365, 180–187 (2010)
[42] Chapman, J., et al.: Period four metal nanoparticles on the inhibition of biofouling. Colloids and Surfaces B: Biointerfaces 78, 208–216 (2010)
[43] Chapman, J.: The development of novel antifouling materials: A multi-disciplinary approach. PhD thesis, Dublin City University (2011)
[44] Sullivan, T.: Investigation of biomimetic and engineered surface textures for biofouling reduction and prevention on environmental sensing platforms. PhD thesis, Dublin City University (2012)
[45] Cleary, J., et al.: Analysis of phosphate in wastewater using an autonomous microfluidics-based analyser. World Academy of Science, Engineering and Technology 28, 196–199 (2009)

Nitrate Anion Sensors: Their Applications and a Case Study of Their Status in Waste Water from Selected Areas of Coastal Guyana via a Spectrophotometric Method

R.C. Jagessar

Department of Chemistry,
University of Guyana, Turkeyen Campus,
P.O. Box, 10-1110, Georgetown, Guyana
raymondjagessar@yahoo.com

Abstract. It is not an easy task to synthesize nitrate anion selective sensors considering that a nitrate anion is trigonally shaped, heavy solvated, has weak basicity and is difficult to form robust hydrogen bonds with ligands. In addition, anion coordination is a difficult aspect of supramolecular chemistry. Many receptors reported to date are halide selective, but few are nitrate selective. Nitrate anion receptors synthesized to date include polyammonium, amide and urea receptors amongst others. In one instance the association constant of a nitrate receptor showing the highest degree of complexation is reported to be $K_a = 110M^{-1}$ in 50% DMSO-d_6/$CDCl_3$ and $20M^{-1}$ in 100% DMSO-d_6 respectively. Globally, the presence of nitrate anions in water beyond the threshold limit can be deleterious to both flora and fauna life. Guyana's waste and domestic water needs monitoring to assess the concentration of toxic anions and cations. High levels of nitrate anions beyond the threshold limit can induce the "blue baby" syndrome amongst other effects. One aspect of this chapter focuses on the determination of nitrate anion concentration from twelve selected areas of coastal Guyana using an ultraviolet spectrophotometric method. Areas monitored in Guyana were 58 Livelihood Village, Rose Hall Town, Skeldon GUYSUCO Estate, Good Hope, Ogle, Stabroek, Parika, Supenaam, Spring Garden in one instance. In another research endeavor, Georgetown Seawall Kingston, LBI, GUYSUCO Estate and the Canje river water at 20 and 40ft from the effluent were monitored. The results showed that the concentrations of nitrates were not as high as expected and are below the internationally accepted threshold values. In the first instance, the average concentration was measured to be 0.030 mg/L (± 0.039 mg/L), 0.064 mg/L (± 0.00292 mg/L), 0.203 mg/L (± 0.00738 mg/L), 1.772 mg/L (± 0.00 mg/L), 2.363 mg/L (± 0.839mg/L), 0.33 mg/L (± 0.156 mg/L), 0.168 mg/L (± 0.043mg/L), 0.142 mg/L (± 0.043 mg/L) and 0.178 mg/L (± 0.039 mg/L) respectively. In the second instance the measured results were 0.45 mg/L (± 0.039 mg/L), 0.01 mg/L (± 0.0033mg/L) and 0.015 mg/L (± 0.0023 mg/L) NO_3^- for Georgetown Seawall (Kingston area), LBI GUYSUCO Estate and the Canje river, 20 feet from the effluent respectively. The results were accepted at the 95% confidence level using statistical analyses. The US public Health Service designated the safe limit

S.C. Mukhopadhyay & A. Mason (Eds.): *Real-Time Water Quality Monitoring*, SSMI 4, pp. 149–169.
DOI: 10.1007/978-3-642-37006-9_7 © Springer-Verlag Berlin Heidelberg 2013

for nitrate in water as 45mg/L. The applicable range of concentrations using the above method is 0.1-2 mg/L NO_3^-. A maximum level of 45 mg/L is established as worldwide guidance for nitrate concentration in water. In Europe, the maximum permitted levels of nitrate in potable water is 50.0 mg/L, while in the US-EPA has established a guideline for the maximum level of nitrate-nitrogen of 10 mg/L. It can safely be informed that the twelve selected areas of coastal Guyana chosen are not polluted with anions. In an effort to improve water quality, the Government of Guyana has embarked on the construction of sand filtration and water treatment plants along the inhabited coastland of Guyana.

Keywords: nitrate anion selective receptors, association constant, threshold limit, flora and fauna, Spectrophotometric method, waste and domestic water, effluent, 95% confidence level, nitrate ion threshold limit.

1 Introduction

This chapter outlines a survey of some nitrate anion sensors reported to date, their application and a review of the status of nitrate ion concentration in waste water from a total of twelve selected areas of coastal Guyana as determined via a spectrophotometric method [44-46].

1.1 Structure and Properties of the Nitrate Anion

Nitrate (NO_3^-) anions are salts of nitric acid and are triangular in shape and exist as the resonance forms (i) to (iii) [1] as shown in Fig 1.

NO_3^- has a D_{3h} symmetry with equivalent N-O bonds [2,3]. The average N-O bond length in many nitrate structures is 1.231 ± 0.025A and the mean ONO angle is 119.98 ± 2.15. Charge density on the individual atom also varies, depending on the

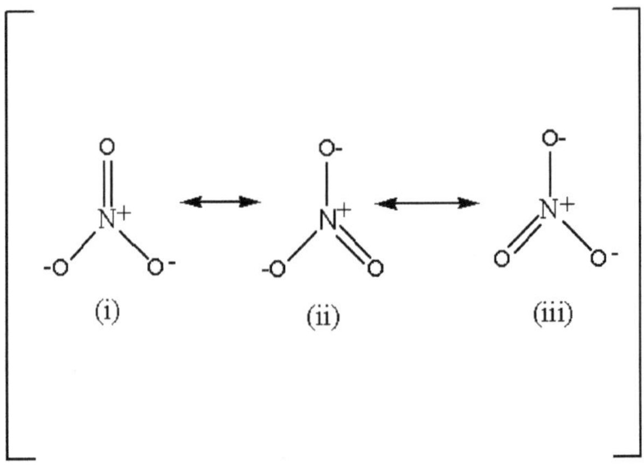

Fig. 1. Resonance in Nitrate ion

environment and the method of calculation. Nitrate is also the conjugate base of nitric acid (pK_a = -1.3). Nitrate ions form water soluble salts with most cations. An extensive hydration shell exists around NO_3^- [4]. Due to its heavy solvation and weak basicity, NO_3^- is weakly coordinative and is thus difficult to form robust hydrogen bonds with ligands [5-8]. Hence, developing NO_3^- selective receptors that use hydrogen bonds is a challenge in supramolecular anion coordination chemistry [9-12].

1.2 Difficulties Encountered in Anion Coordination Chemistry

Anion coordination chemistry is a very difficult area of pursuit as compared with cation coordination chemistry [9-43]. Furthermore, the development of nitrate receptors is more difficult compared with other anions such as chloride, sulphate and phosphate [14-27]. Difficulties in anion coordination chemistry results from the following [9-12]:

(a) **Size:** Compared to their isoelectronic cations, anions are much larger and therefore require much larger arrays for bonding. Their size requires that the macro-cycles developed must be of a larger dimension.
(b) **Geometry:** Unlike cations which are spherical in shape, anions display a variety of geometries. These include spherical, trigonal planar, linear, tetrahedral and octahedral.
(c) **Solvation energies:** Anions display a higher degree of solvation energies than cations which must be overcome by the host species if successful anion complexation is to be realized.
(d) **pH dependence:** Many anions are pH dependent. Below a certain pH range, they become protonated and cease to be anions. These include carboxylates, nitrates, sulphates and phosphates which are protonated to acidic pH.
(e) **Charge:** Anions are negatively charged. Thus, the receptor must contain complementary binding sites such as cationic functions, lewis acids and/or hydrogen bonds for nitrate complexation.

1.3 Synthetic Receptors for Nitrates

As previously mentioned, it is not an easy task to synthesize nitrate anion sensors, considering the fact that the nitrate anion is trigonally shaped and that many receptors reported to date have shown the selectivity trend: $H_2PO_4^- > HSO_4^- > Cl^- > Br^- > NO_3^-$ [14-27]. Thus, designing a host for the nitrate anion is not an easy task for the supramolecular chemist. Synthetic receptors developed for the sensing and complexation of nitrate anion include amide, polyammonium, guanidinium and urea based anion receptors amongst others [38-41]. It is beyond the scope of this chapter to describe them all. Biologically, the amide group, -NHCO-, is used by proteins to complex the nitrate anion [28,29]. Thus, mimicking nature, supramolecular chemists have constructed amide anion receptors for the selective complexation of nitrate anions. These include neutral amide receptors and neutral amide receptors in tandem with a positively charged center [31-37]. For example, Smith and coworkers have synthesized a ditopic receptor (1) for ion pair binding [29]. The receptor (Fig. 2, left) binding motif consists of a cation binding site (crown ether) and an anion (isophthamide) binding site. The crystal structure of (Fig. 2, left) bound to NO_3- with different cations (Li^+, Na^+ and K^+) has been reported [29].

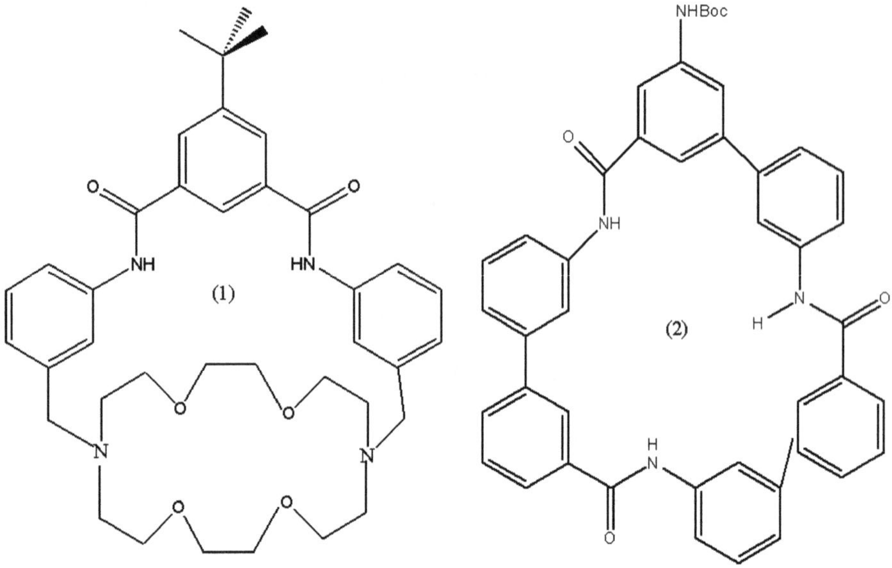

Fig. 2. Amide based macrocyclic nitrate receptors

A cyclic neutral based triamide based anion receptor (Fig. 2, right) has been reported [30]. The association constant for the receptor- NO_3- 1:1 complex in DMSO is one of the largest reported (K_a = 110M^{-1} in 50% DMSO-d_6/$CDCl_3$ and 20M^{-1} in 100% DMSO-d_6).

Positively charged amide anion receptors for the selective complexation of nitrate anion have also been reported [31-33]. For example, the anion receptors α,α,β, β-5,10,15,20- *meso tetrakis* (*ortho* cobaltocenium) amido phenyl porphyrin (Fig. 3, top left) and 5α, 10α-di(*ortho*-amido cobatocenium phenyl)-15,20-diphenyl porphyrin (Fig. 3, top right) receptors have been synthesized, characterized and shown to exhibit the rare selectivity trends: NO_3^- > Cl^- > Br^- . (Fig. 4, top left) and (Fig. 4, top right) are ditopic and monotopic respectively. For (Fig. 3, top left), Log K = 3.58, 3.41 and 3.95 for Cl^-, Br^- and NO_3^- in CD_3CN respectively. For (Fig. 3, top right), the association constant is 5.60×10^2, 1.10×10^3 and 6.76×10^3 for Cl^-, Br^- and NO_3- in CD_3CN respectively. This selectivity trend, which is rare, is in contrast to other porphyrin anion receptors such as (Fig. 3, bottom left) which shows the selectivity trend: Cl^- > Br^- > NO_3^- with association constant of 860, 820 and 160 respectively [31-33]. In contrast, receptor illustrated in Fig. 3, bottom right, shows the nitrate selectivity trend: NO_3- > HSO_4^- > Cl^-.

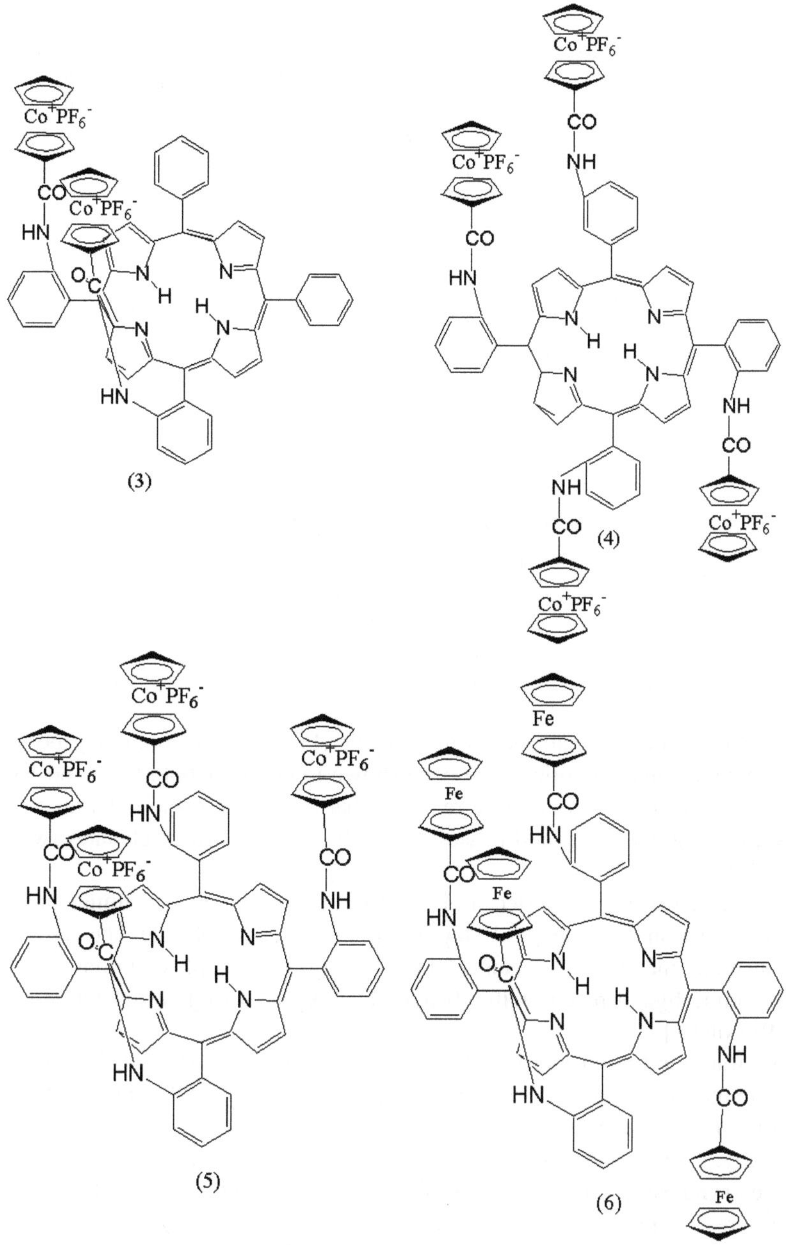

Fig. 3. Macrocyclic porphyrin nitrate selective anion receptors

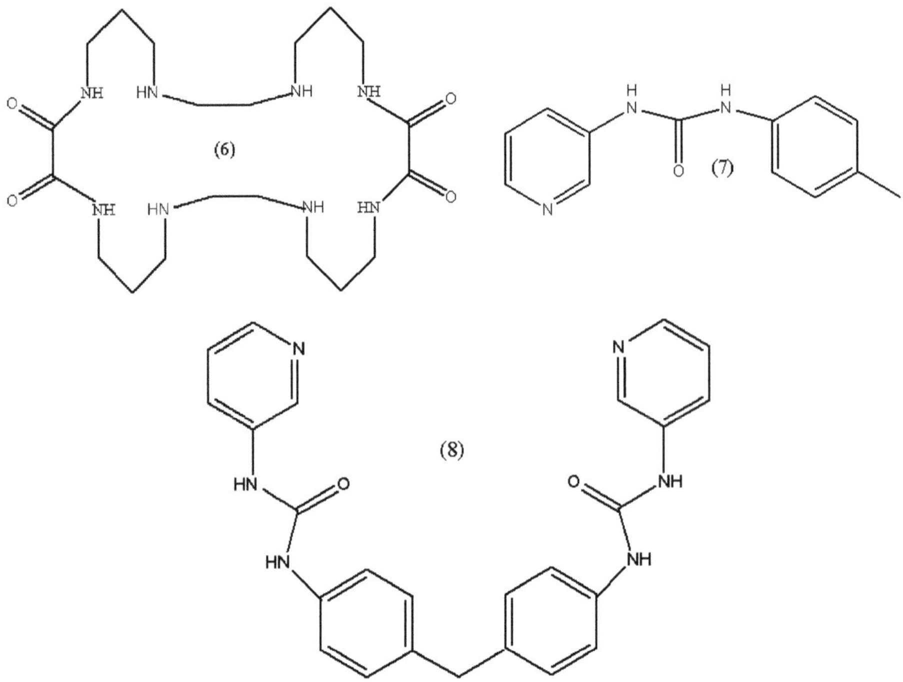

Fig. 4. Macrocyclic and acyclic urea nitrate anion receptors

The 28-membered cyclam based macrocycle (Fig. 4, left), containing both amide and ammonium functionalities have been shown to complex both Cu^{2+} and NO_3^-. X-ray crystal analysis of the structure revealed that the macro-cycle is tetraprotonated , adopts a rectangular shape and possesses a hollow cavity which chelates two nitrate anions [38].

Steed and coworkers reported a class of ditopic receptors containing the pyridyl and urea functionalities (Fig. 4, top right). The supramolecule acts as a ligand for transition metals and retains their ability to complex anions [39]. The ureidopyridine ligand reacts with silver nitrate ($AgNO_3$) to give discrete complexes [Ag $(19)_2$(s)NO_3].

The nitrate binding properties of the bis (ureidopyridine) (Fig. 4, bottom) and its silver complex is yet to be reported. The presence of a metal center in these urea based receptors has enhanced their affinity for nitrate anions.

Recently, a series of neutral, urea appended free base porphyrins and their Zn(II) have been synthesized and characterized [24, 25].. The (α,α,α,α) -5,10,15,20-*tetrakis* (2 –(arylurea)phenyl) porphyrins bind strongly (K (M^{-1}) > 10^2-10^5) to Cl^- anion in DMSO-d_6 and also in the more competitive solvent system DMSO-d_6/D_2O (88: 12, v/v). The porphyrin derivatives exhibited significant binding selectivity since they complexed with the spherical Cl- and Br- to a much greater extent than with the tetrahedral $H_2PO_4^-$ and HSO_4^- and HSO_4^- and trigonal NO_3^- anions in DMSO-d_6 with a selectivity trend of Cl^- > Br^- > H_2PO_4- > HSO_4- > NO_3^-. The latter is novel for any urea anion binding system and is the largest association constant reported for any anion receptor to date. The corresponding metalloporphyrins exhibited a decrease in binding strength and selectivity in DMSO-d_6. X-ray crystallography revealed (α,α,α,α) -5,10.15.20 – *meso tetrakis* (2-(4-chlorophenylurea)phenylporphyrin to be the first coordination complex of an anion (chloride and bromide) bound by a neutral free base porphyrin [24, 25].

1.4 Applications of Nitrate Anion Sensors

Nitrate anion receptors synthesized have found application in chemical sensor technology [40-43]. Only some representative examples will be described here. For example, Poly(vinyl chloride) (PVC)-based membranes of N,N-bis-succinamide-based dendritic molecule with tetrabutyl ammonium bromide (TBAB), as a cation inhibitor and dibutylphthalate (DBP), dioctylphthalate (DOP), dibutyl (butyl) phosphonate (DBBP) and 1-chloronaphthalene (CN) as plasticizing solvent mediators were prepared and used as nitrate ion-selective electrodes. Optimum performance was observed with the membrane having I-PVC-TBAB-DBP in the ratio 1:33:1:65 (w/w). The electrode has been shown to have a linear response to nitrate with a detection limit of $3.9 \times 10^{-5} \pm 0.07$ M and Nernstian compliance (57.0 ± 0.2 mV/decade) between pH 2.8 and 9.6 with a fast response time of about 20s. The selectivity coefficient values of the order of 0.001 for mono-; bi- and trivalent anions, revealed high selectivity for NO_3^- ions over these anions. The electrodes were used for a period of 45 days with good reproducibility of results. The electrodes have been used as good analytical tool for the determination of nitrate ions in waste water samples [41].

A flow through, nitrate-selective, electrochemical sensor developed by electropolymerization (doping) of polypyrrole nanowire (PPy-NWs) electrodes has been reported [43]. The flow through micro-fluidic platform enabled a realistic and controlled testing environment. Cyclic voltammetry, amperometry and flow-through analysis were conducted to evaluate the sensor performance for determination of nitrate ions in two sets of calibration solutions (DI water and IAPSO standard seawater). The sensor response was linear over the range from 10 μM (0.14 ppm Nitrate-N) to 1 mM (14 ppm Nitrate-N), with a sensitivity of 1.17-1.65 nA/μM and detection limit of (4.5 ± 1) μM or (0.063 ± 0.014) ppm Nitrate-N. The electrochemically doped PPy-NWs electrodes show improved selectivity for chloride, sulfate, phosphate and perchlorates ions, without the loss from interfering ions [42].

Zn(II) complex-based potentiometric sensors constructed for the selective determination of nitrate anion present in water samples have been reported [44]. Two Zn(II) complexes coordinated by neutral tetradentate ligands, N,N'-ethylene-bis(N-methyl-(S)-alanine methylamide) and N,N'-ethylene-bis(N-methyl-(S)-alanine dimethylamide), proved efficient as anion-selective carriers, while common phthalocyanine Zn(II) complex rarely responded to any anions. The combination of these novel Zn(II) complexes with dioctylsebacate as a plasticizer offered high sensing selectivity for nitrate anion. They exhibited near-Nernstian slopes in the wide linear concentration range of 5.0×10^{-5} to 1.0×10^{-1} M, and operated well in the wide pH range from 4 to 11 with the response time of less than 25s. The potentiometric selectivity coefficients were evaluated using the fixed interference method, indicating that the two Zn(II) complexes exhibited better selectivity for nitrate anion with respect to a wide variety of inorganic anions. Although the chloride anion proved efficient as an interfering species at a concentration higher than 1.0×10^{-3} M, the novel Zn(II) complex-based sensors were applicable in determination of the nitrate anion after adding silver sulfate to remove the chloride anion [43].

The second aspect of this paper focuses on the monitoring of nitrate anion concentration in waste water from twelve selected areas of coastal Guyana via a spectrophotometric method [44-46]. Guyana, a Third World developing country has an area of 214,970 km^2, and is situated on the north-eastern coast of the South American continent. It is often described as the "Land of Many Waters". The country is divided into five main biophysical divisions: the Coastal Plain, the Savannahs, the Pre-Cambrian Lowland Region, the Southern Upland Region and the Pakaraima Highlands. The coastal zone of Guyana lies between 0.5 to 1.0 m below high spring tide level of the Atlantic Ocean, making it vulnerable to flooding, erosion and salinization. The coastal zone is 430 km long and 26-77 km wide. Approximately 90 percent of the nation's population of 750,000 lives within the coastal zone. Even though it constitutes less than seven percent of the country's total land area, it has the most fertile soils.

1.5 Waste Water

Waste water is one that has been used for washing, flushing, or that which is released from manufacturing processes [44-46]. In Guyana, groundwater provides 90 percent of the potable water supply and is extracted mainly from the coastal artisan basin [47]. However, potable water can be contaminated. The most common examples of resource contamination in Guyana are those arising from water pollution: elements such as mercury, anions: cyanide, phosphates, nitrates, chlorides and cations in calcium and other wastes from mining etc. However, their level of concentration needs to be determined. Others include untreated human and animal wastes in water supplies and wastes from many industries in water tables [48,49]. Guyana's water has become polluted over the years due to industrial activities such as gold, bauxite mining etc. Recently, the Guyana government has resumed the mining of manganese, an element needed for the production of iron ores. Also, the prospect looks good for uranium mining. Whatever it is, our water will become polluted unless correct environmental procedures and practices are put in place and continual monitoring is implemented.

Providing sufficient quantities of high quality water to satisfy our domestic, industrial and agricultural needs is an ongoing global problem. Increasing population size, climate change and pollution will only exacerbate the situation. There is no physical shortage of water on the planet earth. It covers 70% of the globe. However, 97% of the world water is saline and is thus non-drinkable. 2% is locked in glaciers and polar ice caps. This leaves 1% to meet humanities needs [49].

1.6 *Importance of Nitrate Anions*

Some anions are toxic beyond their threshold value because of their mobility in living systems and ability to cross cell membranes. Toxic anions are poisonous and can cause harm or even death via malfunctioning of the organs such as the kidney etc. They enter the body via drinking waters, food, fruits and vegetables, fish and other foods in general that may have been exposed to such waters. Thus, the levels of concentration of anions must be controlled in our water bodies. The U.S. Public Health Service designated safe limits for nitrate in water as 45mg/L or 10mg/L nitrate nitrogen. However, there are deviations from this value in many places on the planet. For example, internationally, the Helmholtz center for environmental research in Germany has shown that Gazan groundwater contains levels of nitrate (NO_3^-) ions up to eight times higher than the World Health Organization (WHO) safe standard. The high level of nitrates could be poisoning many newborn babies. The high levels of nitrates originate mainly from run off from manure used as fertilizer [50].

Biologically, high concentrations of nitrates in drinking water are particularly dangerous to small children, infants and fetuses [51]. High concentrations of nitrate in drinking water can cause blood disorder in babies less than six months of age. In infants' intestine, bacteria, notably *Escherichia coli*, reduce the nitrate ions (NO_3^-) to nitrite ions (NO_2^-). The nitrite ions are absorbed into the bloodstream where they oxidized iron (ii) Fe^{2+} in the hemoglobin to iron (iii) Fe^{3+}. The presence of hemoglobin containing oxidized ions, which is known as Met-hemoglobin, reduces the oxygen carrying capacity of the blood. Babies are more vulnerable to high nitrate levels than adults because their stomachs are less acidic. This allows the *E. coil* to colonize higher up the digestive tract and convert the nitrate ions to nitrite prior to absorption. The use of unsterilized feeding bottles can increase the risk of met-hemoglobin formation. The danger lies in the ability of the bacteria present in the feeding bottle to convert the nitrate in the water to nitrite. A concentration of met-hemoglobin in the blood above 25% causes the skin and lips of the infected infant to take on a bluish hue, the 'Blue Baby Syndrome' or Met-hemoglobinaemia. If untreated, the condition can be fatal. Boiling water contaminated with nitrate increases the nitrate concentration and the potential risk.

Nitrate anions have both beneficial and harmful uses. On the positive side, nitrates (NO_3^-) are essential plant nutrients that are important ultimately for protein synthesis [51-52]. They are responsible for the growth of plants and also nitrogen fixation. Nitrates are found in nature since they are the end product of the aerobic decomposition of organic nitrogenous matter as well as the decomposition of organic micro-organisms via the nitrogen cycle (see Fig. 5).

Other major contributors of nitrate are chemical fertilizers from cultivated land and drainage, from livestock feedlots as well as domestic waste and some industrial waters in the course of leakage. Unpolluted natural waters contain only small amounts of nitrate. In surface water, nitrate is a nutrient which is taken up by plants and assimilated into nucleic acid. Stimulation of plant growth, especially of algae, may cause quality problems associated with eutrophication and consequent death and decay of these algae produces secondary effects on water quality which may be detrimental. Eutrophication results in oxygen deprivation and fish deaths.

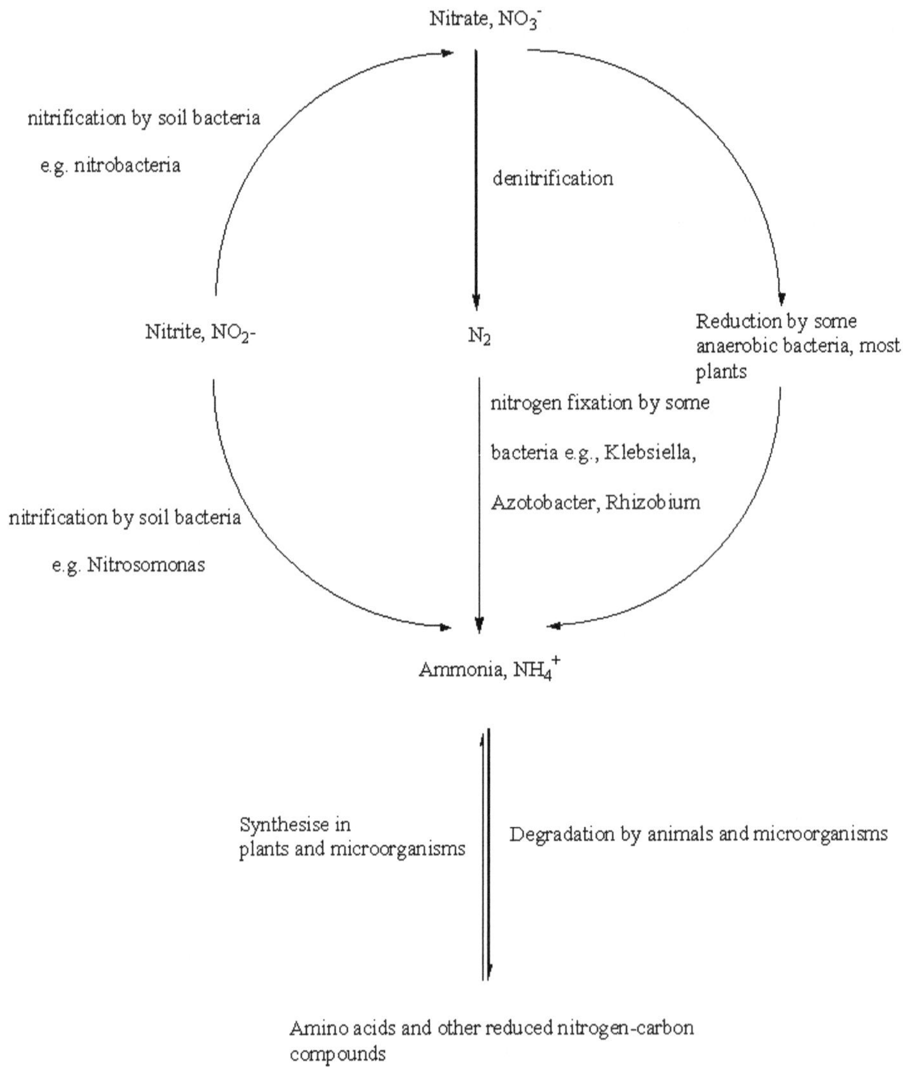

Fig. 5. The nitrogen cycle

Nitrogen, an element of nitrate is part of all living cells and is a necessary part of all proteins, enzymes and metabolic processes involved in the synthesis and transfer of energy. It is also a component of chlorophyll, the green pigment of the plants that is responsible for photosynthesis. Nitrogen also helps plants with rapid growth, increasing seed and fruit production and improving the quality of leaf and forage crops.

Nitrate is a natural constituent of plants and is found in vegetables in varying levels, depending on the amount of fertilizers applied etc. According to the WHO report, most adults ingest about 20-70 milligrams of nitrate (nitrogen) per day.

Foods that are good sources of nitrogen are lettuce, celery, spinach etc. Nitrate is not harmful when foods containing nitrate are eaten as part of a balanced diet.

The excess of nitrates are largely due to the presence of animal manure in the water bodies. This manure is concentrated with ammonia which is not only highly toxic to fish, but can also be converted to dangerous nitrates. Nitrate-nitrogen also comes from the atmosphere and more specifically from snow, fog, or through the decay of material in soil and sediments. Added to their effects to humans they promote the growth of plants and algae in our water bodies thus increasing their populations. With the increase growth in these plants and algae in the soil and water, there is a competition for oxygen in the water body leading to tremendous deaths of these organisms which eventually results in a buildup in decaying matter which may even lead to the filling up of ponds and lakes, thus destroying or creating an imbalance in the entire ecosystem, if not dealt with in time. This process is specifically being referred to as *Eutrophication*, the process of enriching water or algal blooms. Eutrophication is also a cause of the loss of diversity in the sea floor community (including seaweeds, sea grasses, and corals), and amongst planktonic organisms. Since planktonic algae are the basis of marine life, their absence will affect the

Table 1. Sample locations for water samples collected from nine different locations

Key	Location
1a	No. 58 Village
1b	Rose Hall Town
1c	Skeldon GUYSUCO Estate
2a	Good Hope
2b	Ogle
2c	Stabroek
3a	Parika
3b	Supenaam
3c	Spring Garden
4a	Georgetown, Seawall (Kingston area), near effluent
4b	Georgetown Seawall (Kingston area), 20ft from effluent outflow
4c	Georgetown, Seawall (Kingston area), 40ft from effluent outflow
5a	LBI GUYSUCO Estate, near effluent
5b	LBI GUYSUCO Estate, 20ft from effluent
5c	LBIm GUYSUCO Estate, 40ft from effluent
6a	Canje river: Close to effluent
6b	Canje river: 20 ft from effluent
6c	Canje river: 40 ft from effluent

ecological food chain and food webs tremendously. The excess of nitrates contribute to the high levels of eutrophication in most of our drains and trenches in the capital city (Georgetown) in Guyana, and in other towns and villages in Guyana. Sources of nitrate contamination include fertilizers, animal wastes, septic tanks, municipal sewage treatment systems, and decaying plant debris.

Nitrates are an important component of the nitrogen cycle [52, 53]. Most plants withdraw NO_3^- and ammonium $(NH_4)^+$ from the soil and then convert NO_3^- into more NH_4^+ which plants use to produce the amino acid glutamine and other organic nitrogen compounds. Nitrate is the nitrogen form predominantly absorbed by plants. The first step in the cycle is fixation (reduction of nitrate and nitrite) of atmospheric nitrogen by nitrogen fixing bacteria to yield ammonia, NH_3 or NH_4^+. Soil bacteria derive their energy by oxidizing ammonia to nitrite (NO_2^-) and ultimately to nitrate, NO_3^-. These bacteria are abundant and so active that all ammonia reaching the soil is oxidized to nitrate. This process is known as nitrification. Plants and many bacteria take up and readily reduce nitrate and nitrite through the action of nitrate and nitrite reductases. The ammonia that is formed is incorporated into amino acids by plants [9, 10].

All metallic nitrates are soluble in water and some are readily extracted into organic solvents. These are formed when nitrogen from ammonia or other sources combines with oxygenated water. Areas monitored for nitrate in Guyana are in the county of Berbice, Essequibo and Demerara [55, 56]. Those monitored were No. 58 Livehood Village, Rose Hall Town, Skeldon, GUYSUCO Estate, Good Hope, Ogle, Stabroek, Parika, Supenaam, Spring Garden and at three different sites at the Kingston Seawall (Georgetown), the LaBonne Intention, LBI GUYSUCO (Guyana Sugar Corporation) estate and the Canje creek. These are tabulated in Table 1.

2 Determination of Nitrate Anion via a Spectrophotometric Method

Nitrate anion concentration was determined via an ultraviolet spectrophotometric method [44-46]. This is applicable to the analysis of drinking water, surface waters, domestic and industrial waters. The method can be modified to compensate for turbidity, color, salinity and dissolved organic compounds in the sample. The determination of nitrate (NO_3^-) is difficult because of the relatively complex procedures involved, the high probability that interfering constituents will be present and the limited concentration ranges of the various techniques. To determine nitrates, an ultra violet technique that measures the absorbance of nitrates at 220 nm is suitable for screening uncontaminated water (low in organic matter). This method is applicable to the analysis of drinking, surface and saline waters. The minimum quantity required is 100 ml. The NO_3^- calibration curve follows Beer's law up to 11 mg N/L.

2.1 Procedure

The procedure for the determination of nitrate anion can be found in previously published works [44-46].

2.2 Calculations

Some of the formulas used for the determination of nitrate anions are as follows [56, 57]. The corrected or true absorbance of the sample, A, can be calculated using (1).

$$A = T_A - B_A \tag{1}$$

Where T_A is the absorbance of the sample and B_A is the absorbance of a blank. Next, nitrate ion concentration (mg/L), N_{conc}, is calculated via (2), where 4.43 is a constant value and V is the sample volume (mL).

$$N_{Conc} = \frac{A \times 4.43}{V} \tag{2}$$

Statistical data were analyzed in accordance with literature [56-57]. Those investigated were the *standard deviation, variance and confidence limit.* Variance is the square of the standard deviations. It is represented as σ^2, where σ is the standard deviation. The confidence limit at the 95% level was calculated using (3).

$$CL(\mu) = \bar{X} + \frac{Z\sigma}{\sqrt{N}} \tag{3}$$

Where $\bar{X}$ is the measured mean, Z is a statistical factor related to the probability limit required at the 95% level and N is the number of entries. The confidence interval indicates that the true mean, μ, is likely to lie within a certain distance from $\bar{X}$.

2.3 Summary of Results

Using the spectrophotometric method and the statistical analyses, the nitrate ion concentration (mg/L) obtained for the twelve selected areas (as described in Table 1) are shown in Table 2.

Table 2. Nitrate ion concentration (mg/L) obtained for the twelve selected areas

Key	NO_3^- concentration with standard deviation
1a	0.030 mg/L ± 0.00624 mg/L
1b	0.064 mg/L ± 0.00292 mg/L
1c	0.203 mg/L ± 0.00738 mg/L
2a	1.772 mg/L ± 0.000 mg/L
2b	2.363 mg/L ± 0.839 mg/L
2c	0.33 mg/L ± 0.156 mg/L
3a	0.168 mg/L ± 0.043 mg/L
3b	0.142 mg/L ± 0.043 mg/L
3c	0.178 mg/L ± 0.039 mg/L
4b	0.45 mg/L ± 0.039 mg/L
5b	0.01 mg/L ± 0.0033 mg/L
6b	0.015 mg/L ± 0.0023 mg/L

3 Status of Nitrate Anion in Waste Water from Selected Areas of Coastal Guyana via an Ultra Violet Spectroscopy Method

Due to increase agricultural activities and industrialization in Guyana, it is necessary that its water be monitored for toxic anions and cations. Also, it has been found recently that high level of nitrate polluted Gaza wells of Palestinian water to a magnitude of 8 times the limit set internationally. Thus, twelve selected areas of coastal Guyana were monitored for nitrate ion concentration, as previously described in Table 1.

The applicable (permissible range) for the detection of $NO3^-$ concentrations using the above spectrophotometric method is 0.1-2 mg/L NO_3^-. A maximum level of 45 mg/L in water is established internationally.

In Europe, the maximum permitted levels of nitrate in potable water is 50 mg/L, while in the US-EPA (Environmental Protection Agency) has established a guideline for the maximum level of nitrate-nitrogen of 10 mg/L. Nitrate forms a component of total dissolved solids and is widely used as an indicator of water quality. The high level of nitrates originate mainly from run off from manure, manure piles (either human or animal) used as fertilizer which is accentuated by the soil's high permeability [51]. High levels of nitrate can also lead to algae blooms, and when nutrients become limiting such as potassium, phosphate or nitrate, then eutrophication can occur which means an excess of nutrients resulting in oxygen deprivation and fish deaths. While nitrate is much less toxic than ammonia or nitrite levels over 30 ppm of nitrate can inhibit growth, impair the immune system and cause stress in some aquatic species.

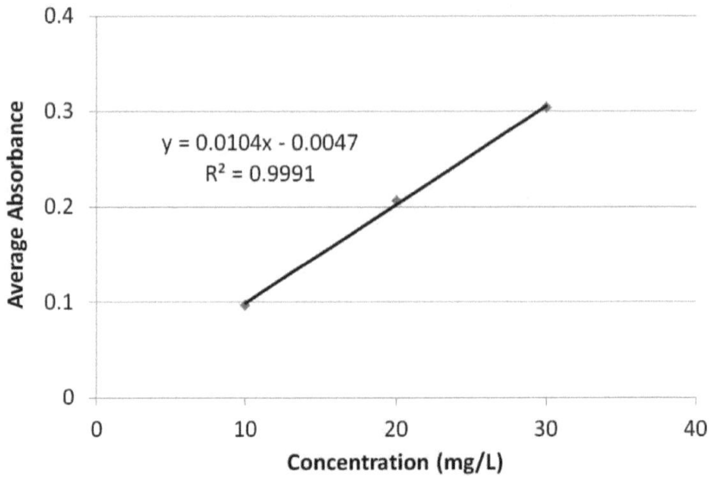

Fig. 6. Plot of the average absorbance of the nitrate standard against its concentration at 10, 20 and 30 mg/L nitrate for areas (1a)-(3c)

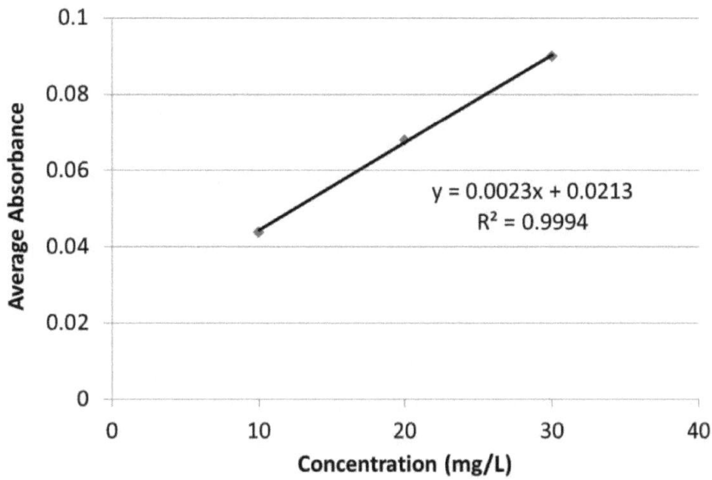

Fig. 7. Plot of the absorbance versus concentration of the nitrate standard (potassium nitrate) at concentrations of 10, 20, and 30 ppm of nitrate for the second set of experiment for areas (4a)-(6c)

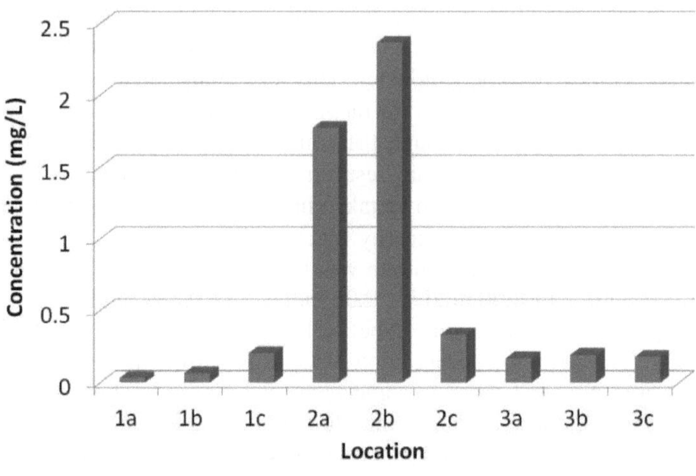

Fig. 8. Results obtained for the analysis of nitrate at the three counties of Guyana: (1) Demerara, (2) Essequibo and (3) Berbice

In terms of sampling, water samples were collected from the twelve selected areas during the non-rainfall period so as to eliminate contamination resulting from rainfall run off. For comparative purposes, KNO_3 standards at concentrations of 10 ppm, 20 ppm, 30 ppm and 40 ppm were prepared and the absorbance at these concentrations was taken. Experimentation was performed in triplicates. A plot of average absorbance versus concentration (in mg/L) is shown as the standard calibration curve in Fig. 6. Accordingly, the absorbance taken for the different concentrations of standards is found to be proportional to concentration. Thus, as concentration is increased, absorbance also increased linearly. This standard graph can also be used to estimate the nitrate concentration of samples at any given absorbance. The NO_3^- calibration curve follows Beer's law up to 11 mg N/L. Two sets of experiments were conducted over a two year period and hence two standard calibration curves are presented in Fig. 6 and Fig. 7. The first sets of experiments were conducted for areas (1a) – (3c), whereas the second sets of experiments were conducted for areas (4a) – (6c).

Fig. 8, which is a plot of concentration of nitrate in mg/L versus locations, indicates that the lowest concentration of nitrate was found in the county of Berbice, whereas the highest is in Demerara. Nitrate ion concentration per county decreases according to the trend: Demerara > Essequibo > Berbice. Considering the county of Berbice, values of 0.030 mg/L (± 0.0064 mg/L), 0.064 mg/L (± 0.00292 mg/L) and 0.20 mg/L (± 0.00738 mg/L) were obtained for (1a) Number 58 Livelihood Village, (1b) Rose Hall Town and (1c) Skeldon Sugar Estate respectively. The highest concentration at Skeldon GUYSUCO (Guyana Sugar Corporation) waste water is due to the use of nitrate fertilizers used by sugar cane farmers for the cultivation of sugar cane and the permeability of the soil. Interestingly, the lowest concentration was obtained for Number 58 Livelihood Village, a village where predominantly cash crop cultivation is performed but which has seen a decline in the past decade due to migration of residents overseas. This low value may be ascribed to the increased rate of evaporation and transpiration that caused the plants to use up the nitrates almost as immediately as they become available. This also contributed to the soil being very dry, hence there was little downward movement of water in the soil, and thus fewer nitrates were being leached through the soil to the nearby water system where the samples were collected. Also, it may result from the low permeability of the soil. It is anticipated that if samples were to be taken in the rainy season, a larger concentration of nitrates would be found in the nearby water ways.

The highest concentration of nitrate was found in the county of Demerara. These were 1.772 mg/L (± 0.00 mg/L), 2.363 mg/L (± 0.839 mg/L) and 0.33 mg/L (± 0.156 mg/L) for (2a) Good Hope, (2b) Ogle and the (2c) Stabroek areas respectively. The highest concentration at Ogle is surprising considering that it's a residential area now but there is some form of crop cultivation done by residents there. The highest concentration of nitrate may be due to run off in trenches, canals from sugar cane lands from nearby sugar cane estate from other regions. It may also be due to nitrogen from industrialized waste from these sugar cane factories and also to the high permeability of the soil. The soil probably has high nitrate content due to past agricultural activities but has now become residential. Nitrate is also a by-product of the septic tank systems. It is a naturally occurring chemical that results from the breakdown or decomposition of animal or human waste. Water quality is also affected through ground water resources that have a high number of septic systems in a watershed. Septics leach down into ground water resources or aquifers and supply nearby bodies of water. The high level of nitrate at Ogle may also result from septic tanks leaching to the nearby water trench. The lowest concentration of nitrate at Stabroek is expected considering that it is the capital city of Georgetown and is non-agricultural.

The second highest concentration of nitrate was found in the county of Essequibo which is predominantly a farming county for cash crops and large scale rice production. Values of 0.168 mg/L (± 0.046 mg/L), 0.192 mg/L (± 0.043 mg/L) and 0.178 mg/L (± 0.039 mg/L) were registered for (3a) Parika, (3b) Supernaam and (3c) Spring Garden respectively.

For the second set of analyses, areas (4a) – (6c), the average concentration of NO_3^- for each of the three locations: Kingston seawall, GUYSUCO LBI estate and the Canje creek at 20ft from the effluent was 0.45 mg/L (± 0.0039 mg/L), 0.01 mg/L (± 0.0033 mg/L) and 0.015 mg/L (± 0.00234 mg/L) respectively. These values are well below the internationally accepted value of 45mg/L. The highest concentration of nitrate anion at the Kingston seawall is probably due to the presence of rocks containing nitrate anions. The lowest concentration of nitrates was found in the Canje creek even though extensive farming is done in the area. There is a lot of farming done in the Canje basin area and at the GUYSUCO LBI estate. Sugar cane planting is done in both areas. In addition, the Canje basin is a source of other agricultural produce. Nitrates in water usually arise from the application of excess fertilisers to crops. In such cases, when the rain falls, the excess is carried or washed away to the nearest river where they are responsible for eutrophication. It is thus reasonable to say that nitrates were applied to crops during the dry season to prevent run off.

The overall low concentration of nitrate for the twelve (12) selected areas from the three counties indicate that they are below the threshold limit of 45mg/L as established internationally. Thus, it can be concluded that the twelve selected areas of coastal Guyana are not contaminated with nitrate above the threshold value. However, further testing of other water ways in Guyana needs to be done to safely conclude that Guyana's water is free from nitrate pollution beyond the threshold limit. No cases of "Blue Baby" syndrome have been reported here in Guyana to the very best of knowledge. The presence of other toxic anions is currently being investigated.

4 Conclusion

It can be concluded from the analysis of nitrate ion, the concentration was found to be 0.030 mg/L (± 0.0064 mg/L), 0.064 mg/L (± 0.00292 mg/L), 0.20 mg/L (± 0.00738 mg/L), 1.772 mg/L (± 0.00 mg/L), 2.363 mg/L (± 0.839 mg/L) and 0.33 mg/L (± 0.156 mg/L), 0.45 mg/L (± 0.0039 mg/L), 0.01 mg/L (± 0.0033 mg/L) and 0.015 mg/L (± 0.00234 mg/L) at 58 Livelihood Village, Rose Hall, Skeldon Guysuco Estate, Good Hope, Ogle, Stabroek, Parika, Supenaam and at Spring Garden, Kingston Seawall, GUYSUCO LBI Estate and at the Camje creek respectively. These values are well below the internationally accepted value of 45mg/L indicator for nitrate ion contamination and thus waste water from the selected twelve areas are not polluted with nitrate beyond the threshold value. However, other water systems must be monitored in Guyana.

Acknowledgements. This project was financially supported by a research grant to Dr. R. C. Jagessar. Gratitude is extended to the University of Guyana for the provision of Chemicals. Analytical services were done at the Institute of Applied Science and Technology, IAST, Turkeyen.

References

[1] Bell, C.F., Lott, K.A.K.: Modern Approach to Inorganic Chemistry, 2nd edn., p. 260. Butterworths, London (1966)
[2] Tongraar, A., Tangkawanwanit, P., Rode, B.M.: A combined QM/MM molecular dynamics simulations study of nitrate anion (NO_3^-) in aqueous solution. J. Phys. Chem. A 110, 12908 (2006)
[3] Oberhammer, H.J.: Mol. Structure 605, 177 (2002)
[4] Gonzales-Lebrero, M.C., Bikel, D.E., Elola, M.D., Estrin, D.A.: Solvent-induced symmetry breaking of nitrate ion in aqueous clusters, a quantum classical simulation study. J. Chem. Phys. 117, 2718 (2002)
[5] Bianchi, A., Bowman-James, K., Garcia- Espana, E.: Supramolecular Chemistry of Anions. Wiley-VCH, New York
[6] Kang, S.O., Begum, R.A., Bowman-James, K.: Amide based ligands for Anion Coordination. Angew. Chem. Int. Ed. 45, 7882–7894 (2006)
[7] Beer, P.D., Gale, P.A.: Amide based ligands for Anion Coordination. Angew. Chem. Int. Ed. 40, 486 (2001)
[8] Gamez, P., Mooibroek, T.J., Teat, S.J., Reedijk: Anion Binding Involving π-Acidic Heteroaromatic Rings. J. Acc. Chem. Res. 40, 435–444 (2007)
[9] Lehn, J.M.: Supramolecular Chemistry: Concepts and Perspectives, 1st edn., ch.3. Wiley-VCH, New York (1995)
[10] Schmidtchen, F.P., Berger, M.: Artificial Organic Host Molecules for Anions. Chem. Rev. 97, 1609–1646 (1997)
[11] Beer, P.D.: Transition-Metal Receptor Systems for the Selective Recognition and Sensing of Anionic Guest Species. Acc. Chem. Res. 31(2), 71–80 (1998)
[12] Bowman-James, K.: The Coordination Chemistry of Anions. Acc. Chem. Res. 38, 671–678 (2005)
[13] Martinez-Manez, R., Sancenon, F.: Fluorogenic and chromogenicchemosensors and reagents for anions. Chem. Rev. 103, 4419–4476 (2003)
[14] Sancenon, F., Descalze, A.B., Miranda, M.A., Sato, J.: A colorimetric ATP sensor based on 1,3,5-triarylpent-2-en-1,5-diones. Angew. Chem., Int. Ed. 40, 2640 (2001)
[15] Gunnlaugason, T., Davis, A.P., O'Brien, J.E., Glynn, M.: Fluorescent sensing of Pyrophosphate and Bis-carboxylates with charge neutral PET chemosensors. Org. Lett. 4, 244 (2002)
[16] Hennrich, G., Sonnennachein, H., Reach-Genger, L.: Tetrahedron Lett. 42, 2805 (2001)
[17] Choi, K., Hamilton, A.D.: Macrocyclic anion receptors based on directed hydrogen bonding interaction. Coordination Chem. Rev. 240, 101–110 (2003)
[18] Wiskur, S.L., Ait-Hissou, H., Lavigne, J.J., Anslyn, E.V.: Teaching Old Indicators New Tricks. Acc. Chem. Res. 34, 963–972 (2001)
[19] Kuo, L., Liao, J.H., Chen, C.T., Huang, C.H., Chen, C.S., Fang, J.M.: Two Arm Ferrocene Amide Compounds: Synclinal Conformations for Selective Sensing of Dihydrogen Phosphate ion. Organic Letters 5(11), 1821–1824 (2003)

[20] Kubic, S., Goddard, R., Kirchner, R., Nolting, D., Seidel: Recognition of Anions. J. Angew. Chem. Int. Ed. 40, 2648 (2001)

[21] Korendovych, I.V., Cho, M., Butler, L.P., Staples, J.R., Rybak-Akimova, E.V.R.: Anion binding to monotopic and ditopic macrocyclic amides. Organic Letters 8(15), 3171–3174 (2006)

[22] Schearder, J.F., Engbersen, J., Casnati, A., Ungaro, R., Reinhoudt, D.N.: Complexation of halide Anions and Tricarboxylate anions by Neutral Urea Derivatives, p-tert-butylcalix[6]arene. J.Org. Chem. 60, 6448–6454 (1995)

[23] Jagessar, R.C., Shang, M., Scheidt, W.R., Burns, D.H.: Neutral Ligands for selective Chloride Anion Complexation (Cis)-5,10,15,20 -Tetrakis (2-(arylurea) phenyl porphyrins. J. Am. Chem. Soc. 120(45), 11684–11692 (1998)

[24] Jagessar, R.C., Burns, D.H. (Cis)-5,10,15,20 -Tetrakis (2-(arylurea) phenyl porphyrins: Novel neutral ligands for remarkablely selective and exceptionally strong chloride anion complexation in $(CD_3)_2SO$. J. Chem. Soc., Chem. Commun., 1685–1686 (1997)

[25] Amemiya, S., Buhlmann, P., Umezawa, Y., Jagessar, R.C., Burns, D.H.: An ion-selective electrode for acetate based on a urea-functionalized porphyrin as a hydrogen bonding ionophore. Anal. Chem. 71(5), 1049–1054 (1999)

[26] Schearder, J., Engbersen, J.F.J., Casnati, A., Ungaro, R., Reinhoudt, D.N.: Complexation of halide Anions and Tricarboxylate anions by Neutral Urea Derivatives, p-tert-butylcalix[6]arene. J. Org. Chem. 60, 6448–6454 (1995)

[27] Konopatkin, N.M., Pakrasi, H.B., Smith, T.J.: Atomic structure of a nitrate-binding protein crucial for photosynthetic productivity. Proc. Natl. Acad. Sci. 103, 9820–9825 (2006)

[28] Fischer, K., Barbier, G.G., Hecht, H.J., Mendel, R.R., Campbell, W.H., Schwarz, G.: Structural basis of eukaryotic nitrate reduction: Crystal structures of the nitrate reductase active site. Plant Cell 17, 1167–1179 (2005)

[29] Mahoney, J.M., Stucker, K.A., Jiang, H., Carmichael, L., Brinkmann, N.R., Beatty, A.M., Noll, B.C., Smith, B.D.: Molecular Recognition of Trigonal Oxyanions Using a Ditopic Salt Receptor: Evidence for Aniostropic Shielding Surface around Nitrate Anion. J. Am. Chem. Soc. 127, 2922 (2005)

[30] Choi, K., Hamilton, A.D.: J. Chem. Soc. 125, 10241 (2003)

[31] Beer, P.D.: Anion selective recognition and optical/electrochemical sensing by novel transition metal receptor systems, Cobalticinium Porphyrins anion receptors. J. Chem. Soc. Chem. Commun., 689–696 (1996)

[32] Beer, P.D., Drew, M.G.B., Jagessar, R.: Selective anion recognition by Novel 5, 10, 15, 20 - tetrakis (o-ferrocenyl carbonylaminophenyl - substituted) zinc metalloporphyrin receptors. J. Chem. Soc., Dalton Trans., 881–886 (1997)

[33] Jagessar, R.C.: PhD Theses "Anion Recognition by novel functionalised Porphyrins": Design, Syntheses and Molecular recognition of guest substrates by novel functionalised Porphyrins, 1-326, Inorganic Chemistry Laboratory, Radcliffe Science and Bodlean library, University of Oxford (1995)

[34] Beer, P.D., Drew, M.G.B., Hesek, D., Jagessar, R.: Spectral and Electrochemical Anion Sensing by a novel 5,10,15,20-tetrakis(R-substituted)Porphyrin Receptor($R{=}C_6H_4$ $NHC(O)C_5H_4CoC_5H_5{+}PF_6^-$). J. Chem. Soc., Chem. Commun., 1187 (1995)

[35] Beer, P.D., Graydon, A.R., Sutton, L.R.: Luminescent anion recognition: selective induced emission by binding of dihydrogenphosphate. Polyhedron 15(14), 2457–2461 (1996)

[36] Chen, Z., Graydon, A.R., Beer, P.D.: Electrochemical response to anions in acetonitrile by neutral molecular receptors containing ferrocene, amide and amine moieties. J. Chem. Soc., Faraday Trans. 92(1), 97–102 (1996)
[37] Beer, P.D., Drew, M.G.B., Smith, D.K.: Selective electrochemical recognition of bidentate anionic guests in competitive solvents using novel ferrocenyl thiourea and guanidinium receptors. J. Organomet. Chem. 543(1-2), 259–261 (1997)
[38] Cronin, L., McGregor, P.A., Parsons, S., Teat, S., Gould, R.O., White, V.A., Lang, N.J., Roberston, N.: Synthesis, structure, and complexation of a large 28-mer macrocycle containing two binding sites for either anions or metal ions Inorg. Chem. 43, 8023 (2004)
[39] Steed, J.W.: A modular approach to anion binding podands: adaptability in design and synthesis leads to adaptability in properties. Chem. Commun., 2637–2649 (2006)
[40] Applegarth, L., Clarke, N., Richardson, A.C., Parker, A.D.M., Radosavljevic-Evans, I., Goeta, A.E., Howard, J.A.K., Steed, J.W.: Modular Nanometer-Scale Structuring of Gel Fibres by Sequential Self-Organization. Chemical Communications 43, 5423–5425 (2005)
[41] Gupta, V.K., Singh, L.P., Chandra, S., Kumar, S., Singh, R., Sethi, B.: Anion recognition through amide-based dendritic molecule: a poly(vinyl chloride) based sensor for nitrate ion. Talanta. 85(2), 970–974 (2011)
[42] Aravamudhan, S., Bhansali, S.: Development of Micro-fluidic Nitrate Selective Sensor Based on Doped-Polypyrrole Nanowires. Sensors and Actuators B: Chemical 132(2), 623–630 (2008)
[43] Mahajan, R.K., Kaur, R., Miyake, H., Tsukube, H.: Zn(II) complex-based potentiometric sensors for selective determination of nitrate anion. Analytica Chimica Acta 584(1), 89–94 (2007)
[44] Eaton, A.D., Clessicens, S.L., Greenberg, E.A.: Standard methods for the Examination of Water and Wastewater, 19th edn., pp. 49–51. American Public Health Association, Washington, DC (1995)
[45] Eaton, A.D., Chair, A.W.W.A., et al.: Standard methods for the examination of Water and Wastewater, 19th edn., pp. 4–67. United Book Press Inc., Baltimore (1995)
[46] Booth, R.L.: Methods for the Chemical analysis of Water and Wastes, 2nd edn., p. 352. Environmental Monitoring and Support Laboratory, Office of Research and Development, U.S. Environmental Protection Agency, Cincinnati, Ohio 45268 (1983)
[47] Jagdeo, B.: National Development Strategy, Environmental Policy. Guyana Government Newsletter, 40–45 (2006)
[48] Williams, N.: Guyana Times, a News Magazine (11), 113 (2010)
[49] Elliot, S.: "Testing the water", Royal Society of Chemistry, RSC. News Magazine 12(5), 12–13 (2008)
[50] Peplow, M.: Nitrates pollute Gaza's wells. Chemistry World 5(10), 25 (2008)
[51] Tindale, S.L., Nelson, W.L.: Soil fertility and Fertilisers, 2nd edn., pp. 174–179. The Macmillan, New York (1970)
[52] Nelson, C.L., Cox, M.M., Lehninger: Principles of BioChemistry, 4th edn., pp. 834–835. W.H. Freeman and Company, New York (2005)
[53] Baum, R.M.: Too much Carbon Dioxide limits plants, Chemical and Engineering News. American Chemical Society News Magazines 31 (2010)
[54] Jagessar, R.C., Sooknanan, L.: Determination of nitrate anion in waste water from nine selected areas of coastal Guyana via a Spectrophotometric method. International Journal of Research and Reviews in Applied Sciences, IJRRAS 7(2), 204–213 (2011)

[55] Jagessar, R.C., Alleyne, O.: Determination of Nitrate Anion concentration in Waste water from selected areas of Coastal Guyana via a Spectrophotometric method. International Journal of Academic Research 3(1), 443–453 (2011)
[56] Skoog, A.D., Holler, J.F., Nieman, A.T.: Principles of Instrumental Analysis, 5th edn., pp. 329–353. Thomson Learning, Inc. (1998)
[57] Daniel, H.C.: Quantitative Chemical Analysis, 6th edn., pp. 61–79. W.H. Freeman and Company, New York (1998, 2003)
[58] Tindale, S.L., Nelson, W.L.: Soil fertility and Fertilisers, 2nd edn., pp. 174–179. The Macmillan, New York (1970)

The Selection of Novel Planar Electromagnetic Sensors for the Application of Nitrate Contamination Detection

M.A. Md. Yunus[1], S.C. Mukhopadhyay[2,*], M.S.A. Rahman[3], N.S. Zahidin[3], and S. Ibrahim[3]

[1] Infocomm Research Alliance,
Department of Control and Mechatronics Engineering (CMED),
Universiti Teknologi Malaysia
[2] School of Engineering and Advanced Technology,
Massey University, Palmerston North, New Zealand
S.C.Mukhopadhyay@massey.ac.nz
[3] Control and Instrumentation Engineering Department,
Universiti Teknologi Malaysia, Skudai, Malaysia

Abstract. Twenty four novel planar electromagnetic sensors based on the combination of meander and interdigital sensors have been designed and fabricated using the simple PCB technology for the application of nitrate contamination detection. Experiments were conducted to obtain the impedance characterization for each sensor, and the results were used to estimate the important parameters that influence the performance of the sensors based on the equivalent electrical circuits. Furthermore, the best sensors have been tested to detect nitrates contamination in distilled water from two sets of experiments. Firstly, two nitrate forms, namely sodium nitrates ($NaNO_3$) and ammonium nitrates (NH_4NO_3), each of different concentration between 5 mg and 20 mg dissolved in 1 litre of distilled water were used to observe the sensor response. Secondly, $NaNO_3$ and NH_4NO_3 were mixed in several different ratios dissolved in 1 litre of distilled water and the responses of the sensors were observed. The outcomes concluded that a sensor with the combination of interdigital sensor enclosed with a meander sensor can very well detect the presence of nitrate added into distilled water and is capable of distinguishing the concentration level. This paper will discuss the process that was involved to select the best sensor for the application of nitrate contamination detection.

Keywords: Interdigital sensor, electromagnetics, nitrate contamination.

1 Introduction

The increase in nitrate concentration in ground water arises mainly from natural and anthropogenic sources. However extensive areas in some parts of the world have increasing nitrate concentration due to natural leaching and nitrate accumulation

* Corresponding author.

S.C. Mukhopadhyay & A. Mason (Eds.): *Real-Time Water Quality Monitoring*, SSMI 4, pp. 171–195.
DOI: 10.1007/978-3-642-37006-9_8 © Springer-Verlag Berlin Heidelberg 2013

under certain environmental conditions [1]. The increase of agricultural fertilizer, manure and sludge application to land is one of the main contributing factors to the nitrate increase in ground water, where the water level is low; this has led to the most widespread diffuse source of nitrate pollution [2]. Methemoglobinemia, having fatal consequences in infants, is caused by nitrate contamination [3]. Furthermore, livestock losses are occurring frequently due to nitrate poisoning [4].

About three decades ago, nitrate measurement was usually being conducted using absorption spectrophotometry using a specific reagent [5]. This method was developed and established for laboratory standards thus making it expensive due to many components being required. It also often involved laborious measuring steps and inevitably consumes a lot of time. They also require controlled working conditions. Therefore, due to the concern regarding the amount of nitrate contaminations in natural water sources and the need of a low cost, convenient, and suitable methods for in-situ measurements of water quality, this paper reports on the results and conclusions of experiments involving novel planar electromagnetic sensors for this purpose.

2 Recent Progress of Planar Electromagnetic Sensors for Environmental Monitoring Applications

For the past ten years, the research of environmental monitoring based on planar meander and interdigital elements has followed several avenues as summarized in Table 1. The early work (since 2001) involved a serially closed circuit inductor-capacitor (LC) element, a TiO_2 (Titanium Dioxide) coating, and flexible parallel plate capacitor which forms a passive sensor as described in the second row of Table 1. The sensor operates on a wireless remote query basis and has been applied for environmental parameter monitoring based on complex permittivity of a surrounding medium, temperature monitoring, and pressure [6]. Later on, this work evolved into other applications such as bacteria growth monitoring [7, 8], monitoring of electrical properties of biological cell solutions [9], quantifying packaged food quality [10], and real-time monitoring of water content in civil engineering materials [11, 12]. Despite offering a good performance, this system is quite complex, considering the material properties estimation was achieved from impedance spectrum of the sensor measured using a remotely located antenna. The material properties (e.g. complex permittivity) are calculated from the impedance spectrum at the resonant frequency with the inductance and capacitance of the sensor values based on calculation of an analytical model. Moreover, looking at the current technologies, the potential of this system to be a portable or home appliance is hindered as it requires either an expensive impedance analyzer or a lock in amplifier.

Woodard et al. [13] and Woodard [14] employed open-circuited self-resonating planar spiralling pattern of electrically conductive material coated with an active material as explained in the fourth section of Table 1. The results reflected that the sensors are suitable to be used in harsh conditions. Nevertheless, the material used to make the sensor, i.e. Silicon Nitride (Si_3N_4), is relatively expensive and only affordable by programs funded for space exploration. Dickey et al. [15] and Varghese et al. [16] have investigated the effect of different sizes of pore and uniformity of the substrate (metal oxide) which was made as a platform base for and interdigital sensor on the sensitivity and accuracy of NH_3 and relative humidity as described in the second part of Table 1. Metal oxide materials (SnO_2, Al_2O_3, and TiO_2) are relatively low-cost and it is well-known that their structure can be manipulated for gas-sensing enhancement. However, the fabrication of the sensor is a laborious process that requires access to expensive facilities and considerably skilled personnel.

Table 1. Overview of existing environmental monitoring sensing systems based on planar meander and interdigital elements

Sensor's Description	Parameters Monitored	Ref
Inductor and Capacitor circuits (LC) connected in a loop with Ti_2O coating and flexible parallel plate capacitor.	Complex permittivity of a surrounding medium, ε' difference (%): 2.9 to 9.1 % and ε'' difference (%): 14.4 to 37.3 %. Temperature monitoring: linear shift of 6.4 kHz/°C. Humidity: resonant frequency shift of 3.1 MHz (≈14 %) during humidity change between 0-98 %. Pressure: linear resonant frequency shift of 1.28 MHz/psi over the pressure range between 0-5 psi.	[6]
Interdigital sensor with gold electrodes deposited on Metal Oxide films which formed through an anodization process	Detection of NH_3 at room temperature and relative humidity.	[15, 16]
Inductor and Capacitor (LC) wireless sensor with Elecrolyte-Nitrate-Oxide-Silicon (ENOS) structure as an electrochemical potential-to-capacitance transductor fabricated using CMOS technology.	pH measurement with sensitivity of 1.2 % frequency/pH.	[17, 18]
Self-resonating-open-circuited planar spiralling pattern of electrically conducted material, coated with active material such as silicon nitride and polyaniline.	Application in harsh condition. Gases detection such as NH_3, CO, CO_2, O_2, H_2, and CH_4. Temperature measurement. Ambient pressure.	[13, 14]

Several researchers [17, 18] have addressed the application of inductor and capacitor (LC) wireless sensors with an Elecrolyte-Nitrate-Oxide-Silicon (ENOS) structure as an electrochemical potential-to-capacitance transducer fabricated using CMOS technology for pH measurement of liquid material such as water. These studies have made improvements of shorter response times and linear response. The main drawback is the operating frequency had to be adjusted to a proper value to minimize the unwanted response of the sensor to the sample (material under test) conductivity. In this research, our attention and interest has been drawn into to developing an electromagnetic sensor consisting of both inductive and capacitive elements which can be integrated as a low-cost and convenient solution, which is suitable for in-situ measurement in water quality monitoring.

3 Description of the Novel Planar Electromagnetic Sensors

Applications of planar electromagnetic sensors have been reported in [19-48] where the designs can be found. In this chapter, two main types of sensors are introduced as following:

(1) Sensors with meander and interdigital elements placed side by side having parallel and series combinations.
(2) Sensors with meander surrounding the interdigital element, where the interdigital element is placed in the middle having parallel and series combinations.

All the sensors were designed using Altium Designer 9 and have been fabricated using simple printed circuit board (PCB) fabrication technology where the thickness of the PCB is 0.25 mm. All the sensors were sprayed with Wattyl Killrust Incralac to form an acrylic resin-based protective coating. Wattyl Killrust Incralac is an acrylic resin-based protective coating containing special tarnish and corrosion control additives designed especially for the protection of copper, brass and similar alloys. The thickness of the coating was set to be approximately 30 microns.

Fig. 1, 2, 3, and 4 show the inductive-capacitive planar electromagnetic sensors connected in series: SECS1_1, SECS2_1, SECS3_1, and SECS4_1, respectively. In general, each sensor has a dimension of 26 mm × 54 mm (width × length). SECS1_1, SECS2_1, SECS3_1, and SECS4_1 consist of type 1 interdigital, type 2 interdigital, type 3 interdigital, and type 4 interdigital, respectively (each interdigital has the same effective area of 10 mm × 40 mm), placed in the middle of a meander sensor with three turns. The distances between the sensors are kept at 2.5 mm.

The type 1 interdigital sensor, whose electrodes' pattern is + - + - + - +, is an arrangement of alternative positive and negative electrodes separated by a distance of 3.3 mm. It consists in total of 13 electrodes, each 10 mm in length and 0.5 mm wide, 7 connected to exciting/driving electrodes and 5 are connected to ground. SECS2_1 shares the same design as SECS1_1 but it is actually the type 2 interdigital sensor, and the negative electrodes width is 3 mm as shown in Fig. 2. Negative electrodes with wider size enable a better or strong current signal off the sensor.

The type 3 interdigital, in the design configuration of SECS3_1, has electrodes which are arranged in the pattern of + - - + - - + etc, a driving electrode followed by a series of 2 ground electrodes as shown in Fig. 3. The above pattern was repeated with a driving electrode again. Each electrode is separated by a distance of 2.24 mm; each electrode is of 10 mm in length and 0.5 mm in width. The interdigital sensor type 4 in SECS4_1 is similar to SECS3_1, however it employs the electrode pattern of + - - - + - - - + etc, where each electrode is separated by a distance of 1.67 mm as can be seen in Fig. 4. The positive electrodes are driven by an AC electric potential (same source as the meander) to create an AC electric field between the positive and negative electrodes. This field passes through the material under test, thus material dielectric properties as well as the electrode and material property affect the capacitance and the conductance (impedance) between the two electrodes. The penetration depth of the electric field is determined from the spatial wavelength, λ, which is the length between the two adjacent electrodes of same type. The penetration depth is one fourth of the spatial wavelength [43, 45, 47, 49].

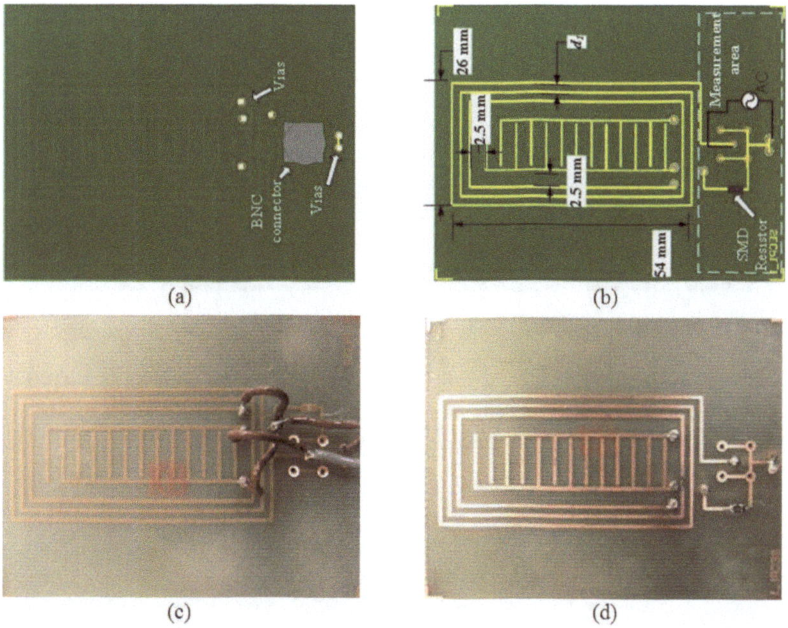

Fig. 1. SECS1_1 schematic diagrams of (a) top layer and (b) bottom layer, and pictures of (c) top layer and (d) bottom layer

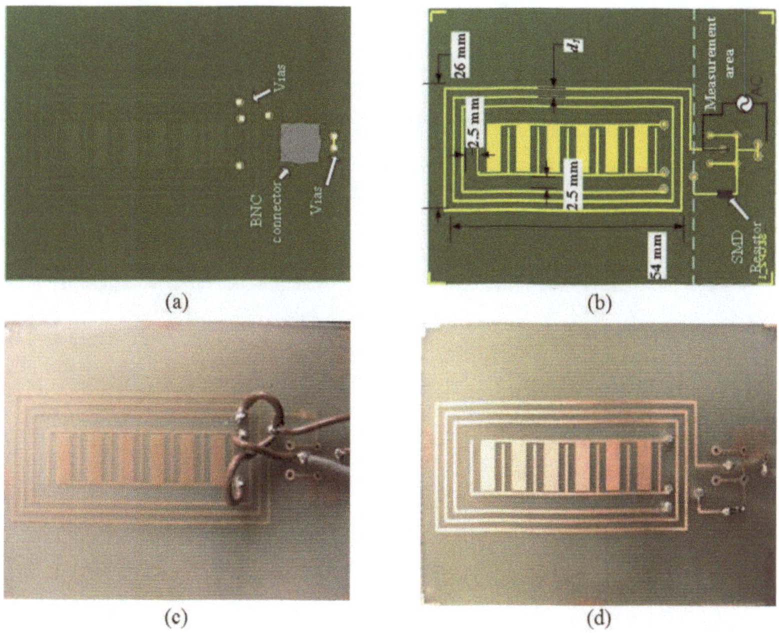

Fig. 2. SECS2_1 schematic diagrams of (a) top layer and (b) bottom layer, and pictures of (c) top layer and (d) bottom layer

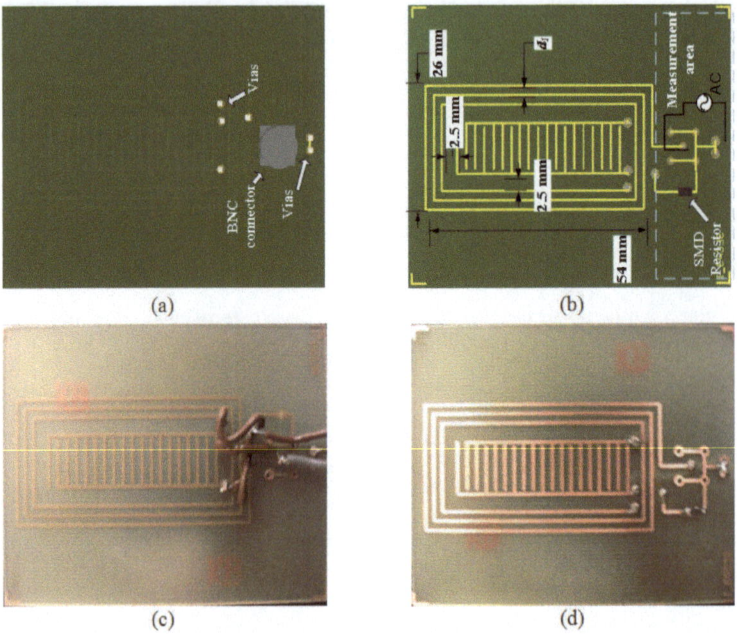

Fig. 3. SECS3_1 schematic diagrams of (a) top layer and (b) bottom layer, and pictures of (c) top layer and (d) bottom layer

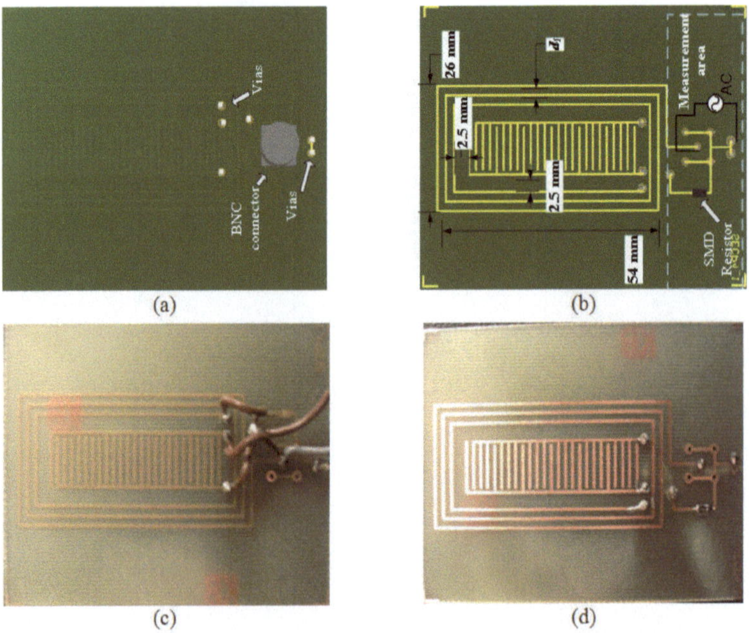

Fig. 4. SECS4_1 schematic diagrams of (a) top layer and (b) bottom layer, and pictures of (c) top layer and (d) bottom layer

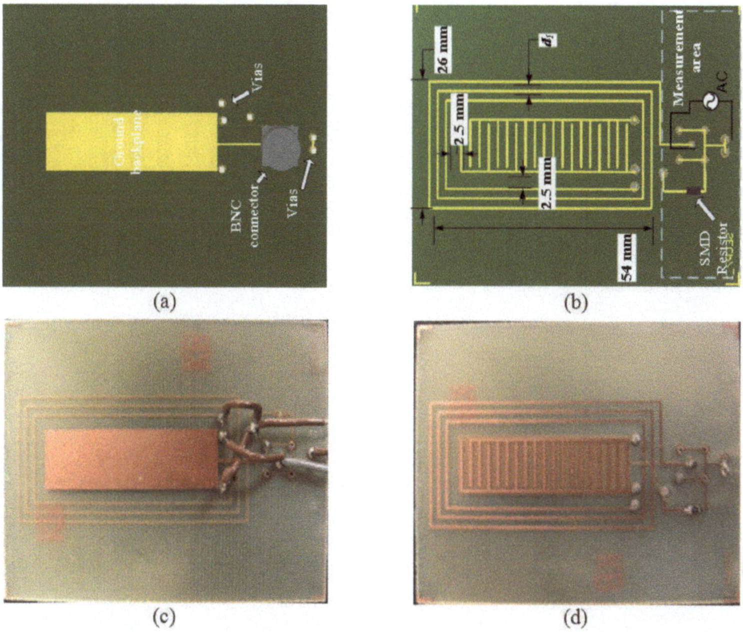

Fig. 5. SECS7_1 schematic diagrams of (a) top layer and (b) bottom layer, and pictures of (c) top layer and (d) bottom layer

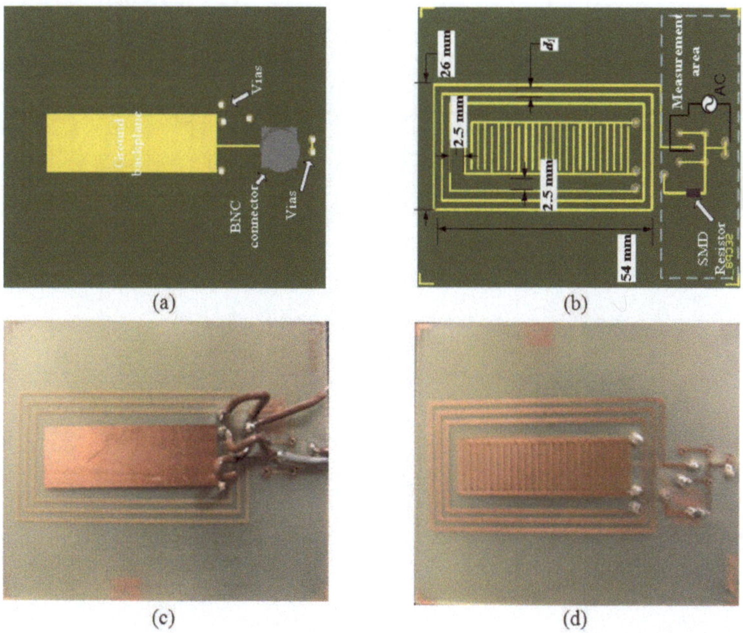

Fig. 6. SECS8_1 schematic diagrams of (a) top layer and (b) bottom layer, and pictures of (c) top layer and (d) bottom layer

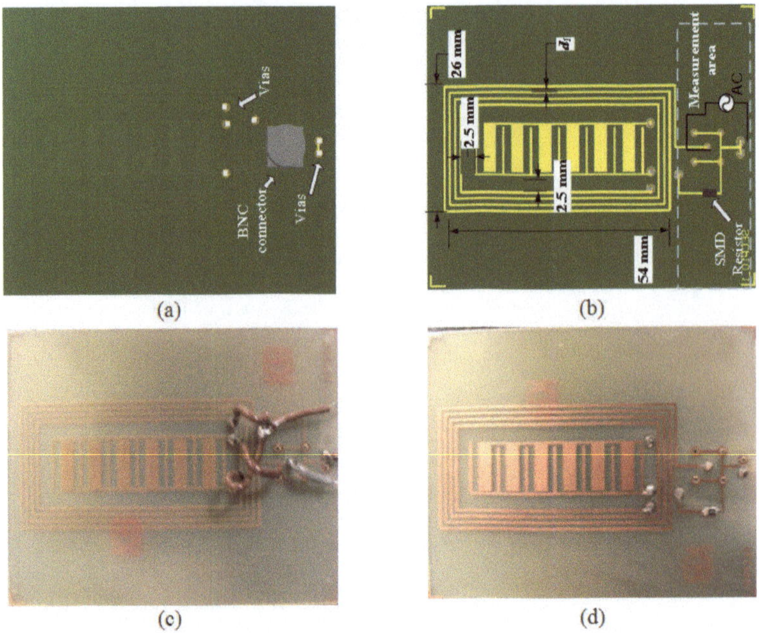

Fig. 7. SECS10_1 schematic diagrams of (a) top layer and (b) bottom layer, and pictures of (c) top layer and (d) bottom layer

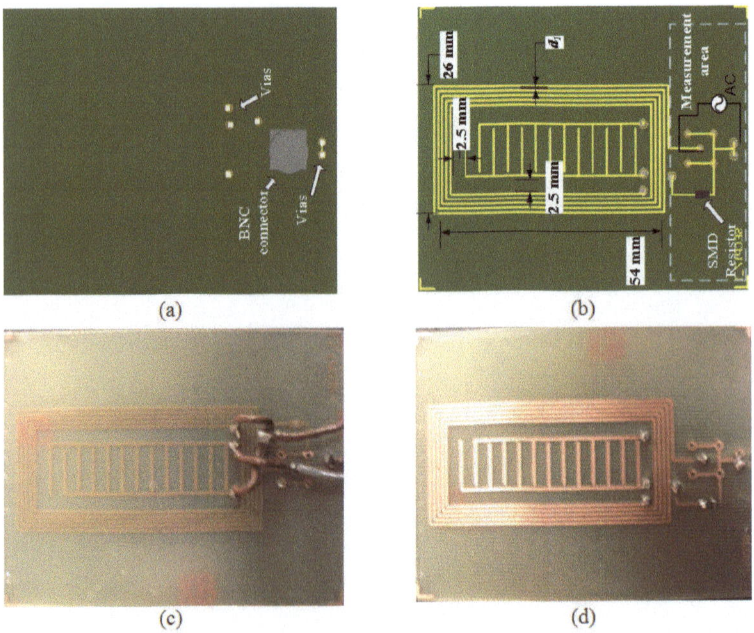

Fig. 8. SECS17_1 schematic diagrams of (a) top layer and (b) bottom layer, and pictures of (c) top layer and (d) bottom layer

Table 2. Summary of SECP1_1 to SECP24_1

Sensor	Interdigital type				Meander	
	Type	Wavelength, λ (mm)	Distance between electrode, *d* (mm)	Negative backplane	Number of turns	Distance between turn, d_l (mm)
SECS1_1	1	6.6	3.3	No	3	2.0
SECS2_1	2	6.6	1.8	No	3	2.0
SECS3_1	3	6.6	2.24	No	3	2.0
SECS4_1	4	6.6	1.67	No	3	2.0
SECS5_1	1	6.6	3.3	Yes	3	2.0
SECS6_1	2	6.6	1.8	Yes	3	2.0
SECS7_1	3	6.6	2.24	Yes	3	2.0
SECS8_1	4	6.6	1.67	Yes	3	2.0
SECS9_1	1	6.6	3.3	No	4	1.35
SECS10_1	2	6.6	1.8	No	4	1.35
SECS11_1	3	6.6	2.24	No	4	1.35
SECS12_1	4	6.6	1.67	No	3	1.35
SECS13_1	1	6.6	3.3	Yes	4	1.35
SECS14_1	2	6.6	1.8	Yes	4	1.35
SECS15_1	3	6.6	2.24	Yes	4	1.35
SECS16_1	4	6.6	1.67	Yes	4	1.35
SECS17_1	1	6.6	3.3	No	5	1
SECS18_1	2	6.6	1.8	No	5	1
SECS19_1	3	6.6	2.24	No	5	1
SECS20_1	4	6.6	1.67	No	5	1
SECS21_1	1	6.6	3.3	Yes	5	1
SECS22_1	2	6.6	1.8	Yes	5	1
SECS23_1	3	6.6	2.24	Yes	5	1
SECS24_1	4	6.6	1.67	Yes	5	1

SECS5_1, SECS6_1, SECS7_1, and SECS8_1 are similar to SECS1_1, SECS2_1, SECS3_1, and SECS4_1, respectively, but have a grounded backplane on the other side of the interdigital sensor (top layer). Fig. 5 and 6 show the design layouts for SECS7_1 and SECS8_1, respectively. The second variation and the third variation have meander sensors with three and five turns, respectively. Fig. 7 and 8 show the designs for SECS10_1 (three turns) and SECS17_1 (five turns), respectively. Table 2 summarizes the designs of sensors in this part. In general, the sensors are denoted by the generic code of *SECSXX_V*. *SECS* denotes the code for the meander sensor that is enclosing the interdigital sensor and the sensors are connected in series, *XX* is the number of the variation from 1 to 24 and *V* is the version number of the sensor (1 in all cases).

4 Principle of Measurement

The electrical equivalent circuit of SECS1_1 to SECS24_1 sensors is shown in Figures 9, where:

R_g is the output resistance of the function generator with a nominal value of 50 Ω.
R_{ms} is the real part resistance of the sensing coil (meander).

X_{ms} is the imaginary part reactance of the sensing coil (meander).
R_{is} is the real part resistance of the interdigital sensor for the sensors connected in series.
X_{is} is the imaginary part reactance of the interdigital sensor for the sensors connected in series.

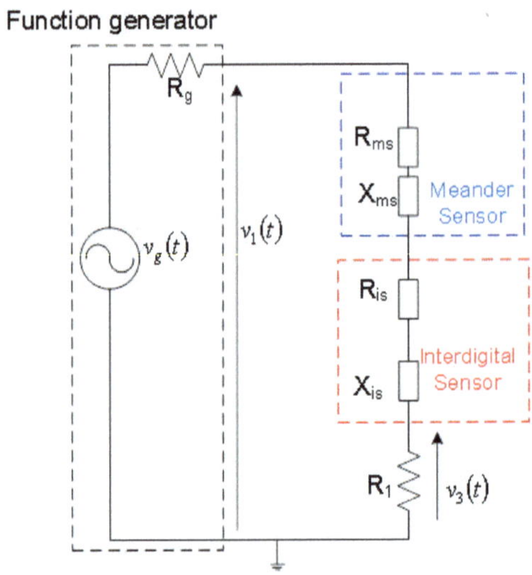

Fig. 9. The equivalent circuits for SECS1_1 to SECS24_1

For the sensors in Fig. 9, R_1 denotes the series surface mount resistor connected to the sensors with the nominal value of 120 kΩ. Hence, R_{total_s} and X_{total_s} can be calculated from (1).

$$I_1 = V_3\angle 0^\circ / R_1 \tag{1}$$

Where I_1 is the RMS value of current through the sensor and V_3 is the voltage across R_1. Introducing (1) into (2), gives Z_{total_s} (the total impedance):

$$Z_{total\_s} = V_1\angle\theta_1 / I_1\angle 0^\circ \tag{2}$$

Where θ_1 is the phase difference between $v_1(t)$ with $v_3(t)$ in degrees, taking $v_3(t)$ as reference. Therefore, R_{total_s} and X_{total_s} are given by (3) and (4).

$$R_{total\_s} = Z_{total\_s} \cdot \cos(\theta) - R_1 \tag{3}$$

$$X_{total\_s} = Z_{total\_s} \cdot \sin(\theta_1) \tag{4}$$

5 Characterization of the Sensors and the Experimental Setup

The characteristics of the sensors was determined by calculating the total impedance (absolute) of the sensors at different frequencies between 100 Hz and 100 MHz when no material is placed near the sensor (air).

5.1 Experimental Setup and Labview Program

The experimental setup involved a frequency waveform generator where a standard sinusoidal waveform with a 10V peak-to-peak value was set as the input signal for the sensors. The Agilent 54622D mixed signal oscilloscope was interfaced to a PC, and then the output signals and the sensors' impedance were recorded and calculated consecutively using developed programs in the *LabView* software. The experimental setup and the caption of the developed *LabView* program are shown in Fig. 10 and 11 respectively.

The *LabView* program consists of five parts: (a) the main controls, (b) the oscilloscope controls, (c) the function generator controls, (d) the displays, and (e) the controls to save the recorded data as can be seen in Fig. 11. Fig. 12 shows the *LabView* block function to calculate the impedance and phase shift.

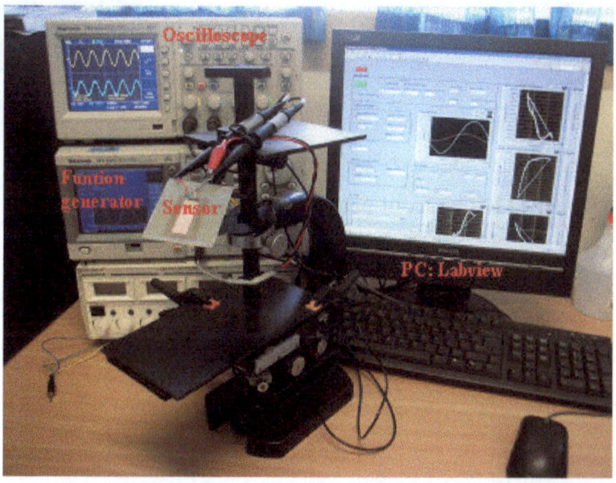

Fig. 10. The experimental setup used for sensor characterization

Given a wave form, the phase angle, θ and V amplitude given in root mean square (*RMS*) of a given frequency are found by calculating the number of samples in a single period of the desired frequency according to (5).

$$\frac{Samples}{cycle} = \frac{1}{t_{samplerate} \times f_{desiredfrequency}} \tag{5}$$

Then, one wave cycle is extracted from the waveform by taking half the number of samples per cycle from each side of the centre. Fourier transform is applied on the result. Since the window size covers exactly one period of the given frequency, it represents the fundamental frequency. This is element 1 in the Fourier transform results, (Element 0 is the DC component of the signal), as described by the following equation (6).

$$F_n = \sum_{k=0}^{N-1} f_k e^{-\frac{2\pi i n k}{N}} \tag{6}$$

Where N is the number of sample and f_k is the data point (where $k = 0,\ldots,N-1$). The result is expressed in amplitude and phase by converting it into polar form, and then scaled by $1/\sqrt{2}\,\mathrm{N}$ to convert them to *RMS*.

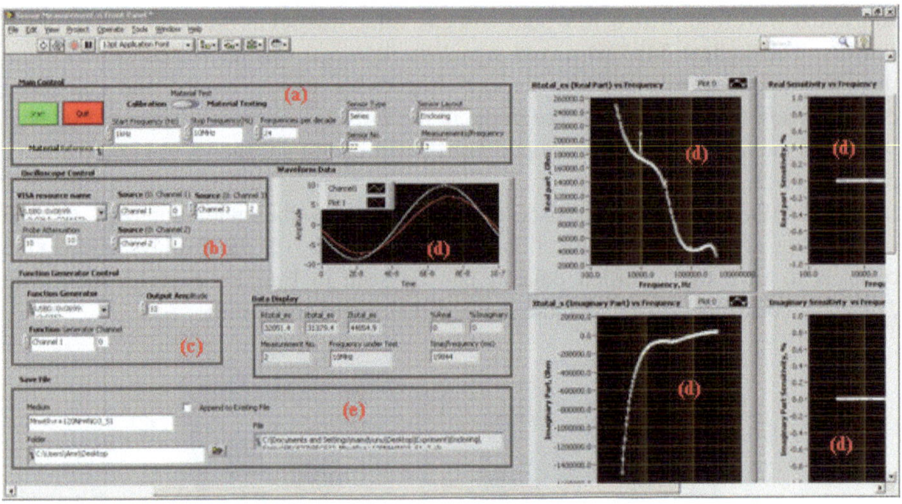

Fig. 11. The front panel of the LabView program

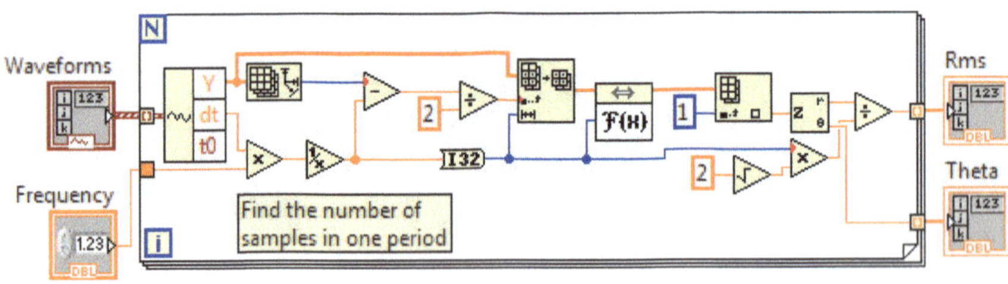

Fig. 12. Block function to calculate the waveform's amplitude in RMS and phase angle in radian

Scaling the result is not actually necessary as it cancels out later because we are only interested in the gain. Once the amplitude and phase angle have been found for the input (V_1) and output (V_2/V_3), the gain and phase shift to calculate the impedance can be found. An array of V amplitudes (RMS) and phase angle (θ) are passed to the *LabView* function to calculate the impedance as shown in Fig. 13. The input voltage, V_1 and output voltage, V_2/V_3 are split up, and the output (V_2/V_3) is scaled with the calibration value found for the particular frequency that is being measured.

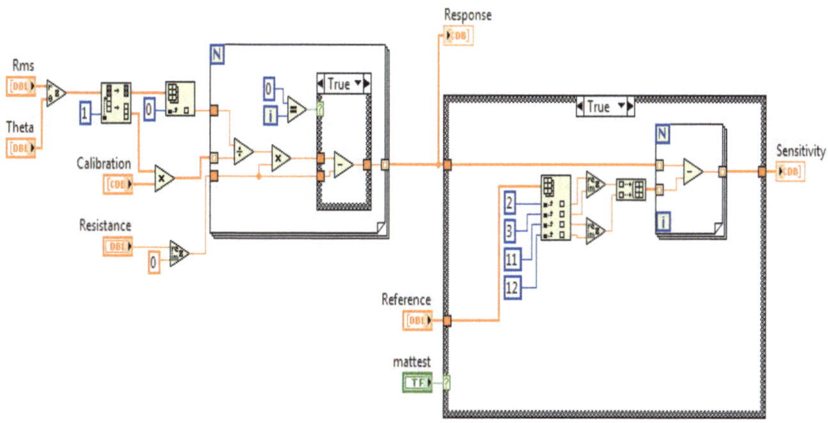

Fig. 13. Block function to calculate the impedance

Finally, the total impedance in polar form can be calculated from (7).

$$Z\angle\theta = \left(V_{input}\angle\theta_{input} / V_{output}\angle\theta_{ouput}\right)\times R \tag{7}$$

Where:

V_{input} is the *RMS* value of the input signal (V_1).
θ_{input} is the phase angle of the input signal.
V_{output} is the *RMS* value of the output signal measured across the series resistor (V_2/V_3).
θ_{output} is the phase angle of the output signal measured across the series resistor.
R is the resistance value of the series resistor.

5.2 Characteristics of the Sensors

Fig. 14 depicts the impedance characterization of SECS1_1 to SECS24_1. In general, for most of the meander and interdigital sensors connected in series, a U-shaped impedance characteristic was obtained as shown in Fig. 14. The right-half and left–half curves show that most of the sensors are capacitive at a low-frequency range and inductive at a high-frequency range, respectively. It is seen that, within a certain frequency range, the sensor shows the combined effect of inductive and capacitive response.

The frequency range, which represents the combined effect of inductive and capacitive response, can be found at between around 100 kHz to 30 MHz. The introduction of a negative backplane basically causes the frequency range of inductive-capacitive state to be narrower, and the impedance values to be increased as can be seen from Fig. 14 (a), (b), and (c).

The SECS sensors are modeled as an electrical circuit as can be seen in Fig. 15 [50, 51]. The circuit consists of total inductance (*L*), total capacitance (*C*), total resistance (R_1) and total resistive losses (R_2). The values of electrical parameters in Fig. 15 are also estimated for each sensor from the real part and imaginary part responses using EIS spectrum analyzer software [52].

Table 3 summarizes the electrical parameter values of SECS. For all SECS sensors, the average total resistive losses, R_2 value is relatively high, i.e. between around $2.0\times10^{7}\ \Omega$ to $2.0\times10^{18}\ \Omega$. The total resistance average value (R_1) basically increases when the negative backplane is introduced as summarized in Table 3. Moreover, there is no significant improvement of capacitance and inductance values except for the sensors of SECS21_1 to SECS24_1. Therefore, the best series connected planar electromagnetic sensors are SECS21_1 to SECS24_1, as can be seen in Table 3, where the capacitance and inductance values are the highest as compared to other planar electromagnetic sensors connected in series.

Table 3. Electrical parameters values of SECS1_1 to SECS24_1

		Meander: One turn		Meander: Three turns		Meander: Five turns	
		No negative backplane	With negative backplane	No negative backplane	With negative backplane	No negative backplane	With negative backplane
Interdigatal: type 1		**SECS1_1**	**SECS5_1**	**SECS9_1**	**SECS13_1**	**SECS17_1**	**SECS21_1**
	R_1, Ω	306140	1.55E+06	501430	1.68E+06	432790	1.89E+06
	R_2, Ω	1.66E+10	1.00E+08	4.98E+10	1.00E+08	4.69E+10	3.14E+10
	C, F	1.50E-12	2.06E-12	3.78E-12	1.76E-12	4.42E-12	5.95E-12
	L, H	1.03E-04	1.54E-03	3.77E-04	3.27E-03	3.17E-04	1.67E-02
Interdigatal: type 2		**SECS2_1**	**SECS6_1**	**SECS10_1**	**SECS14_1**	**SECS18_1**	**SECS22_1**
	R_1, Ω	320510	1.59E+06	3.03E+06	4.59E+06	323630	1.41E+06
	R_2, Ω	1.88E+10	8.54E+08	4.02E+10	1.18E+09	6.39E+08	1.82E+06
	C, F	5.40E-12	2.43E-12	7.03E-13	2.59E-12	2.64E-12	7.90E-12
	L, H	2.36E-04	1.46E-03	1.97E-03	1.12E-03	1.97E-04	6.54E-03
Interdigatal: type 3		**SECS3_1**	**SECS7_1**	**SECS11_1**	**SECS15_1**	**SECS19_1**	**SECS23_1**
	R_1, Ω	384540	1.54E+06	384760	1.69E+06	375450	1.68E+06
	R_2, Ω	2.91E+10	1.15E+09	5.18E+10	3.92E+10	1.25E+11	3.93E+10
	C, F	4.76E-12	2.26E-12	4.94E-12	2.13E-12	1.94E-12	5.51E-12
	L, H	2.24E-04	1.46E-03	3.19E-04	1.65E-03	2.45E-04	2.12E-03
Interdigatal: type 4		**SECS4_1**	**SECS8_1**	**SECS12_1**	**SECS16_1**	**SECS20_1**	**SECS24_1**
	R_1, Ω	325100	1.35E+06	335280	1.63E+00	325570	1.34E+06
	R_2, Ω	3.90E+10	1.00E+08	7.38E+08	7.84E+08	2.74E+10	7.19E+08
	C, F	2.93E-12	2.93E-12	5.41E-12	2.39E-12	5.65E-12	5.62E-12
	L, H	2.45E-04	2.14E-03	1.73E-04	2.07E-03	1.98E-04	4.75E-03

6 Experimental Results and Discussion

In this section, the best sensors are tested to detect nitrate contamination in distilled water from two set of experiments. Firstly, two nitrates, namely sodium nitrates ($NaNO_3$) and ammonium nitrates (NH_4NO_3), each of different concentration between 5 mg and 20 mg dissolved in 1 liter of distilled water were used to observe the sensor response. Secondly, $NaNO_3$ and NH_4NO_3 were mixed in several different ratios dissolved in 1 liter of distilled water and the responses of the sensors were observed. The best sensors for the series combination sensors comprise of SECS17_1, SECS21_1, and SECS22_1. Each sensor consists of a meander coil spiraling inwards in rectangular form and the inner end is connected to an interdigital sensor completing a serial connection. The overall best of these sensors is determined based on the interpretation from both nitrates experiment results.

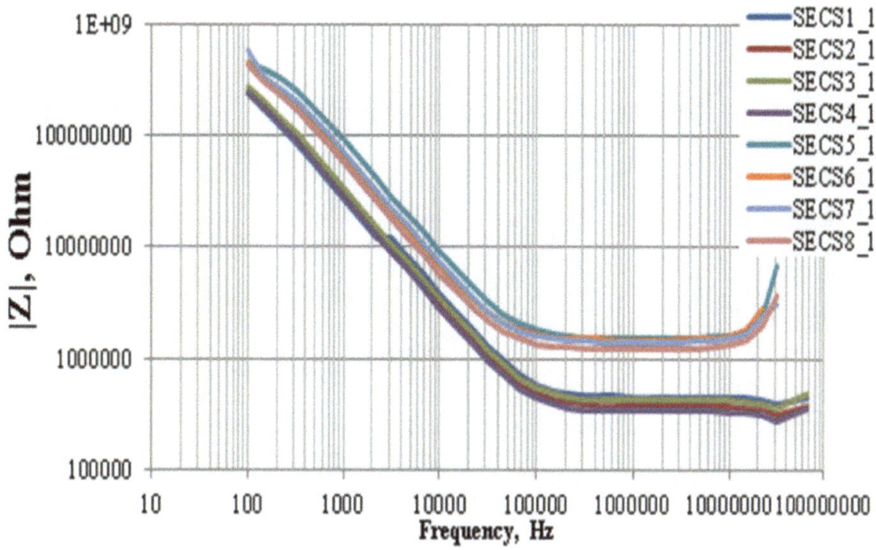

(a)

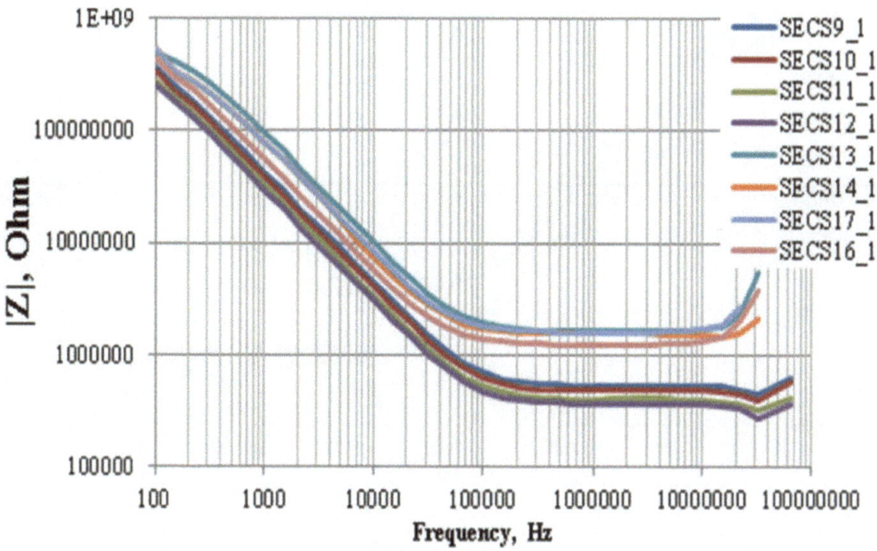

(b)

Fig. 14. Impedance characterization of (a) SECS1_1 to SECS8_1, (b) SECS9_1 to SECS16_1, (c) SECS17_1 to SECS24_1, and (d) SECS1_1 to SECS24_1

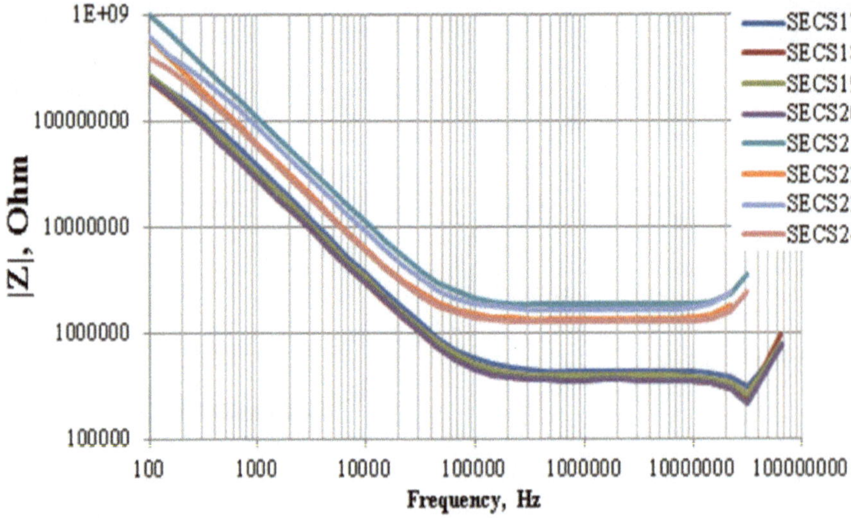

(c)

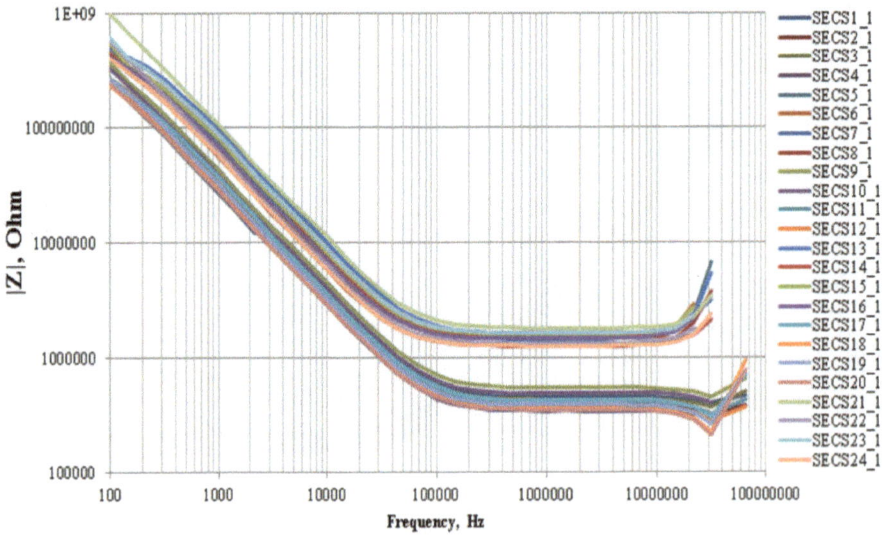

(d)

Fig. 14. (*continued*)

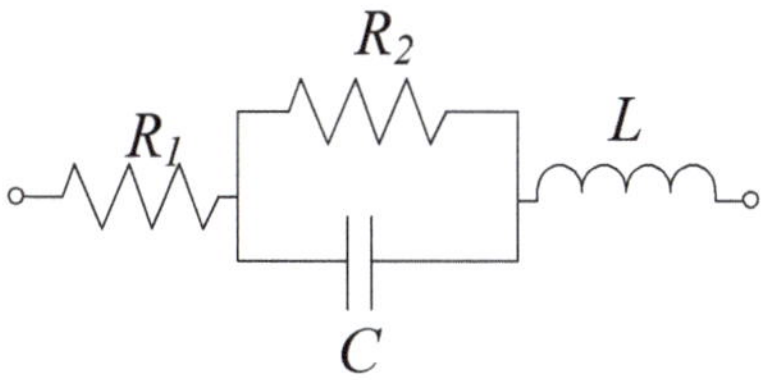

Fig. 15. Circuit representation of the sensors combined in series

6.1 Experimental Setup

The experimental setup for the water sample experiments is shown in Fig. 16. The setup has a frequency waveform generator where a standard sinusoidal waveform with a 10V peak-to-peak value was set as the input signal for the sensors. The sensors were partially immersed into the water sample. The Agilent 54622D mixed signal oscilloscope was interfaced to a PC where the output signals and the sensor's impedance was recorded and calculated using *LabView*.

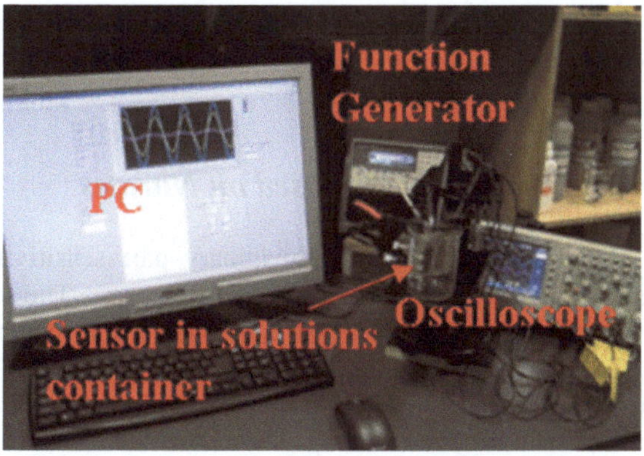

Fig. 16. Experimental setup nitrate experimental work

6.2 Experiment with Chemical Based on Nitrates and Water Samples: Single Frequency Operation

The operating frequencies for all series sensors during water sample experiments were set at 500 kHz. Two sets of experiments were conducted to find the best sensor for further application. Each of the sensors was sprayed with Wattyl Killrust Incralac to form an acrylic resin-based protective coating. The effect of the samples on the sensor's impedance was recorded and the sensitivities of the sensors are calculated from the real part and imaginary part as governed equations (8) and (9).

$$\text{Real part sensitivity, } \%R = \frac{\left((R_{total})_{sample} - (R_{total})_{distilled}\right)}{(R_{total})_{distilled}} \times 100 \tag{8}$$

$$\text{Imaginary part sensitivity, } \%X = \frac{\left((X_{total})_{sample} - (X_{total})_{distilled}\right)}{(X_{total})_{distilled}} \times 100 \tag{9}$$

Where:

$(R_{total})_{distilled}$ is the real part of the impedance value[1].
$(R_{total})_{sample}$ is the real part of the impedance value[1].
$(X_{total})_{distilled}$ is the imaginary part of the impedance value[1].
$(X_{total})_{sample}$ is the imaginary part of the impedance value[1].

In this section, the first experiment involved the response of the sensors with different concentrations of nitrates in the form of sodium nitrate ($NaNO_3$) and ammonium nitrate (NH_4NO_3) mixtures. Then, the sensors with the best sensitivities were chosen from each combination for a second experiment. Five different ratio mixes were prepared where the total weight of every single mix is given in 20 mg. Then, each mix is diluted in 1 liter of distilled water. The ratios of $NaNO_3$ to NH_4NO_3 were 1:1(Ratio_1), 1:2(Ratio_2), 2:1(Ratio_3), 1:3(Ratio_4) and 3:1(Ratio_5).

The second experiment objective was to test the chosen sensors with different concentration of nitrates samples (5 mg, 10 mg, 15 mg and 20 mg) in the form of sodium nitrate ($NaNO_3$) and ammonium nitrate (NH_4NO_3) diluted in 1 liter of distilled water.

6.3 Experiment with Chemical Based on Nitrates

Fig. 17 and 18 depict the real part and imaginary part sensitivities, respectively calculated for SECS17_1, SECS21_1, and SECS22_1 sensors. The sensitivities of the real part values stay in the negative region and this is due to the increase of the conductivity level of the mixed solution. It is clear that SECS22_1 has the best response amongst the series sensors and will be used for further applications.

6.4 Experiment with Chemical Based on Nitrates

Fig. 19 shows the sensitivity values of SECS22_1 when tested with a solution based on $NaNO_3$ and NH_4NO_3. A similar response can be observed for both solution types. The imaginary values are in the range around 100% to around 200% and the values are independent of the total concentration showing that SECS22_1 is already able to detect contamination in the water sample. Moreover, the real part negative values progressively decrease with the total concentration of the chemicals. This is true since the electrical conductivity (σ) of the solution is highly dependent on its concentration of dissolved salts where $NaNO_3$ and NH_4NO_3 were added in the distilled water. The purer the water samples, the lower the conductivity (the higher the resistivity) and vice versa. Therefore, the real part sensitivities give the indirect relationship of total

[1] When the sensor is immersed in distilled water.

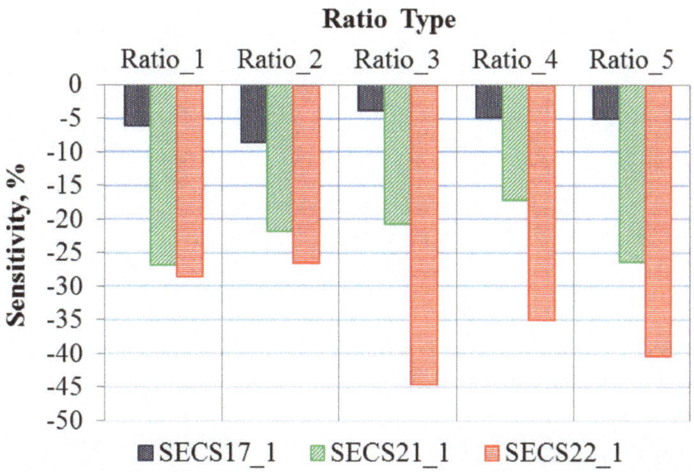

Fig. 17. Series sensors real part sensitivity from experiment with nitrate mixes of varying ratio

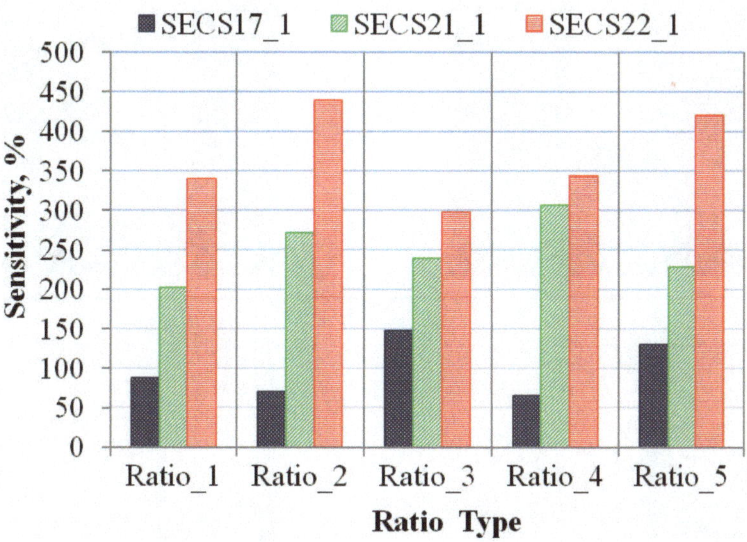

Fig. 18. Series sensors imaginary part sensitivity from experiment with nitrate mixes of varying ratio

amount of chemical concentrations with the quality level of the water samples. In previous work, it has been shown that with more negative electrodes, the field distribution becomes more uniform and that improves the sensitivity [45]. SECS22_1 comprises a thick negative plane as shown in Fig. 20 and this is similar to many negative electrodes. Fig. 21 illustrates the re-plotted real part sensitivities using SECS22_1 for both chemicals with the equivalent linear equation. The graphs in Fig. 21 show very good correlation with R^2=0.9948 and R^2=0.9652, between SECS22_1's real part sensitivities with the chemical concentration and contamination level.

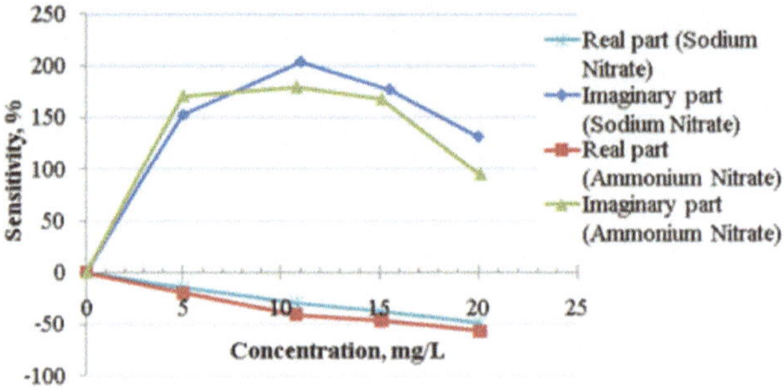

Fig. 19. SESS22_1 sensitivity from experiment with different concentrations of sodium nitrate ($NaNO_3$) and ammonium nitrate (NH_4NO_3)

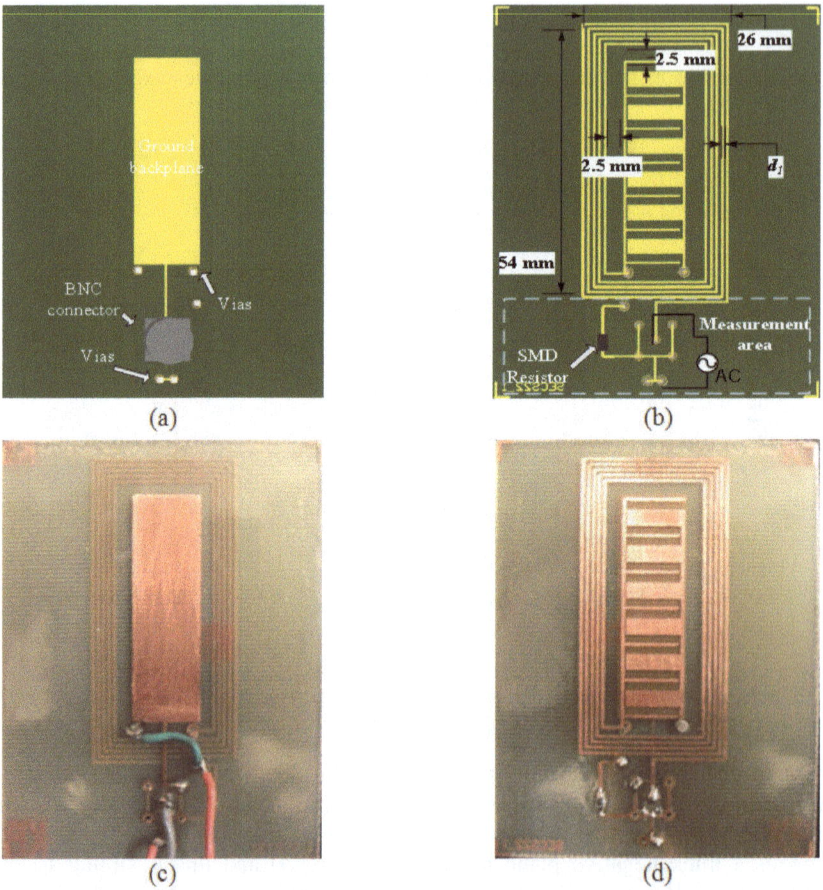

Fig. 20. SECS22_1 schematic diagrams of (a) top layer and (b) bottom layer, and pictures of (c) top layer and (d) bottom layer

6.5 *Stability Test of SECS22_1*

Long-term response stability has been identified as one of the important factors of a chemical sensor performance. In order to evaluate the response stability of the SECS22_1, one reference response value of distilled water was recorded. Then, the experiments involving solutions with different ratio mixes as well as the water samples were repeated separately where the sensors output response to each solution was recorded at an interval of approximately 550s to determine the real part and imaginary part sensitivities. The outcome from Fig. 22 is the extensions of those that have been obtained in Fig. 18. As a whole, the previous results are still valid and significant. Moreover, the real part sensitivities remained constant as shown in Fig. 22 (left) while the imaginary part sensitivities moderately varied as can be seen in Fig. 22 (right).

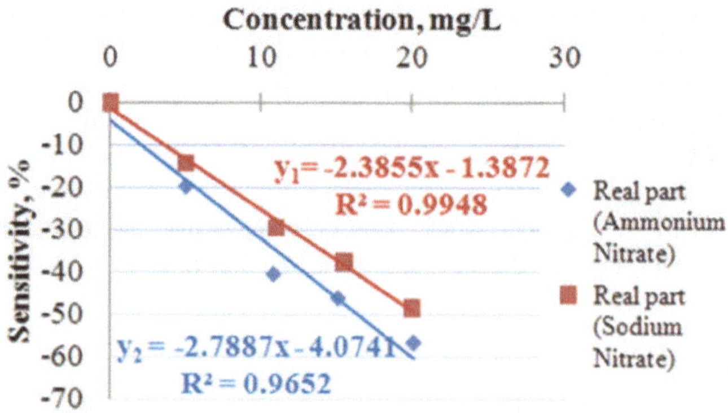

Fig. 21. Graphs of SECS22_1 sensitivity from experimentation with different concentrations of sodium nitrate ($NaNO_3$) and ammonium nitrate (NH_4NO_3)

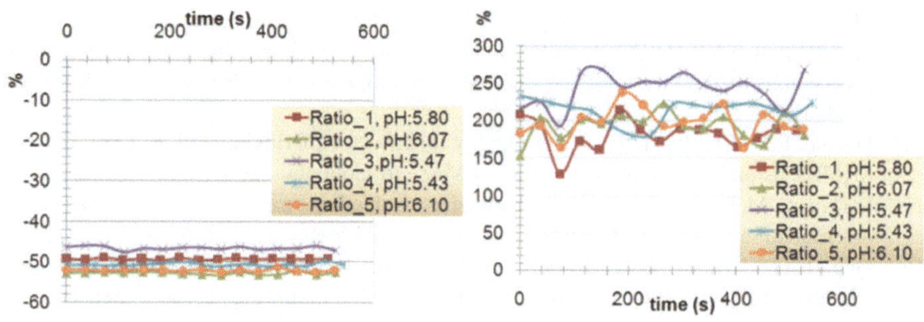

Fig. 22. Sensitivity of SECS22_1 to nitrate solutions: (left) real part and (right) imaginary part

7 Conclusion

In total, 24 planar electromagnetic sensors have been fabricated in this research resulting from the different designs of the sensors (meander and interdigital) and the combination of the sensors (meander and interdigital). Design, basic operation, and principle of measurements have been discussed for all sensors. The printed circuit board technology has proven to be a simple and reliable method to construct the sensors. The used of Incralac coating not only provides protection to the sensors but also opens the way for the investigation of planar electromagnetic sensors for environmental monitoring. In this chapter, the characteristics of all the planar electromagnetic sensors and the effects of different designs to the electrical parameters of the sensors were discussed. The best series connected planar electromagnetic sensors (SECS21_1 to SECS24_1) were determined based on the highest capacitance and inductance compared to other similar sensors. The studies show that a very distinct differentiation of nitrates concentration can be achieved using a planar electromagnetic sensor with meander sensor and an interdigital sensor combined in series (SECS22_1). Furthermore, it can be used as a tool for water sources monitoring on farms where the nitrate level should not exceed 100 mg·$(NO3)^{-}$–N/L.

Acknowledgement. The authors would like to acknowledge the Universiti Teknologi Malaysia's Research University Grant Scheme (Vote No. 4P049).

References

[1] Terblanche, A.P.S.: Health hazards of nitrate in drinking water. Water Sa 17, 77–82 (1991)
[2] Di, H.J., Cameron, K.C.: Nitrate leaching losses and pasture yields as affected by different rates of animal urine nitrogen returns and application of a nitrification inhibitor - a lysimeter study. Nutrient Cycling in Agroecosystems 79, 281–290 (2007)
[3] Metcalf, N.F., Metcalf, W.K., Wang, X.: The Differing Sensitivities of the Hemoglobin of Fetal and Adult Red-Cells to Oxidation by Nitrites in Man - the Role of Plasma. In: Proceedings of the Physiological Society, Cambridge Meeting: Poster Communications, p. 44 (1987)
[4] Donald, L.P., Charles, D.F.: Water Quality for Livestock Drinking, vol. EQ 381, pp. 1–4. MU Extension Publication, University of Missouri-Columbia (2001)
[5] Downes, M.T.: An improved hydrazine reduction method for the automated determination of low nitrate levels in freshwater. Water Research 12, 673–675 (1978)
[6] Ong, K.G., Grimes, C.A., Robbins, C.L., Singh, R.S.: Design and application of a wireless, passive, resonant-circuit environmental monitoring sensor. Sensors and Actuators a-Physical 93, 33–43 (2001)
[7] Ong, K.G., Wang, J., Singh, R.S., Bachas, L.G., Grimes, C.A.: Monitoring of bacteria growth using a wireless, remote query resonant-circuit sensor: application to environmental sensing. Biosensors & Bioelectronics 16, 305–312 (2001)

[8] Ong, K.G., Bitler, J.S., Grimes, C.A., Puckett, L.G., Bachas, L.G.: Remote query resonant-circuit sensors for monitoring of bacteria growth: Application to food quality control. Sensors 2, 219–232 (2002)
[9] Hofmann, M.C., Kensy, F., Buchs, J., Mokwa, W., Schnakenberg, U.: Transponder-based sensor for monitoring electrical properties of biological cell solutions. Journal of Bioscience and Bioengineering 100, 172–177 (2005)
[10] Tan, E.L., Ng, W.N., Shao, R., Pereles, B.D., Ong, K.G.: A wireless, passive sensor for quantifying packaged food quality. Sensors 7, 1747–1756 (2007)
[11] Ong, J.B., You, Z.P., Mills-Beale, J., Tan, E.L., Pereles, B.D., Ong, K.G.: A Wireless, Passive Embedded Sensor for Real-Time Monitoring of Water Content in Civil Engineering Materials. IEEE Sensors Journal 8, 2053–2058 (2008)
[12] Sharavanan, B., Jung-Rae, P., Tarisha, M., Niwat, A., Alkim, A., Tenneti, R., Nagarajan, N.: Conformal Passive Sensors for Wireless Structural Health Monitoring. In: Material and Devices for Smart Systems III, Boston, pp. 341–348 (2009)
[13] Stanley, E.W., et al.: Wireless temperature sensing using temperature-sensitive dielectrics within responding electric fields of open-circuit sensors having no electrical connections. Measurement Science and Technology 21, 075201 (2010)
[14] Woodard, S.E.: SansEC sensing technology - A new tool for designing space systems and components. In: 2011 IEEE Aerospace Conference, pp. 1–11 (2011)
[15] Dickey, E., Varghese, O., Ong, K., Gong, D., Paulose, M., Grimes, C.: Room Temperature Ammonia and Humidity Sensing Using Highly Ordered Nanoporous Alumina Films. Sensors 2, 91–110 (2002)
[16] Oomman, K.V., Craig, A.G.: Metal Oxide Nanoarchitectures for Environmental Sensing. Journal of Nanoscience and Nanotechnology 3, 277–293 (2003)
[17] Garcia-Canton, J., Merlos, A., Baldi, A.: High-Quality Factor Electrolyte Insulator Silicon Capacitor for Wireless Chemical Sensing. IEEE Electron Device Letters 28, 27–29 (2007)
[18] García-Cantón, J., Merlos, A., Baldi, A.: A wireless LC chemical sensor based on a high quality factor EIS capacitor. Sensors and Actuators B: Chemical 126, 648–654 (2007)
[19] Greenhou, H.M.: Design of Planar Rectangular Microelectronic Inductors. IEEE Transactions on Parts Hybrids and Packaging 10, 101–109 (1974)
[20] Hwang, H.Y., Yun, S.W., Chang, I.S.: A design of planar elliptic bandpass filter using SMD type partially metallized rectangular dielectric resonators. In: 2001 IEEE Mtt-S International Microwave Symposium Digest, vols. 1-3, pp. 1483–1486 (2001)
[21] Um, K., An, B.: Design of rectangular printed planar antenna via input impedance for supporting mobile wireless communications. In: Proceeedings IEEE 55th Vehicular Technology Conference, Vtc Spring 2002, vols. 1-4, pp. 948–951 (2002)
[22] Yunas, J., Rahman, N.A., Chai, L.T., Majlis, B.Y.: Study of coreless planar inductor at high operating frequency. In: Proceedings 2004 IEEE International Conference on Semiconductor Electronics, pp. 606–610 (2004)
[23] Fava, J., Ruch, M.: Design, construction and characterisation of ECT sensors with rectangular planar coils. Insight 46, 268–274 (2004)
[24] Quirarte, J.L., Silva, M.J.M., Palacios, M.S.R.: Software for analysis and design of rectangular planar arrays. In: 2004 1st International Conference on Electrical and Electronics Engineering (ICEEE), pp. 90–95 (2004)
[25] Valderas, D., Melendez, J., Sancho, I.: Some design criteria for UWB planar monopole antennas: Application to a slotted rectangular monopole. Microwave and Optical Technology Letters 46, 6–11 (2005)

[26] Li, L.W., Li, Y.N., Mosig, J.R.: Design of a novel rectangular patch antenna with planar metamaterial patterned substrate. In: 2008 IEEE International Workshop on Antenna Technology: Small Antennas and Novel Metamaterials - Conference Proceedings, pp. 119–122 (2008)
[27] Oraizi, H., Noghani, M.T.: Design and Optimization of Linear and Planar slot Arrays on Rectangular Waveguides. In: 2008 European Microwave Conference, vols. 1-3, pp. 1468–1471 (2008)
[28] Fava, J.O., Lanzani, L., Ruch, M.C.: Multilayer planar rectangular coils for eddy current testing: Design considerations. Ndt & E International 42, 713–720 (2009)
[29] Goldfine, N.J.: Magnetometers for Improved Materials Characterization in Aerospace Applications. Materials Evaluation 51, 396–405 (1993)
[30] Goldfine, N.J., Clark, D.: Near-surface material property profiling for determination of SCC susceptibility. Presented at the in 4th EPRI Balance-of-Plant Heat Exchanger NDE Symp. (1996)
[31] Yamada, S., Fujiki, H., Iwahara, M., Mukhopadhyay, S.C., Dawson, F.P.: Investigation of printed wiring board testing by using planar coil type ECT probe. IEEE Transactions on Magnetics 33, 3376–3378 (1997)
[32] Mukhopadhyay, S.C., Yamada, S., Iwahara, M.: Experimental determination of optimum coil pitch for a planar mesh-type micromagnetic sensor. IEEE Transactions on Magnetics 38, 3380–3382 (2002)
[33] Mukhopadhyay, S.C.: Quality inspection of electroplated materials using planar type micro-magnetic sensors with post-processing from neural network model. IEEE Proceedings-Science Measurement and Technology 149, 165–171 (2002)
[34] Mukhopadhyay, S.C.: A novel planar mesh-type microelectromagnetic sensor - Part II: Estimation of system properties. IEEE Sensors Journal 4, 308–312 (2004)
[35] Mukhopadhyay, S.C.: High Performance Planar Electromagnetic Sensors-A Review of Few Applications. In: New Zealand National Conference on Non-Destructive Testing, pp. 33–41 (2004)
[36] Mukhopadhyay, S.C.: Novel planar electromagnetic sensors: Modeling and performance evaluation. Sensors 5, 546–579 (2005)
[37] Sheiretov, Y., Grundy, D., Zilberstein, V., Goldfine, N., Maley, S.: MWM-Array Sensors for In Situ Monitoring of High-Temperature Components in Power Plants. IEEE Sensors Journal 9, 1527–1536 (2009)
[38] AnimAppiah, K.D., Riad, S.M.: Analysis and design of ferrite cores for eddy-current-killed oscillator inductive proximity sensors. IEEE Transactions on Magnetics 33, 2274–2281 (1997)
[39] Karunanayaka, D., Gooneratne, C.P., Mukhopadhyay, S.C., Sen Gupta, G.: A Planar Electromagnetic Sensors Aided Non-destructive Testing of Currency Coins. NDT.net 11, 1–12 (2006)
[40] Radke, S.M., Alocilja, E.C.: A microfabricated biosensor for detecting foodborne bioterrorism agents. IEEE Sensors Journal 5, 744–750 (2005)
[41] Syaifudin, A.R.M., Jayasundera, K.P., Mukhopadhyay, S.C.: A low cost novel sensing system for detection of dangerous marine biotoxins in seafood. Sensors and Actuators B-Chemical 137, 67–75 (2009)
[42] Syaifudin, A.R.M., Yunus, M.A., Mukhopadhyay, S.C., Jayasundera, K.P.: A novel planar interdigital sensor for environmental monitoring. In: 2009 IEEE Sensors, Christchurch, New Zealand, pp. 105–110 (2009)

[43] Mukhopadhyay, S.C., Gooneratne, C.P., Demidenko, S., Gupta, G.S.: Low Cost Sensing System for Dairy Products Quality Monitoring. In: Proceedings of the IEEE Instrumentation and Measurement Technology Conference, IMTC 2005, vol. 1, pp. 244–249 (May 2005)

[44] Mukhopadhyay, S.C., Sen Gupta, G., Woolley, J.D., Demidenko, S.N.: Saxophone reed inspection employing planar electromagnetic sensors. IEEE Transactions on Instrumentation and Measurement 56, 2492–2503 (2007)

[45] Mukhopadhyay, S.C., Gooneratne, C.P.: A novel planar-type biosensor for noninvasive meat inspection. IEEE Sensors Journal 7, 1340–1346 (2007)

[46] Mukhopadhyay, S.C., Choudhury, S.D., Allsop, T., Kasturi, V., Norris, G.E.: Assessment of pelt quality in leather making using a novel non-invasive sensing approach. Journal of Biochemical and Biophysical Methods 70, 809–815 (2008)

[47] Yunus, M.A.M., Kasturi, V., Mukhopadhyay, S.C., Sen Gupta, G.: Sheep skin property estimation using a low-cost planar sensor. In: IEEE Instrumentation and Measurement Technology Conference, I2MTC 2009, pp. 482–486 (2009)

[48] Yunus, M.A.M., Mukhopadhyay, S.C.: Novel Planar Electromagnetic Sensors for Detection of Nitrates and Contamination in Natural Water Sources. IEEE Sensors Journal 11, 1440–1447 (2011)

[49] Mamishev, A.V., Sundara-Rajan, K., Yang, F., Du, Y.Q., Zahn, M.: Interdigital sensors and transducers. In: Proceedings of the 2006 IEEE Sensors Applications Symposium, vol. 92, pp. 808–845 (May 2004)

[50] Paul, C.R.: Analysis of Linear Circuits. In: Analysis of Linear Circuits B2 - Analysis of Linear Circuits. McGraw-Hill, New York (1989)

[51] Cano, E., Lafuente, D., Bastidas, D.M.: Use of EIS for the evaluation of the protective properties of coatings for metallic cultural heritage: a review. Journal of Solid State Electrochemistry 14, 381–391 (2010)

[52] Bondarenko, A.S., Ragoisha, G.A.: EIS Spectrum Analyser, `http://www.abc.chemistry.bsu.by/vi/analyser/`

[4] [illegible]

Microwave Sensors for Real-Time Nutrients Detection in Water

A. Mason*, O. Korostynska, and A.I. Al-Shamma'a

Built Environment and Sustainable Technologies (BEST) Research Institute,
Liverpool John Moores University,
Byrom Street, Liverpool, L3 3AF, UK
A.Mason1@ljmu.ac.uk

Abstract. Current wastewater monitoring techniques rely on the use of nutrients detection as the result of some chemical reaction, which is undesirable for long-term use in real-time applications. In addition, new legislation may render such systems obsolete if they cannot reliably determine the amount of nutrients in wastewater relative to allowable levels. This chapter attempts to address this issue by considering the use of microwave sensing techniques as an alternative real-time approach that has the potential to monitor wastewater nutrients such as phosphate and nitrate. The method utilizes a broad range of microwave frequencies (1-15 GHz) and is demonstrated with two different types of structure for this purpose, namely a traditional resonant cavity and a flexible interdigitated electrode structure. A variety of experimental results are shown that validate the applicability of the microwave sensing for detecting phosphates and nitrates in the solutions. LabView software used for analysis of captured data and for easy user interpretation of this data is also demonstrated. Future work to be undertaken is discussed in relation to improving the performance of the sensor further, as well as adding the capability to automatically determine both the type and concentration of nutrients in water solutions.

Keywords: Water quality monitoring, wastewater, nitrate, phosphate, microwave sensor, interdigitated electrode, flexible sensor.

1 Introduction

Urban wastewater is defined by the Council of the European Communities [1] as "*Domestic wastewater or mixture of domestic wastewater with industrial wastewater and/or run-off rain water*". Domestic wastewater is defined as "*Wastewater from residential settlements and services which originates predominantly from the human metabolism and from household activities*". On the other hand, industrial wastewater is "*Any wastewater which is discharged from premises used for carrying on any trade or industry, other than domestic wastewater and run-off rain water*". Therefore we can say that wastewater refers to a broad spectrum of contaminated water.

---

* Corresponding author.

S.C. Mukhopadhyay & A. Mason (Eds.): *Real-Time Water Quality Monitoring*, SSMI 4, pp. 197–216.
DOI: 10.1007/978-3-642-37006-9_9 © Springer-Verlag Berlin Heidelberg 2013

In order to maintain a healthy environment and to control the spread of disease, wastewater has to be collected and treated prior to discharge back to the environment. UK directives have been implemented through the urban wastewater treatment regulations since 1994 [2]. The collection and treatment of wastewater plays a vital part in the protection of public health, water resources and wildlife; directives set the standards to be used for its collection and treatment. As part of these standards, a sewerage system is provided for all urban areas above a specified population size, and the collected sewage receives at least secondary (biological) treatment before it is discharged to the environment. Discharges from the sewerage systems are only allowed under storm conditions. Areas where sewage requires extra treatment before discharge are identified by the Council of the European Communities as "sensitive areas" [1]. One example is the atrophic waters, where additional nutrients, mainly nitrogen or phosphorus, stimulate the growth of algae and other plants, damaging the natural environment. In these areas, larger sewage discharges must be treated to reduce their load of nutrients.

Wastewater treatment plants (WWTP) and industrial sites that discharge more than 1 m^3 of effluent on a daily basis have to provide periodic reports for the quality and quantity of their effluent before discharging. They also need to pay fines for pollution events. Charges could be reduced if continuous monitoring is available to control and solve problems before discharging. As mentioned before, wastewater treatment involves removing nutrients from wastewater before it is discharged to the water course (e.g. canals and rivers). Many technologies have been developed for this purpose, such as the activated sludge system EBPR (enhanced biological phosphorus removal). These technologies made the detection of such nutrients more difficult because it reduces the nutrient levels significantly compared with previous practice. To solve these difficulties, many types of sensor and analyzer systems have been developed to detect and monitor the wastewater treatment process. However, most of these analyzers are based on off-line measurements which imply low frequency data sampling and significant delays between sampling and availability of results.

Effective monitoring for the quality of effluent can be achieved by obtaining representative samples for lab analysis, or by installing an on-line analyzer. Current on-line technologies for monitoring the limits have a high capital cost, are unreliable and incur high maintenance costs.

2 The Wastewater Treatment Process

Wastewater, also known as sewage, contains more than 99% water and is characterized by volume or rate of flow, physical condition, chemical constituents and the bacteriological organisms that it contains. In general it contains pathogens such as bacteria, viruses, and parasitic worms, as well as organic particles such as plant material, humus, and paper fiber. The soluble organic material that also can be found in sewage could come from urea, fruit sugar and soluble proteins. Sand, grits and metal particles are considered as inorganic particles, and sewage also contains soluble inorganic material such as ammonia, road salt and hydrogen sulphide. In the UK, over

350,000 kilometers of sewers collect over 11 billion liters of wastewater a day, and this is treated at about 9000 sewage treatment works [3] before the treated effluent is discharged to inland water, estuaries and the sea. Wastewater treatment is an important component in the water cycle [4], as it ensures that the environmental impact of human usage of water is significantly reduced.

WWTPs use a series of treatment stages to clean up the contaminated water so that it is safely released into lakes, rivers, or streams. Wastewater treatment consists of several processes (physical, biological and chemical) that aim to reduce nitrogen, phosphorous, organic matter and suspended solids content. To reduce the amount of these substances, WWTPs consist of (in general) five treatment stages: (1) a mechanical pre-treatment stage, (2) a primary treatment stage, (3) a secondary treatment stage, (4) a tertiary treatment stage and (5) a final sedimentation stage. Note that not all of these stages can be found in every wastewater treatment plant, depending on the size of the plant and where the treated wastewater is discharged. Fig. 1 shows the five treatment stages in a typical WWTP [5].

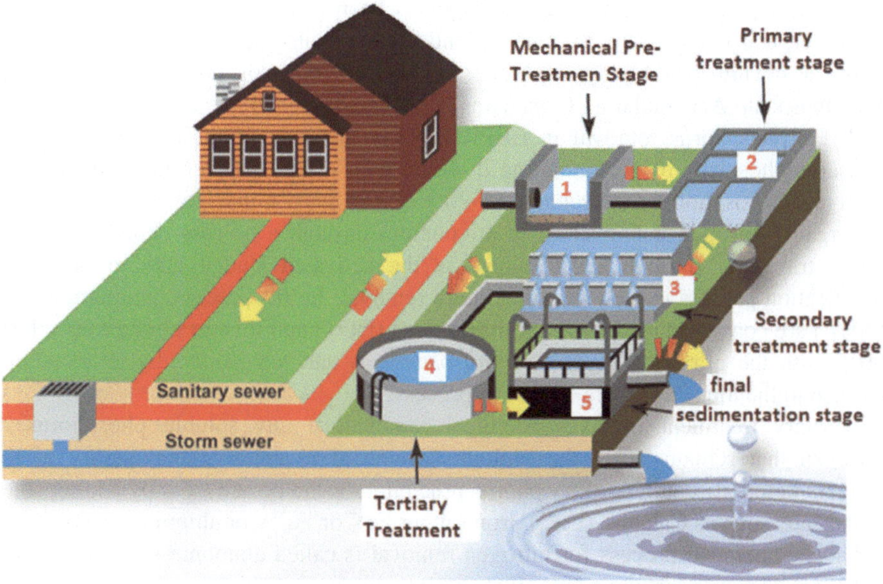

Fig. 1. A simplified overview of the wastewater treatment process

The mechanical pre-treatment stage is essential to remove different types of suspended solids from the incoming wastewater. It also adjusts the pH level to avoid damage to equipment, underwater sewer pipes, and microorganisms used as part of the treatment process [6]. This physical treatment is meant to protect the next stages from different types of grits and larger particles. Practically, this stage consists of multiple screening grids that remove larger objects found in wastewater, an aerated sand filter that removes sand, and a sedimentation unit that reduces the content of suspended solids by means of sedimentation. Since October 2007, the non-hazardous

UK waste produced by the pre-treatment stage such as sewage screenings and grits must be treated before it can be land filled [7].

The primary treatment stage consists of a large settling tank, and aims to settle down the suspended solid and grits that are still in wastewater after the pre-treatment stage. The settled matter (sediment), known as sludge, is pushed into hoppers and carried out to be treated. Also the settling tank will allow the grease and oil to float on the surface so that it can be skimmed off. The primary treatment stage reduces the amount of solids by 50%, and also BOD (Biochemical Oxygen Demand) by at least 20% before it is discharged [8].

Once all noticeable solids have been removed, a biological treatment that targets the organic matter takes place during the secondary treatment stage. The most common type of biological treatment is called 'Activated sludge' where WW is mixed with air to provide oxygen for bacteria to grow and then consume the organic matter. Turbines and surface aerators can be used as air diffusers. Microorganisms degrade the content of the organic matter in the WW aerobically, i.e. when air is supplied to the biological reactor.

The biological treatment stage has originally been solely to remove organic matter. However, many wastewater treatment plants today are also designed for the biological removal of nitrogen and phosphorous. A unique type of microorganism called Polyphosphate Accumulating Organisms (PAO) are enriched into the activated sludge tank for phosphorus removal in EBPR process The role of these organisms is to consume phosphate by accumulating it within their cells. PAO can incorporate up to 0.38 mg/L phosphorous and remove 15-20% of it in many municipal wastewaters [9].

Organic nitrogen is converted to ammonium through a process called hydrolysis while travelling through sewer pipes. In the activated sludge tank, a biological nitrification process is used for ammonium removal. In this process, ammonium ions (NH_4^+) are converted or oxidized to nitrite ions (NO_2^-) and then to nitrate ions (NO_3^-) [10], with the nitrifying organisms (*nitrosomonas* and *nitrobacter* bacteria) adding oxygen to the ions during the oxidation process.

Tertiary treatment involves chemical removal of any soluble phosphorus or nitrogen that remains after the biological removal to enhance the quality of the effluent before discharging. Phosphorus precipitation can be achieved by adding some metal salts such as Calcium (Ca^{2+}), iron (either Fe^{2+} or Fe^{3+}), or aluminum ($Al_2(SO_4)_3$) [11]. The chemical process for nitrogen removal is called ammonia stripping, where pH is raised to convert the ammonium ion into ammonia, which can be stripped from the water by passing large quantities of air through the water [12]. Also, Chlorine could be added to oxidize ammonia-nitrogen into nitrogen gas; 9-10 mg/L of chlorine is required for every 1 mg/L of ammonia-nitrogen [13].

At the end of the treatment process, a secondary settling tank is used to settle down the remaining precipitates that resulted from either the biological or the chemical nutrients removal in the activated sludge stage. The final settled sludge is then carried out to be treated. Sludge can be recycled to produce an organic-based fertilizer and soil conditioner for use in agriculture. It also may be used for energy generation, large tanks called digesters are used to transform the organic solids of the sludge into gaseous end products with the absence of oxygen in a process called anaerobic digestion. The final effluent is discharged into ship canals or coastal waters, and it is here that initially this work is aimed; i.e. to detect nutrients remaining in water after treatment. Such information is important in many cases, but particularly (for example) in cases where water is discharged into coastal waters close to shellfish habitats, since contaminated coastal waters impact negatively on both the marine life and agriculture, which in turn harms people in the area of wastewater discharge. Ultimately however it is envisaged that the technique proposed in this work could be utilized at many stages during the treatment process in order to determine the effectiveness of the process and give some real-time alert when nutrient levels unexpectedly vary.

3 Microwave Sensing

Microwave analysis (or microwave spectroscopy) can be applied to suit a broad range of requirements [14-19] and has a number of advantages over competing technologies for wastewater sensing applications, some of which were highlighted by the current authors in a prior paper [20]. The primary advantages in this case are twofold:

(1) **True real-time sensing**, the analyte material flows past or through the sensor and an instantaneous measurement is acquired without the need for significant pre-processing.
(2) **Direct sample measurement**, a feature which is currently unavailable in many competing technologies which rely on some chemical reaction to induce a change (most commonly a color change) in the analyte material.

Further work by the authors considers a variety of sensing techniques and compares them to that of microwave sensors [21]. The microwave sensor (or cavity) is constructed from closed sections of metallic materials, such as aluminum or copper, which are often selected due to their excellent conductive properties. A microwave sensor may have multiple inputs and outputs, which are commonly referred to as *ports*. In this work no more than two-ports are utilized for the sake of simplicity. Fig. 2 illustrates a schematic overview of a two-port device.

Microwave analysis can provide unique signal spectrum signatures which consist of a reflected signal, S_{11}, and/or a transmitted signal, S_{21}, based on parameters such as conductivity and permittivity [22]. Conductivity is a measure of a material's ability to conduct an electric current. Permittivity is a measure of how an electric field is

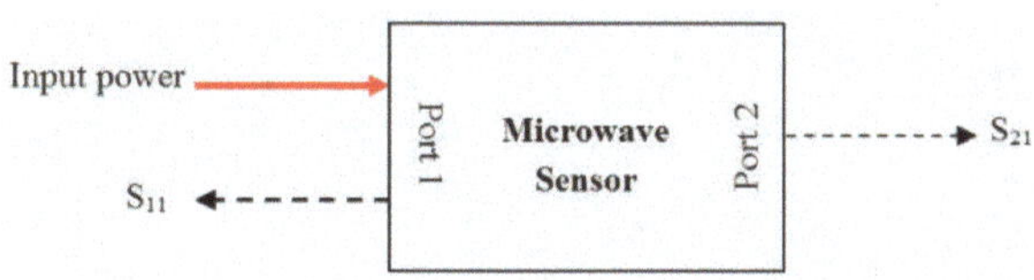

Fig. 2. Illustration of a microwave structure, showing where measurements of S_{11} and S_{21} may be acquired

affected by a dielectric medium, which is determined by the ability of a material to polarize in response to the field, and reduce the total electric field inside the material. Therefore, permittivity relates to a material's ability to transmit an electric field and is a complex value which varies with frequency, and accounts for both the energy stored by a material (ε') as well as any losses of energy (ε") which might occur. The dielectric properties of materials are closely related to their molecular structure. As a result, any change in these structures can be detected by means of a microwave sensor provided that the microwave response directly depends on the permittivity of the material with which the electromagnetic waves interact [23].

For much of the work conducted so far, a cylindrical cavity has been utilized. When one excites a cylindrical cavity with microwave energy, at certain frequencies predictable arrangements of the electric and magnetic fields will form inside, typically referred to as *resonant modes*. These modes are based upon such parameters as the relative permeability (μ_r), the relative permittivity (ε_r), the radius (a) of the cavity and also its depth (d). Using work detailed by Pozar [24], it is possible to predict the frequencies at which these modes will occur from (1).

$$f_{nml} = \frac{c}{2\pi\sqrt{\mu_r \varepsilon_r}} \left[\left(\frac{p_{nm}}{a} \right)^2 + \left(\frac{l\pi}{d} \right)^2 \right]^{1/2} \tag{1}$$

here c is the speed of light and p_{nm} is the mth root of the of Bessel function of the nth order for TM modes or the mth root of the first derivative of Bessel function of the nth order for TE modes. It is also possible to visualize these modes using 3D modeling software such as the Ansys High Frequency Structure Simulation (HFSS) package [25]. For the cavity shown in Fig. 3, a number of resonant modes are shown in Fig. 4 by way of example.

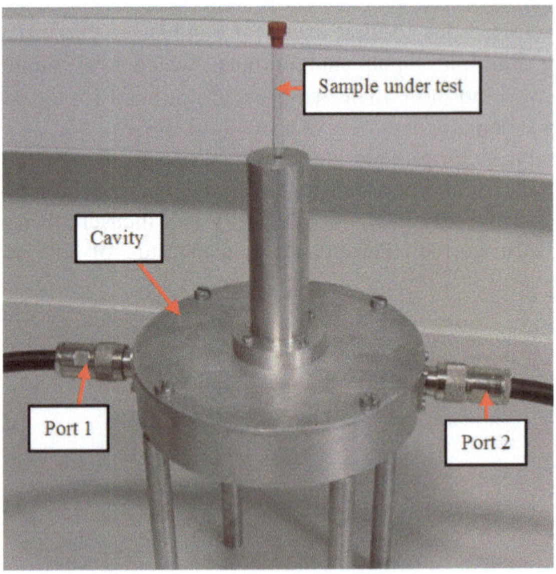

Fig. 3. Example of a cylindrical microwave cavity used by authors in other works [26-28]

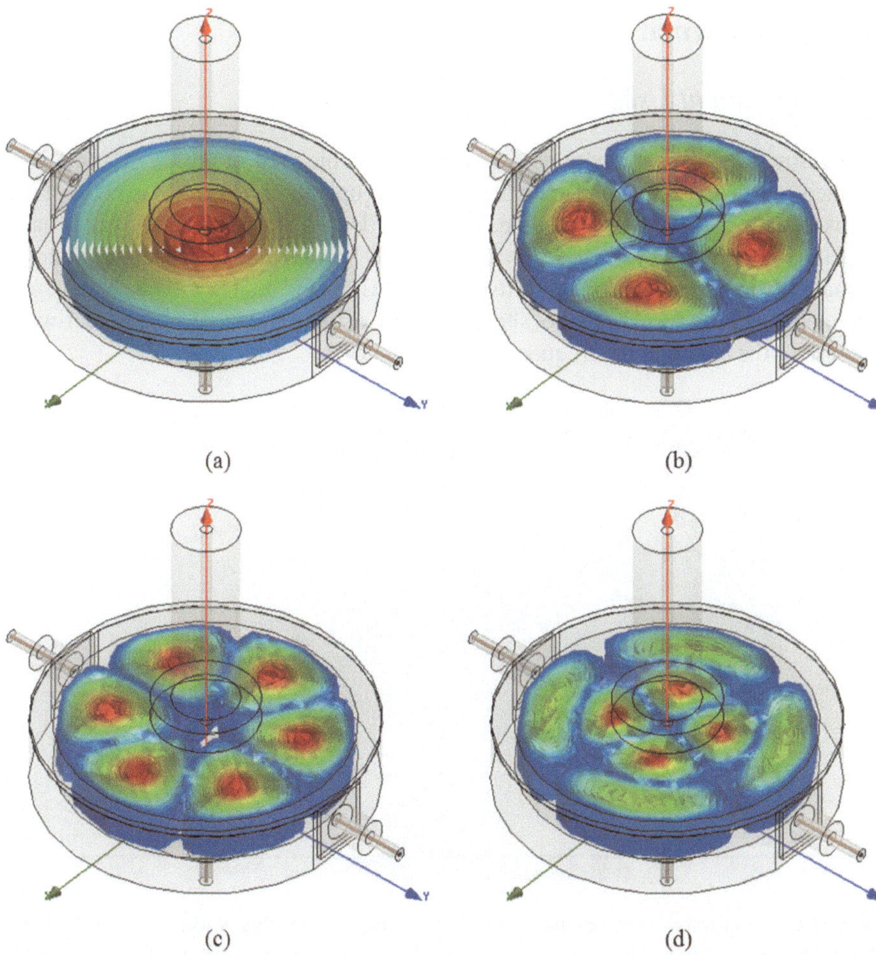

Fig. 4. An example of the electromagnetic fields formed inside the cavity shown in Fig. 3. at increasing frequencies, namely showing the modes (a) TM_{010}, (b) TM_{210}, (c) TM_{310} and (d) TM_{220}

This work considers, for the moment, *narrow band analysis* to enable primarily the optimization of system sensitivity to nutrients in wastewater. This means that the work is dominantly based around the response which occurs at the TM_{010} mode of a cylindrical cavity when different analytes (i.e. phosphates and nitrates at varying concentrations) are placed within the cavity. The multi-parameter nature of broadband microwave analysis is discussed as a topic for future work toward the end of this paper, and is the mechanism by which the issue of nutrient *specificity* will be explored.

4 Experimental Setup

4.1 Equipment Used

The microwave resonant cavity used in this work is pictured in Fig. 5. The cavity is designed such that its fundamental mode of operation (i.e. TM_{010}) occurs at approximately 2.5 GHz [29] when the central PTFE pipe is water filled. This is based upon internal dimensions where $a = 36$ mm and $d = 30$ mm. When the water is evacuated, the operating mode shifts to approximately 3 GHz. This behavior is explained generally by (1). Fig. 6. shows a comparison of the modeled and real-world response of the cavity in the frequency range 1-6 GHz.

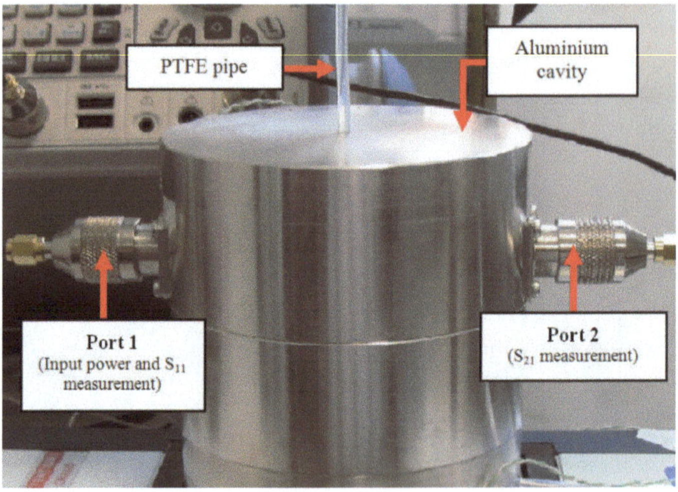

Fig. 5. The 2-port microwave resonant cavity sensor utilized for this work, showing the PTFE tubing passing through the center of the aluminum construction thus allowing nutrient concentration measurements via interaction with the electric field formed inside the cavity

The experimental system, shown in Fig. 7, is constructed from a number of components which are briefly detailed as follows:

(1) **Four Port HLPC Pump.** This allows sample material (i.e. synthetic wastewater) to be mixed and passed through the sensor to characterize and determine its response to differing nutrients and nutrient concentrations.

(2) **Vector Network Analyzer.** This device is responsible for providing input power to the microwave cavity in addition to measuring the S_{11} and S_{21} response to differing nutrients and nutrient concentrations. The analyzer is a Rohde and Schwarz ZVL6 unit, and is calibrated prior to measurements in order to ensure negligible external impact on measurements which is particularly important considering the length of cable required for this experimental work (i.e. 0.5 m).

(3) **Heating Element.** The permittivity of water is known to be affected heavily by temperature [30]; if temperature increases then the permittivity decreases [22, 31].

Therefore the system is designed to alleviate ambient temperature changes by heating the nutrients to a fixed temperature.

(4) **Microwave Cavity Sensor.** As described earlier, this is the device which responds to the changing nutrient type and concentration.

(5) **Synthetic Wastewater (SWW).** A synthetic wastewater is created in this case from a simple combination of deionized water and either KH_2PO_4 (Potassium Phosphate) or KNO_3 (Potassium Nitrate).

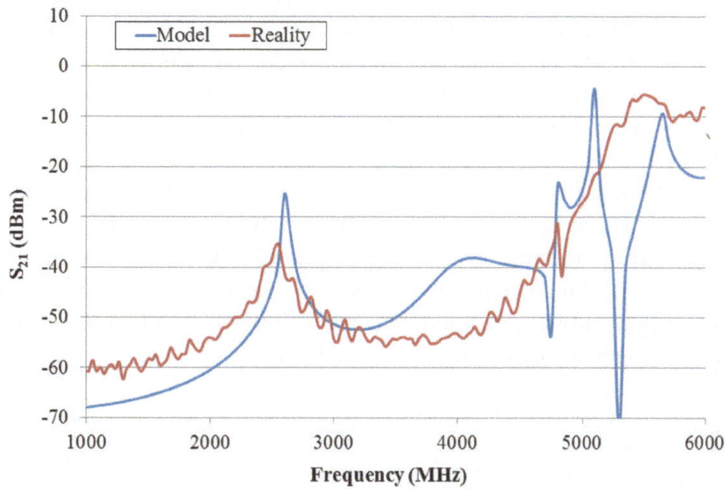

Fig. 6. Shows the microwave cavity S_{21} response during simulated and real world measurements

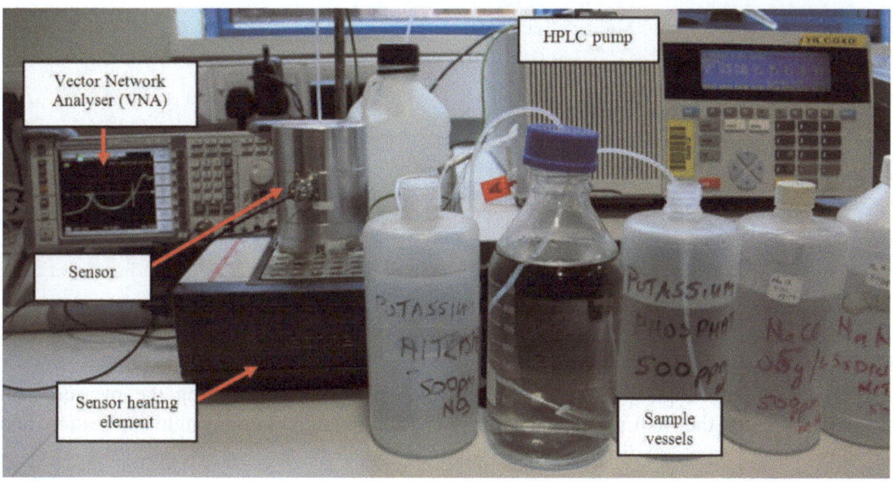

Fig. 7. Shows the experimental setup used for this work, with heating elements being employed to negate ambient temperature variation

4.2 *Experimental Procedure*

Beginning with the SWW, this is created to emulate a simplistic wastewater, and is envisaged to be similar to the final effluent of a wastewater treatment plant (i.e. relatively free of solids). This removes the need to incorporate filtration into the system, and allows better assessment of the sensor performance in relation to water soluble nutrients.

Two varieties of SWW were prepared, one utilizing Potassium Phosphate (KH_2PO_4) and the other Potassium Nitrate (KNO_3) – both are used as agricultural fertilizer and are therefore one of the main sources for such nutrients being present in actual wastewater. The phosphate solution was prepared by diluting 2.2 g of KH_2PO_4 salt in 1L of deionized water to obtain 500 mg/L of PO_4-P (or 1535 mg/L of PO_4). The nitrate was prepared by diluting 3.6 g of KNO_3 salt in 1L of deionized water to obtain 500 mg/L of NO_3-N (or 2200 mg/L of NO_3).

Using the 500 mg/L solutions, it is possible to have numerous concentration levels mixed by the HLPC pump in real-time ready for introduction to the sensor. For this preliminary study, concentrations of 500 mg/L down to 0 mg/L of PO_4-P and NO_3-N are considered in 100 mg/L intervals. This is intended to consider the limitations of the current design, and suggest improvements which will lead to a more fine-grained approach to testing and characterization in the future.

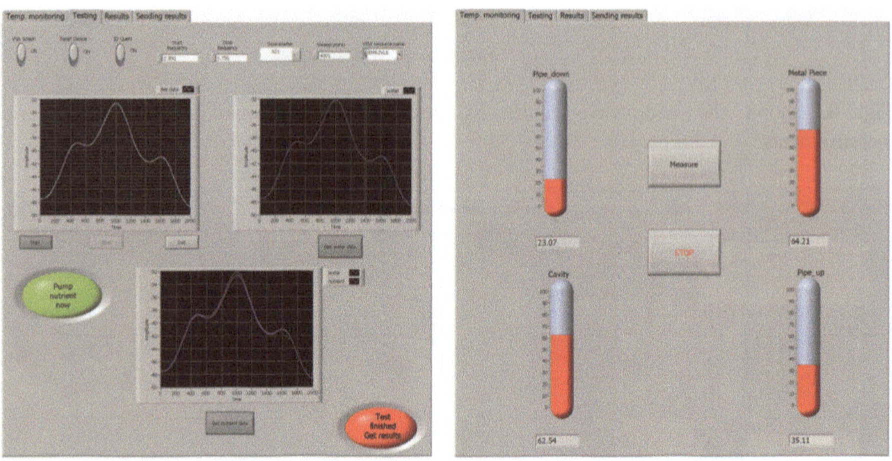

(a) (b)

Fig. 8. LabView front panel window for capturing data from (a) the Rohde and Schwarz VNA and (b) the thermocouple probes used to monitor sample testing during the measurement process. Note that the front panels shown here are all integrated into a single application, and the above images represent different "tabs" within the application.

The output of the HPLC pump is connected to the PTFE piping that runs through the center of the sensor and feeds to a waste vessel – samples are not recirculated to prevent algae build up in the piping and also to prevent unwanted sample contamination or dilution. A heating block heats the cavity to approximately 63°C, which heats the SWW to a maximum of 35°C when it leaves the cavity. The stability of the cavity and SWW is monitored via thermal couple measurements which are incorporated into to a LabView interface (see Fig. 8) for monitoring.

Measurements from the system are acquired in two stages. In the first stage water only is pumped through the sensor to provide a reference level, and the LabView interface acquires an average of the stabilized microwave spectrum between 2480 and 2620 MHz over the course of a 1 minute period. Secondly, the SWW is pumped through the system and a similar procedure is applied. Between each stage sufficient time is allowed for the water or SWW sample to propagate through the tubing. When pumping at 10 ml/min, this waiting time is approximately 45 seconds.

The method used here for acquiring data has proven to give considerable measurement reliability, and along with the heating block, helps to further negate external impacts on measurements to ensure that the system provides both short-term and long-term repeatability. This is particularly important when the system is used as a predictive tool, as discussed later in this chapter.

5 Results

Data acquired for phosphate and nitrate concentrations are shown in Fig. 9. Table 1 notes the prominent features of the results – i.e. resonant peak amplitude and frequency – for each nutrient. This data is also represented in Fig. 10, showing clearly the linear relationship between nutrient concentration and peak amplitude. The linearity of the peak frequency is not quite so clear, although there is a general trend of increasing frequency with increasing nutrient concentration.

Table 1. Peak amplitude and frequency data for PO_4-P and NO_3-N concentrations, as shown in Fig. 9 and 10

Concentration (mg/L)	Peak Amplitude (dBm)		Peak Frequency (MHz)	
	PO_4-P	**NO_3-N**	**PO_4-P**	**NO_3-N**
0	-30.04	-30.04	2547.4	2547.41
100	-30.36	-30.56	2547.83	2547.83
200	-30.77	-30.99	2548.11	2548.11
300	-31.20	-31.40	2548.11	2548.11
400	-31.59	-31.79	2548.25	2548.25
500	-31.89	-32.09	2548.39	2548.39

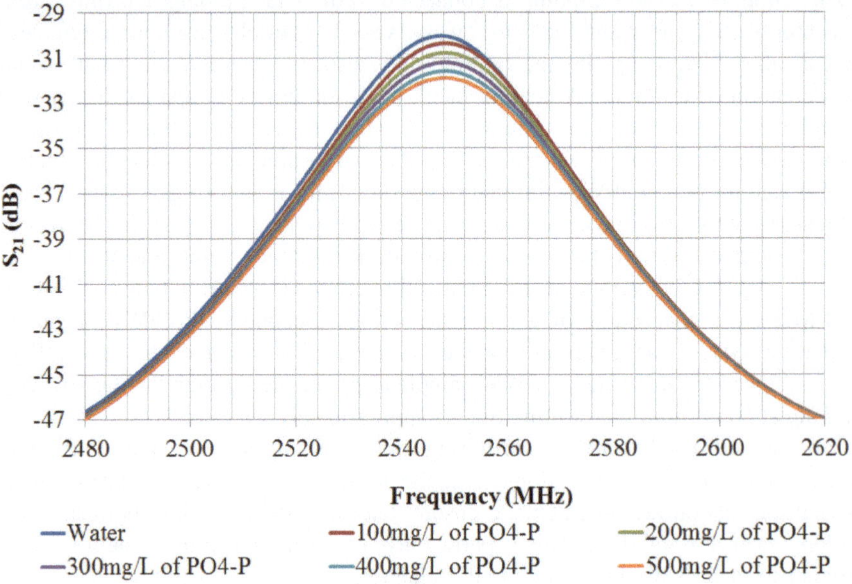

(a)

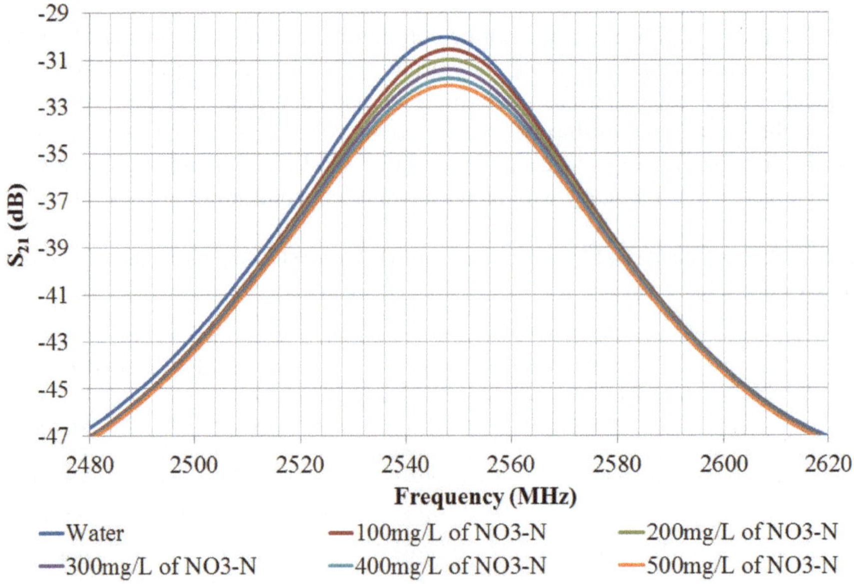

(b)

Fig. 9. S_{21} results for (a) PO_4-P and (b) NO_3-N concentrations from 0-500 mg/L in the frequency range 2480 – 2620 MHz

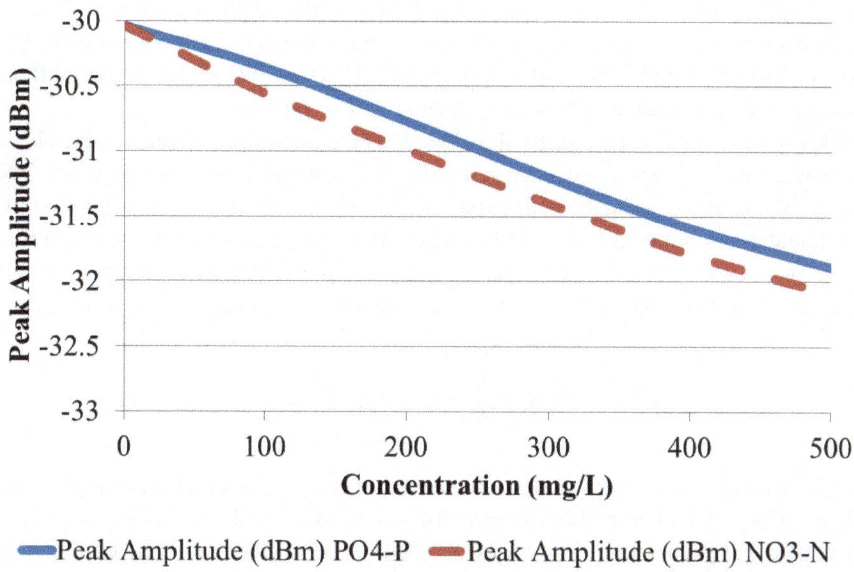

(a)

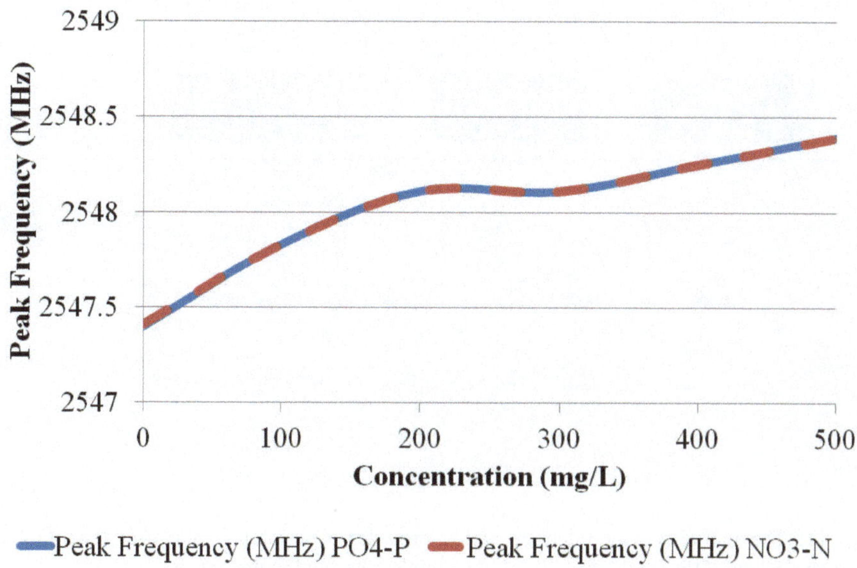

(b)

Fig. 10. Indicators of linearity for the sensor peak (a) amplitude and (b) frequency

From the results displayed it is possible to determine the concentration of PO_4-P and NO_3-N in a SWW solution. Since these results show a great deal of repeatability, it has been possible to develop a semi-automated user interface for the sensor which can reliably determine the concentration of PO_4-P or NO_3-N in water, although currently the user must specify which nutrient will be present.

This user interface has been developed using National Instrumental LabView software, which is capable of interfacing directly with the Rohde and Schwarz VNA. Using this interface, a database of expected data for each nutrient has been compiled. This database contains differential variance values for each nutrient concentration in the range of 0-500 mg/L. In order to calculate the differential variance, first the variance of the two captured spectra (i.e. water and SWW) is calculated using (2).

$$\sigma^2 = \frac{1}{N}\sum_{i}^{N}(x_i - \mu)^2 \tag{2}$$

Where N is the number of data points acquired in each spectra and μ is the population mean. Upon calculating the variance for water (σ^2_W) and the SWW (σ^2_{SWW}), the difference between them (i.e. $\sigma^2_{SWW} - \sigma^2_W$) is calculated and utilized for future predictions. The LabView interface for achieving this prediction is shown in Fig. 11.

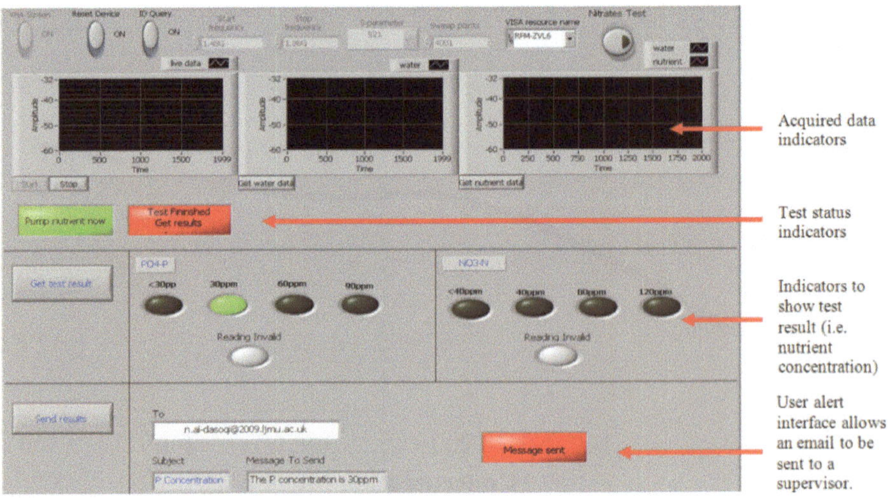

Fig. 11. The LabView user interface which provides a decision to the user regarding the nutrient content of the synthetic waste water sample presented to the sensor

6 Discussion and Future Progression

This work has shown great promise of the microwave technique in detecting nutrients such as phosphate and nitrates in wastewater. However, microwave sensing is not limited to the use of cavity structures, which are seen by many as bulky and expensive due to the use of reasonably expensive high conductivity metals (e.g. aluminum or copper). Thus the authors have been considering new ways in which to achieve the benefits of microwave sensor capability but in a variety of formats. The use of microstrip printed antennas operating as microwave sensors in the GHz frequency range with a Ag planar pattern printed on flexible substrate was recently reported [32]. These sensors are conformable to planar and non-planar surfaces, simple and cost-effective to manufacture using modern printed-circuit technology, mechanically robust when mounted on rigid surfaces, and when the particular patch shape and mode are selected they are very versatile in terms of resonant frequency, polarization and impedance.

As an example, an interdigitated electrode (IDE) structure shown in Fig. 12 operating at microwave frequencies was chosen for its versatile design that combines ease of manufacturing with desired functionality [32]. A distinct feature of IDE type sensors is their superior sensitivity to change close to the sensor surface, with this sensitivity decaying rapidly with distance away from the surface. This is advantageous as it reduces significantly the chance of undesirable factors influencing sensor response. Thin flexible substrate provides not only structural benefits for a wide range of applications, but also plays a pivotal role in controlling the strength of a microwave signal fed into the sensor. Thicker substrates are known to be prone to the following effect: as the height increases, surface waves are introduced which usually is not desirable because they extract power from the total available for interaction with the analyte material. The surface waves travel within the substrate and they are scattered at bends and surface discontinuities and degrade the antenna pattern and polarization characteristics [33].

Silver was used as a metal material for both the bottom layer, which acted as a ground plane, and top IDE pattern to maintain chemical neutrality when the device is placed in contact with water. Fig. 13 (a) illustrates optical images of the manufactured prototype microwave sensor, which is bent to illustrate the flexibility of the substrate.

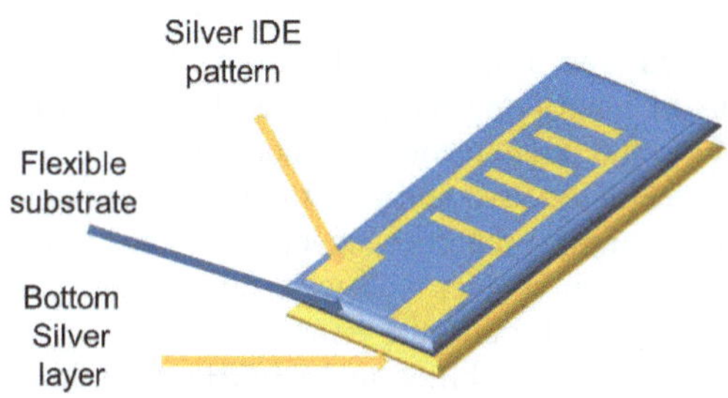

Fig. 12. Schematic of a microwave sensor with interdigitated electrodes

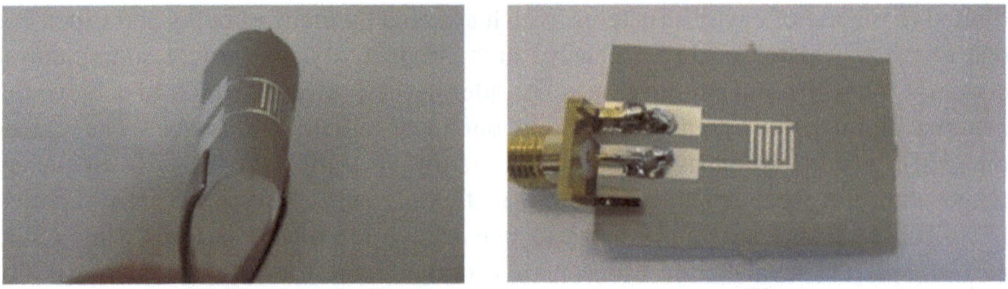

(a) (b)

Fig. 13. Images of the microwave sensor: (a) sensor bent to illustrate the flexibility of the substrate and (b) sensor with SMA connector soldered [32]

To give a feeling of the performance of sensors such as that shown in Fig. 13, preliminary S_{11} measurements where obtained using a VNA (as described in Section 4). These results are shown for 0-500 ppm NO_3 in Fig. 14 (a). There is a clear distinction between the samples, and similar principles can also be applied to smaller concentrations. In terms of sensitivity, the design of the sensor, particularly its geometry, plays a crucial role and the authors are currently investigating a variety of structures which provide enhanced sensitivity. While current legislation requires detection of mg/L quantities of nutrients such as nitrates, in the future it is likely that this will be pushed lower and thus will require sensors which can detect μg/L levels.

Also under investigation is the notion of *selectivity*, such that the sensor will respond in a different manner to the variety of nutrients that might be present in water or wastewater. Although other works suggest the use of coatings (e.g. polymers [34-36], sol-gels [37-39] or metal oxides [40-42]), it is also noteworthy that broadband microwave analysis provides us with a unique spectrum for different analytes [28] and in some cases this may be sufficient to determine the analyte present in some media. Fig. 14 (b) demonstrates this with different media by way of an example.

Fig. 14 (b) illustrates the power S_{11} distribution measured with the flexible microwave sensor in the range 1-15 GHz when in contact with air, deionized water and tap water samples. One may note the significant number of resonant peaks available with the IDE sensor, which indicates that the various sensing elements each influence the obtained spectrum. It is believed that this will give significant advantages in terms of identifying the presence of water contaminants with greater sensitivity, selectivity and high resolution. Focusing on 1-4 GHz and 7-10 GHz ranges, one may notice distinctive shifts in the resonant frequencies of the spectra, corresponding to the sample properties under test. Thus, there is no peak for the sensor response when not in contact with water at around 2 GHz, whereas once the deionized water was placed onto the sensing pattern, a well-pronounced peak occurred at 2.11 GHz and it has shifted to 1.93 GHz when in contact with the water sample taken from the tap. Similarly, at higher frequencies there is a change in the resonant peaks for all the samples and, having maintained all other experimental parameters constant, the only explanation to these shifts is that they are connected with the properties of the water, namely its composition. This particular feature makes the developed microwave sensor an attractive option for real-time monitoring of water purity.

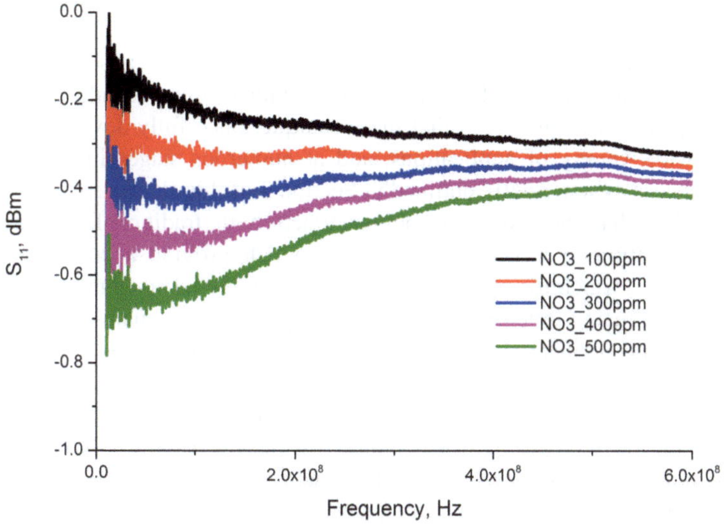

(a)

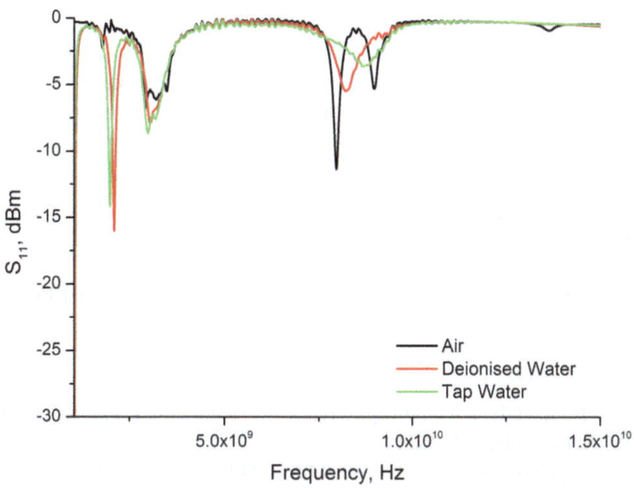

(b)

Fig. 14. Illustrating the response of planar microwave structures to (a) different concentrations of NO_3 and (b) to different analyte materials; air, deionized water and tap water [32]

7 Conclusion

This research was driven by the industrial need for a novel real-time monitoring method of water purity that would be able to meet strict regulatory demands and yet be versatile, sensitive and cost-efficient. It has been demonstrated that the method can provide adequate sensitivity for measurement of different nutrients which would be found in wastewater, namely phosphate (PO_4) and nitrate (NO_3). There is still work to be done in enhancing this sensitivity further, however this area is still under active investigation by the authors and the subject of a number on-going industrially focused research projects. It is envisaged that the technique, once fully developed, could be used for a wide variety of applications, and at numerous stages in the wastewater treatment process as was briefly outlined in Section 2 of this chapter.

It is notable that the work has demonstrated both traditional resonant cavity methods in addition to novel IDE structures which provide a cost effective and flexible means by which to apply the suggested technique. This highlights one of the key benefits of using microwave sensors; they can take many different forms depending on the application and its requirements. It is believed that the small flexible sensor shown in Section 6 would provide for long-term usage since this configuration is less prone to failure due to mechanical damage than perhaps a rigid substrate planar device. The sensor response was tested using a VNA in 1-15 GHz frequency range. It was clearly seen that the resonant peaks have shifted once deionized and tap water samples were placed in contact with the antenna pattern. Notably, the sensor's response returned to its original position, namely the air spectrum, after each water sample measurement, confirming that the developed microwave sensor is reliable, re-usable and thus a sustainable solution for water monitoring.

Acknowledgements. This work is financially supported by the European Community's Seventh Framework Programme through the FP7-PEOPLE-2010-IEF Marie-Curie Action project 275201, Water-Spotcheck.

References

[1] Concerning urban waste water treatment, Council Directive (91/271/EEC) (1991)
[2] E. Agency. Urban Waste Water Treatment Directive, (April 2, 2012), `http://www.environment-agency.gov.uk/business/regulation/31907.aspx`
[3] D. o. E. F. a. R. A. (DEFRA), Sewage Treatment in the UK, London (2002)
[4] Vowell, P.: Aqua cycle, vol. 2011 (2010)
[5] Gilmore, A., Tighe, C., Donato, T., Gill, M.L.: D.O. Home (2010)
[6] Guyer, H.H.: Industrial processes and waste stream management. John wiley and sons, Canada (1998)
[7] E. Agency, Treatment of non-hazardous wastes for landfill, Bristol, UK (2007)
[8] E. Agency. Urban waste water treatment (June 15, 2011), `http://www.environment-agency.gov.uk/homeandleisure/37809.aspx`
[9] Henze, M., Loosdrecht, M., Ekama, G., Brdjanovic, D.: Biological wastewater treatment: principle, modeling and design. IWA Publishing (2008)
[10] Gerardi, M.H.: Nitrification and denitrification in the activated sludge process. John Wiley and Sons, New York (2002)

[11] Droste, R.L.: Theory and practice of Water and wastewater treatment. John Wieley & Sons (1997)

[12] Davis, M.L., Cornwell, D.A.: Introduction to Enviromental Engineering, 2nd edn. McGraw-Hill (1991)

[13] Halling-Sorensen, B., Jorgensen, S.E.: The removal of nitrogen compounds from wastewater. Elsevier Science (1993)

[14] Bababjanyan, A., Melikyan, H., Kim, S., Kim, J., Lee, K., Friedman, B.: Real-Time Noninvasive Measurement of Glucose Concentration Using a Microwave Biosensor. Journal of Sensors 2010 (2010)

[15] Cataldo, A., Piuzzi, E., Cannazza, G., De Benedetto, E., Tarricone, L.: Quality and anti-adulteration control of vegetable oils through microwave dielectric spectroscopy. Measurement 43, 1031–1039 (2010)

[16] Choi, J., Cho, J., Lee, Y., Yim, J., Kang, B., Oh, K., Jung, W., Kim, H., Cheon, C., Lee, H.: Microwave Detection of Metastasized Breast Cancer Cells in the Lymph Node; Potential Application for Sentinel Lymphadenectomy. Breast Cancer Research and Treatment 86, 107–115 (2004)

[17] Nyfors, E., Vainikainen, P.: Industrial microwave sensors. In: IEEE MTT-S International Microwave Symposium Digest, vol. 3, pp. 1009–1012 (1991)

[18] Goh, J.H., Mason, A., Al-Shamma'a, A.I., Field, M., Shackloth, M., Browning, P.: Non Invasive Microwave Sensor for the Detection of Lactic Acid in Cerebrospinal Fluid (CSF). Presented at the Sensors and their Applications XVI, Cork, Ireland (2011)

[19] Mason, A., Wylie, S., Thomas, A., Keele, H., Shaw, A., Al-Shamma'a, A.I.: HEPA Filter Material Load Detection Using a Microwave Cavity Sensor. International Journal on Smart Sensing and Intelligent Systems 3, 322–337 (2010)

[20] Al-Dasoqi, N., Mason, A., Alkhaddar, R., Al-Shamma'a, A.I.: Use of Sensors in Wastewater Quality Monitoring - A Review of Available Technologies. Presented at the 2011 World Environmental & Water Resources Congress, Palm Springs, California (2011)

[21] Korostynska, O., Mason, A., Al-Shamma'a, A.I.: Monitoring of Nitrates and Phosphates in Wastewater: Current Technologies and Further Challenges. International Journal on Smart Sensing and Intelligent Systems 5, 149–176 (2012)

[22] Kajfez, D.: Temperature characterization of dielectric-resonator materials. Journal of the European Ceramic Society 21, 2663–2667 (2001)

[23] Afsar, M.N., Birch, J.R., Clarke, R.N., Chantry, G.W.: The measurement of the properties of materials. Proceedings of the IEEE 74, 183–199 (1986)

[24] Pozar, D.M.: Circular Waveguide. In: Microwave Engineering, 3rd edn., pp. 119–120. John Wiley and Sons, New York (2005)

[25] Ansys, Ansys HFSS (2012), `http://www.ansoft.com/products/hf/hfss`

[26] Goh, J.H., Mason, A., Al-Shamma'a, A.I., Field, M., Browning, P.: Lactate Detection Using Microwave Spectroscopy for In-Situ Medical Applications. International Journal on Smart Sensing and Intelligent Systems 4, 338–352 (2011)

[27] Blakey, R.T., Mason, A., Al-Shamma'a, A.I., Rolph, C.E., Bond, G.: Dielectric Characteristics of Lipid Droplet Suspensions Using the Small Perturbation Technique. Presented at the IMPI 46, Las Vegas, USA (2012)

[28] Goh, J.H., Mason, A., Al-Shamma'a, A.I., Wylie, S., Field, M., Brown, P.: Lactate Detection Using a Microwave Cavity Sensor for Biomedical Applications. Presented at the Fifth International Conference on Sensing Technology, Palmerston North, New Zealand (2011)

[29] Pozar, D.M.: TM Modes. In: Microwave Engineering, 3rd edn., pp. 121–123. John Wiley and Sons, New York (2005)
[30] Catenaccio, A., Daruich, Y., Magallanes, C.: Temperature dependence of the permittivity of water. Chemical Physics Letters 367, 669–671 (2003)
[31] Tamura, H., Matsumoto, H., Wakino, K.: Low temperature properties of microwave dielectrics. Japanese Journal of Applied Physics 28, 21–23 (1989)
[32] Korostynska, O., Mason, A., Al-Shamma'a, A.I.: Proof-of-Concept Microwave Sensor on Flexible Substrate for Real-Time Water Composition Analysis. Presented at the ICST 2012: 6th International Conference on Sensing Technology, Kolkata, India (2012)
[33] Balanis, C.A.: Antenna Theory: Analysis and Design, 3rd edn. Wiley-Blackwell, United States (2005)
[34] Levit, N., Pestov, D., Tepper, G.: High surface area polymer coatings for SAW-based chemical sensor applications. Sensors and Actuators B: Chemical 82, 241–249 (2002)
[35] Cornila, C., Hierlemann, A., Lenggenhager, R., Malcovati, P., Baltes, H., Noetzel, G., Weimar, U., Göpel, W.: Capacitive sensors in CMOS technology with polymer coating. Sensors and Actuators B: Chemical 25, 357–361 (1995)
[36] Tan, Y., Yin, J., Liang, C., Peng, H., Nie, L., Yao, S.: A study of a new TSM bio-mimetic sensor using a molecularly imprinted polymer coating and its application for the determination of nicotine in human serum and urine. Bioelectrochemistry 53, 141–148 (2001)
[37] Pandey, P.C., Upadhyay, S., Pathak, H.C.: A new glucose sensor based on encapsulated glucose oxidase within organically modified sol–gel glass. Sensors and Actuators B: Chemical 60, 83–89 (1999)
[38] Gupta, R., Mozumdar, S., Chaudhury, N.K.: Effect of ethanol variation on the internal environment of sol–gel bulk and thin films with aging. Biosensors and Bioelectronics 21, 549–556 (2005)
[39] Calabria, J.A., Vasconcelos, W.L., Daniel, D.J., Chater, R., McPhail, D., Boccaccini, A.R.: Synthesis of sol–gel titania bactericide coatings on adobe brick. Construction and Building Materials 24, 384–389 (2010)
[40] Stoycheva, T., Vallejos, S., Blackman, C., Moniz, S.J.A., Calderer, J., Correig, X.: Important considerations for effective gas sensors based on metal oxide nanoneedles films. Sensors and Actuators B: Chemica 161, 406–413 (2012)
[41] Meixner, H., Gerblinger, J., Lampe, U., Fleischer, M.: Thin-film gas sensors based on semiconducting metal oxides. Sensors and Actuators B: Chemical 23, 119–125 (1995)
[42] Tomchenko, A.A., Harmer, G.P., Marquis, B.T., Allen, J.W.: Semiconducting metal oxide sensor array for the selective detection of combustion gases. Sensors and Actuators B: Chemical 93, 126–134 (2003)

Remote Monitoring of Water Quality for Intensive Fish Culture

D. Li* and S. Liu

College of Information and Electrical Engineering,
China Agricultural University,
Beijing 100083,
P.R. China
dliangl@cau.edu.cn

Abstract. Water quality monitoring and forecasting plays an important role in modern intensive fish farming management. This paper describes an online water quality monitoring system for intensive fish culture in China, which is combined with web-server-embedded and mobile telecommunication technology. Based on historical data, this system is designed to forecast water quality with artificial neural networks (ANNs) and control the water quality in time to reduce catastrophic losses. The forecasting model for dissolved oxygen half an hour ahead has been validated with experimental data. The results demonstrate that multi-parametric, long-distance and online monitoring for water quality information can be accurately acquired and predicted by using this established monitoring system.

Keywords: water quality monitoring, intensive fish culture, wireless sensor network, LSSVR.

1 Introduction

Aquaculture is the fastest growing food-producing sector in the world, with an average annual growth rate of 8.9% since 1970 [1]. China is one of the most important contributors to world aquaculture production. 41.3 million tons, or 69.6% of the world production, was produced in China [2]. As a result of a significant shift from wild fishing to aquaculture in the 1980s, aquaculture development has accelerated throughout the country. The production of intensive fish culture has been increased rapidly in China from 1.6million tons in 1990 to 13.5million tons in 2005 [2,3].

Automatic remote monitoring and computer-controlled intensive culture is the future trend in aquaculture. In modern aquaculture management, water quality monitoring plays an important role. Appropriate control of water quality to keep the concentration of the water environment parameters in the optimal range can

* Corresponding author.

S.C. Mukhopadhyay & A. Mason (Eds.): *Real-Time Water Quality Monitoring*, SSMI 4, pp. 217–238.
DOI: 10.1007/978-3-642-37006-9_10 © Springer-Verlag Berlin Heidelberg 2013

enhance the fish growth rate, impact dietary utilization and reduce the occurrence of large-scale fish diseases [4,5]. Without gathering information regarding physical and chemical parameters of water quality together with the related ecological factors it is almost impossible to perform the appropriate water quality control at the right time and in the right place.

However, there are a few applications of systems which could carry out real-time water quality monitoring continuously in China. According to the conventional methods of water quality monitoring, samples of water are taken and transported to a chemical laboratory to analyze the hazardous substances. On the one hand, the maintenance of the measurements and control process is manual and influenced by the personal experience. On the other hand, the process of forecasting is time-consuming and some contamination episodes might be missed [5]. For example, fish mortality occurred overnight in one incident and was only detected the next morning, after huge losses had already been caused.

With the advent of new sensor technologies, data telemetry and wireless communication technology, various equipment has been developed to monitor remote areas in real-time [6-9]. At present, continuous monitoring of drinking water and wastewater quality at most treatment plants is applied in Europe, North America and Japan [10,11]. In China, online monitoring installations have been constructed for several large rivers, such as the Huanghe River and the Huaihe River, to provide real-time information to support environmental protection decision-makers [12]. However, the financial burden for building the fundamental hardware of these high-tech facilities may only be affordable to governments. Realizing real-time data collection in a secure, robust, manageable and low-cost manner, without long-distance cable connections, will likely become a bottleneck in the development of information monitoring in fish culture. Therefore, using web-server-embedded and next generation telecommunication technologies will become increasingly important in sensing networks.

In recent years, some researchers investigated integrated water quality remote monitoring systems [13,14] and management systems based on culture knowledge models and forecasting models [15-18], but these systems are not aimed at the present needs to develop aquaculture and not connected with any online monitoring system. Moreover, these installations cannot achieve real-time communication between data collection and control terminals, which is not yet a fully viable alternative for high-density, open, and dynamic fish breed circumstances.

In this work, water quality remote monitoring systems using a GPRS service combined with IPsec-based virtual private networking (VPN) functionality were developed for constructing a wireless sensing network on a countrywide scale. Integrated with a forecasting model on the basis of artificial neural networks (ANN), the system is able to provide real-time information and the dynamic trend of the water quality at different monitoring sites. The detected data can be collected and analyzed at any time via the Internet so as to know the status and changes of the system.

2 Aquaculture and Water Quality Requirements

Aquaculture is defined as the high-density production of fish and plant forms in a controlled environment. Water quality for aquaculturists refers to the quality of water that enables successful reproduction of the desired organisms. The required water quality is determined by the specific organisms to be cultured and has many components that are interwoven. Aquaculture obeys a set of physical, chemical and biological principles. Since these principles compose the subject of water quality, in Section 2.1 we describe common water quality parameters related to these principles which have been used as indicators of water quality on fish culture, as well as the respective classification of these parameters by monitoring importance. In Section 2.2, we present a classification of the parameters based on their impact level in an ecosystem.

2.1 Physical, Chemical and Biological Analysis

The monitoring of environmental parameters in fish aquaculture allows the control and good management of water quality in fish ponds, avoiding the occurrence of unfavorable conditions that can be harmful for organisms [19,20] .

Water quality is based on the results of toxicity tests. These tests measure the responses of aquatic organisms to defined quantities of specific pollutants [21]. The aquatic species have different tolerances for a specific toxic compound; in this paper the characteristics of the fish are analyzed to evaluate the performance of the model.

Table 1. Water quality parameters classified by monitoring frequency

Monitored daily	*Monitored Weekly*	*Monitored by request*
Temperature (Temp)	Total ammonia (NH)	Alkalinity (Ak)
Dissolved oxygen (DO)	Nitrate (NO3)	Phosphorus (P)
Salinity (Sal)	Nitrite (NO2)	Hydrogen sulfide (H2S)
pH	Non ionized ammonia (NH3)	Non ionized hydrogen sulphide (HS-)
	Turbidity (Tb)	Dioxide of carbon (CO2)
		Suspended solids (Ss)
		Potential redox (Px)
		Silicate (Si)
		Chlorophyll A (ChA)
		Total inorganic nitrogen (N)
		Total marine bacteria (Tmb)
		Vibrio (Vb)
		Fecal coliforms (Fc)

In extensive aquaculture systems on china, the water quality parameters are monitored in different frequencies. Dissolved oxygen, temperature, pH and salinity are monitored daily while ammonia, nitrates, turbidity and algae counts are analyzed weekly. Chemical analyses are not taken into consideration for water quality management on a routine bases, they are only monitored by request [22]. Table 1 lists common water quality parameters used as indicators of water quality on fish marine culture and their respective classification by monitoring frequency.

In order to understand the effects of these water quality parameters, Table 2 and Table 3 show the optimal and harmful ranges (reported in the literature) for daily, weekly and by request parameters which will be considered for the assessment of water quality in our work.

Table 2. Daily and weekly measured water quality parameters and their importance to fish farming

Parameters	*Importance on marine fish culture*
Temperature	The temperature of water plays an important role in both environmental and intensive aquaculture processes. First, it affects the ability of living organisms to resist certain pollutants. Some organisms cannot survive when the water temperature takes a value beyond a specific range. Changes in temperature rates can stress fish and consequently high mortality rates can be present in the population [23]. Second, it controls solubility of gases, chemical reactions and toxicity of the ammonia. The demand of dissolved oxygen increases when temperature is high [24]. Temperature can be considered as normal from 28 to 32 °C [22].
Dissolved oxygen	The dissolved oxygen is breathed by fish and zooplankton and is necessary for their survival. Fluctuation of dissolved oxygen, hypoxia and anoxia crisis are events that can be normally presented in aquaculture systems. Dissolved oxygen is considered the most critical quality parameter, since fish in low dissolved oxygen concentrations are more susceptible to disease. The minimum levels recommended by authors oscillate between 4 and 5 ppm. It is recommended that
Salinity	Salinity is the saltiness or dissolved salt content of a body of water. The salt content of most natural lakes, rivers, and streams is so small that these waters are termed fresh or even sweet water. The actual amount of salt in fresh water is, by definition, less than 0.05%. The water is regarded as brackish, or defined as saline if it contains 3 to 5% salt. The ocean is naturally saline and contains approximately 3.5% salt. High salinity concentrations reduce dissolved oxygen in water ponds, and it can be dangerous for fish cultivation [21]. The optimal concentrations of salinity are from 15 to 23 ppt [21].

Table 2. (*continued*)

pH	pH is a measure of the relative amount of free hydrogen and hydroxyl ions in the water. Water that has more free hydrogen ions is acidic, whereas water that has more free hydroxyl ions is basic. The values of pH range from 0 to 14 (this is a logarithmic scale), with 7 indicating neutral. Values less than 7 indicate acidity, whereas values greater than 7 indicate a base. Extremely low or high pH stresses fish and causes soft shell and poor survival [25]. Water bodies with 6.5 to 9.0 pH concentrations are appropriate for aquaculture production. Reproduction decreases outside of this range. Acid death appears with values below than 4.0 and an alkaline death in values above 11 [22,24]. The presence of chemicals in the water affects its pH, which in turn can harm the animals and plants that live there. For example, even a mildly acidulous seawater environment can harm shell cultivation. This renders pH an important water quality indicator.
Ammonia	Ammonia is the main end product of protein catabolism in crustaceans. Ammonia increases tissue oxygen consumption, damages gills and reduces the ability of blood to transport oxygen. Ammonia exists in water in both ionized (NH_4^+) and unionized (NH_3) forms. Unionized ammonia is the most toxic form of ammonia due to its ability to diffuse readily across a cell membrane [26]. The safe level for unionized ammonia, recommended by Chien (1992) and Wickins (1976) [27], is less than 0.1 mg/l and for total ammonia is under 1.0 mg/l.
Water nitrogen	Inorganic nitrogen in water is chiefly present as ammonia, nitrate and nitrite. In fish, the respiratory pigment is hemocynanin, which can still bind oxygen in the presence of oxidizing agents such as nitrite [28]. The safe concentration of NO_2 is from 0.4 to 0.8 mg/l. Nitrates are nitrogenous compounds which can be toxic when their levels rise. According to Clifford (1994) [29], the optimal level for nitrates is from 400 to 800 µg/l. The expected total inorganic nitrogen recommended for crops is from 2.0 to 4.0 mg/l. [21,25].
Turbidity	A high concentration of suspended solids can cause high turbidity in water, preventing the penetration of light and affecting photosynthesis. The amount of suspended solids can be determined indirectly by measuring the turbidity. The accepted range for suspended solids is from 50 to 150 mg/l; or turbidity from 35 to 45 cm depth [24].

2.2 *Environmental Classification*

Water quality parameters can be classified in different impact levels, depending on the toxicity and harmful situations the parameters introduce to the ecosystem. In order to classify the behavior of a water quality parameter, it is necessary to define levels and allowed deviations for optimal or harmful concentrations. These deviations are useful to determine the ranges where values are considered closer to or farther from

Table 3. Water quality parameters measured by request and their importance to fish farming

Parameters	*Importance on marine fish culture*
Alkalinity	Related to important factors in fish culture such as buffer effect on daily variation of pH in the pond, setting the soluble iron precipitated, and in ecdysis (molting) and growth [19,20].
Phosphorus	Nutritive element, mainly appearing as orthophosphate, essential to aquatic life. From Esteves [30], phosphorus acts particularly in metabolic processes of living beings, such as energy storage and in the structure of the cell membrane [20].
Hydrogen sulfide	In water, hydrogen sulfide exists in unionized (H_2S) and ionized forms (HS– and S_2). Only the unionized form is considered toxic to aquatic organisms. Unionized H_2S concentration is dependent on pH, temperature and salinity, and it is mainly affected by pH [25].
Dioxide of carbon	When dissolved oxygen concentrations are low, carbon dioxide prevents oxygen penetration. According to Boyd (2001) [31], the normal range of carbon dioxide is from 1 to 10 mg/l
Potential redox	This is an indicator of substance oxidation or reduction levels. Low values are indicators of strong reduction of sediment, which is associated with toxic metabolite formation, hypoxic or anoxic conditions and low pH values. In a pond, optimal ranges of potential redox are from 500 to 700 mV for water and from 400 to 500 mV for sediment [29].
Silicate	In water, silicate is a composite of high importance particularly for diatoms. Optimal levels for silicate are established from 0.1 to 0.3 mg/l. [20,30].
Chlorophyll A	Phytoplankton biomass represents the primary consumer feed, and indirectly determines the feed availability of the next trophic system level. The ideal concentrations for fish ponds are from 50 to 70 μg/l [29].
Total marine bacteria	Microorganisms, particularly bacteria, play a vital role in pond ecosystems. Both beneficial (nutrients recycling, organic matter degrading etc.) and harmful (such as parasites) issues are caused by bacteria in the pond ecosystem. The optimal range for total bacteria counts should be below 10,000 UFC/ml [20,31].
Vibrio	Vibriosis is a bacterial disease responsible for mortality of cultured fish worldwide [32]. Vibrio related infections frequently occur in hatcheries, but epizootics are also commonly in pond reared fish species. Optimal ranges are defined as being below 1000 UFC/ml
Fecal Coliforms	Fecal Coliforms in water come from the feces of warm-blooded animals and they are an indicator of water pollution. The optimal range of fecal coliforms is below 1000 MPN/ml and for crop it should not exceed 1400 MPN/ml [33].

specified levels. In this study, tolerance thresholds were chosen using minimal changes in water parameters [33]. The levels for classification of the water quality parameters were defined by taking into account the levels and limits reported in the literature (see Table 2 and 3). In Table 4, 5 and 6 we show the classification, in different impact levels, for the water quality parameters from Table 1. The deviation column in these tables represents the tolerance for each level.

Table 4. Classification levels for daily monitored parameters

Water quality parameters	*Deviation*	*Levels*				
		Hypoxia acid	*Low*	*Normal*	*High*	*Alkaline*
Temperature (°C)	1.0		0–20	20–30	< 30	
Dissolved Oxygen (mg/l)	0.5	0–2	2–5	< 5		
Salinity (ppt)	1.0		0–15	15–23	< 23	
pH	0.5	0–4	4–6.5	6.5–9.5	9.5–11	11–14

Table 5. Classification levels for weekly monitored parameters

Water quality parameters	*Deviation*	*Levels*		
		Low	*Normal*	*High*
Total ammonia (mg/l)	0.10	0–0.1	0.1–1.0	< 1.0
Nitrites (μg/l)	100	0–400	400–800	<800
Nitrates (mg/l)	0.10		0–0.5	< 0.5
Non ionized ammonia (mg/l)	0.01		0–0.1	< 0.1
Turbidity (cm)	1.00	4–35	35–45	< 45

The importance of water quality management, the correct interpretation of water parameters and the appropriate techniques for integrating these parameters are problems studied in the aquaculture field. This research deals with one of the most important objective of the aquaculture management: it proposes a new way to join dissimilar parameters for getting an accurate assessment of water quality, increasing the effectiveness of the proposed system over traditional methodologies. In this sense, we hypothesize that different effects and levels of parameter concentrations degenerate in different water quality, thus, an appropriate join of these parameters

Table 6. Classification levels for monitored by request parameters

Water quality parameters	*Deviation*	*Levels*		
		Low	*Medium*	*High*
Alkalinity (mg/l)	10	0–100	100–140	< 140
Phosphorus (mg/l)	0.01	0–0.1	0.1–0.3	< 0.3
Hydrogen sulfide (mg/l)	0.01	0–0.05	0.05–0.1	< 0.1
Non ionized hydrogen sulfide (mg/l)	0.001	0–0.002	0.002–0.005	< 0.005
Carbon dioxide (mg/l)	2	0–10	10–20	< 20
Suspension solids (mg/l)	5	0–50	50–150	< 150
Potential redox (mV)	10	0–400	400–500	< 500
Silicate (mg/l)	0.2	0–2.0	2.0–4.0	< 4.0
Chlorophyll A (μg/l)	5	0–50	50–75	< 75
Total inorganic nitrogen (mg/l)	0.2	0–2	2–4	< 4
Total marine bacteria (UFC/ ml)	1000	0–5000	5000–10,000	< 10,000
Vibrio (UFC/ ml)	100	0–500	500–1000	< 1000
Fecal coliforms (MPN/ml)	100	0–500	500–1000	< 1000

could determine a better assessment of water quality. This assessment could be achieved using a fuzzy inference system, which involves different situations generated by water quality parameters.

3 System Design

3.1 System Architecture

Accuracy, reliability, real-time and expandability are essential in the remote monitoring system. Therefore, the sensors of high sensitivity should be chosen and rationally distributed for data accuracy. Since some locations have no access to any cable network (telephone line) and harsh production environment could damage cable connections, wireless devices would be necessary. Accordingly, each station is designed to communicate with the server via wireless communication technology. In order to offer a better expandability, an intelligent "plug and play" sensor technology has been used.

As shown in Fig. 1, the basic structure of the system can be divided into three major parts: the data layer, the transport layer and the application layer for data acquisition, reliable transmission, intelligent information processing and logical operation, respectively. These three parts communicate with each other through the telecommunication system.

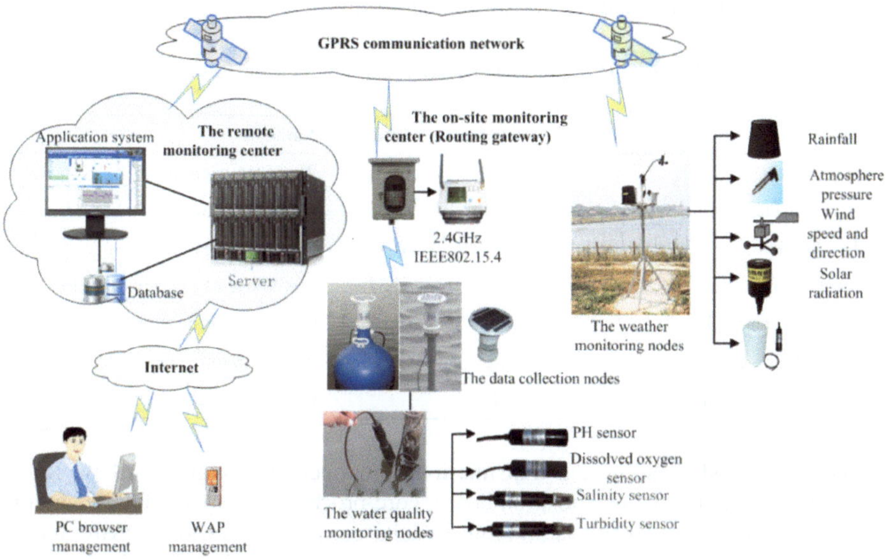

Fig. 1. The structure diagram of the digital remote wireless monitoring system

3.2 *Remote Monitoring Platform*

The remote monitoring platform (RMP) contains three parts: data acquisition, data transformation and transmission, and water quality control components. The system architecture of the remote monitoring platform is shown in Fig. 2.

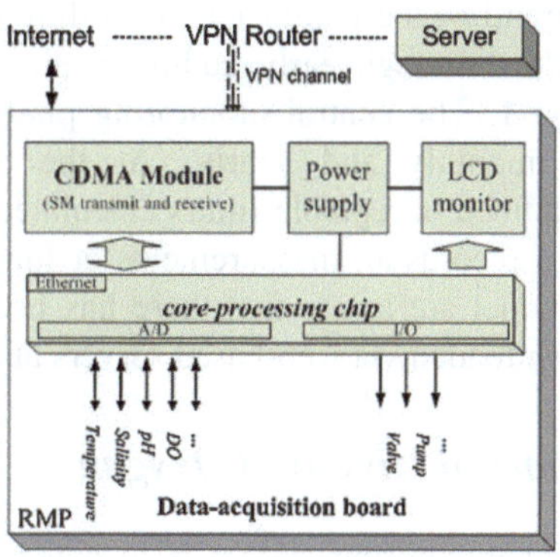

Fig. 2. System architecture of the remote monitoring platform

The function of the data-acquisition component is to obtain signals of the most important environmental factors by using various sensors. The main variables that can be monitored are reported in Table 7. With the current measurement methods, pH value is measured by the glass electrode method, dissolved oxygen by the membrane electrode technique, and temperature by thermometer sensing technology. A method of measuring the conductivity and transforming it to salinity has been adopted to replace the common method for measuring the salinity.

Table 7. The water quality variables of data acquisition

Name	Variable	Units	Name
Water temperature	T_w	(°C)	Water temperature
Indoor temperature	T_a	(°C)	Indoor temperature
Solar radiation	S	(W/m^2)	Solar radiation
Percentage of oxygen saturation	DO_S	(%)	Percentage of oxygen saturation
Oxygen concentration	DO	(mg/l)	Oxygen concentration
Pouvoir Hydrogène	pH		Pouvoir Hydrogène
Electrical conductivity	EC		Electrical conductivity

The data transformation and transmission component is composed of the signal conditioning circuits, data-acquisition board, core-processing chip and GPRS module. The sensors and the signal conditioning circuits convert the various environmental factors to electrical voltage standard signals in the range of 0–5V. The signal is transmitted to the Web-based monitoring chip, and then is converted into the digital signal through A/D conversion. Onsite data-acquisition nodes compose a wireless LAN, and the GPRS module enables the RMP to receive the data and transmit them to a PC for further analysis.

3.3 Central Monitoring Platform

The central monitoring platform receives, pre-processes and analyzes the data from the RMP, predicates the trend of the parameters according to historical information, and then warns stakeholders through early audio warning or early short message warning, as shown in Fig. 1. The central monitoring platform stores the data to a database daily, weekly, monthly and yearly. At the same time it compares measurements to the predefined acceptable limits calculated by the expert empirical knowledge. Furthermore, it records all measurements or functional errors in different log files, so that the personnel are aware that there has been an alarm in a specific tank. Real-time data is downloaded via web-based servers at scheduled intervals.

3.4 Forecasting Model of Dissolved Oxygen

One purpose of the current monitoring system is to detect a trend of water quality fluctuation using historical data. Most current models for prediction that focus on pollutants in a river or lake are not applicable for intensive aquaculture. In this study, the stored water quality data is analyzed for temporal trends focusing on the dissolved oxygen half an hour after measurement. Due to their ease of development, decreased reliance on expert knowledge of the system under investigation and non-linear modeling capabilities, Least Squares Support Vector Regression (LSSVR) was selected as the modeling tool. In spite of this, the LSSVR performance heavily depends on the choice of several hyper parameters, which are necessary to define the optimization problem and the final LSSVR model. To design a LSSVR, one must choose a kernel function, set hyper parameters such as the kernel parameters and determine a regularization parameter γ. The hyper parameters that should be optimized include the regularization parameter γ and the kernel function parameters such as the gamma (σ) for the radial basis function (RBF) kernel. Thus, selecting appropriate model parameters has a crucial impact on the prediction accuracy [34]. Unfortunately, there no exact method to obtain the optimal set of LSSVR hyper-parameters, so a search algorithm must be applied to obtain the parameters.

In general, the search algorithms used to obtain LSSVR hyper-parameters can be summarized in two categories. One is based on analytical techniques, and the other is based on heuristic searches. The first kind of techniques determines the hyper parameters with gradients of some generalized error measures [35-39]. This procedure is time-consuming and can't converge at the global optimum. The second kind of techniques determines the hyper parameters with modern heuristic algorithms including simulated annealing algorithms, differential evolution, genetic algorithms, particle swarm optimization algorithm and other evolutionary strategies [40-44], which are applied to implement a robust research on the hyper parameters search space. Compared with other heuristic algorithms, for example, genetic algorithm, particle swarm optimization (PSO) does not need evolutionary operators such as crossover and mutation. Furthermore, the advantages of PSO are that PSO possesses the capability to escape from local optima, is easy to be implemented, and has fewer parameters to be set [45-47]. Thus in this study, the novel prediction method based on the combination of least squares support vector regression (LSSVR) and improved particle swarm optimization (IPSO) is proposed to the water quality prediction in the intensive aquaculture of river crab, which IPSO is applied to optimize the hyper-parameters. Traditional LSSVR model and BP neural network are used as comparison basis. The experiments results show that the predictive accuracy and capability of generalization is greatly improved by our proposed approach.

For the nonlinear LSSVR, its generalization performance depends on a good setting of parameters γ and the kernel parameters σ. Inappropriate hyper-parameters combination in LSSVR lead to over-fitting or under-fitting, so IPSO is used to optimize the parameters of SVR: γ and kernel parameter σ of RBF-kernel function, which are two attributes of each particle. In solving the hyper-parameter selection, each particle represents a potential solution, comprised of a vector $d = (\gamma, \sigma)$. The performance of each particle is measured according to the fitness function. In the training and testing process of LS-SVR, the objective is to minimize the errors between the actual values and prediction values of the testing samples. Therefore, the fitness function of IPSO is defined as:

$$Fitness() = \sqrt{\frac{1}{z}\sum_{j=1}^{z}(\hat{y}_{ij} - y_{ij})^2} \tag{1}$$

where z is the number of each subset as validation, y_{ij} represents the actual values, and $\hat{y}_{ij}$ represents the prediction values. The goal is to minimize the fitness, so the particle with the minimal fitness value will outperform others and should be reserved during the optimization process. Accordingly, also the optimal combinational parameters values of γ and σ are obtained.

The implementation process of water quality prediction based on IPSO-LSSVR is described in steps as follows:

(1) Input data of water quality, construct training sample set and test sample set. Initialize the original water quality data by normalization and then form training patterns.
(2) Set algorithm parameters, select the kernel function, the regularization parameter γ and the kernel function parameter σ.
(3) Train LSSVR on the training set, solve the optimization problem and obtain the parameters of LSSVR by IPSO, get IPSO-LSSVR prediction model. Test the performance of the prediction model with test sample.
(4) For a new application of the prediction task, extract water quality index and form a set of input variables x.

4 System Implementation

4.1 Testing Environment

In the testing environment set up for the purposes of this work, one central monitoring platform with an IPsec based VPN (Virtual Private Network) router (BV-601, NESCO Co., China) is deployed in China Agricultural University located in Beijing. The remote monitoring system is deployed in an intensive fish farm culture site, Fengze Corporation, located in Shandong province, which is a typical recycling aquaculture system [48]. Each fish tank is approximately 6.77m × 6.55m × 0.55m. The average fish stock density is 30–40 kg/m^3.

4.2 Remote Monitoring Platform

Two prototypes of the remote monitoring platform have been installed in a practical fish farm in Shandong to verify the performance of the system, as shown in Fig. 3.

The probes of the sensor in the present study for temperature, DO, pH, salinity are made by the Nantu Company (China) with accuracies of 0.1 °C, 0.1mg/L, 0.1, 0.1 ppt, respectively. The HQ 40d18 (HACH, USA) is chosen as a contrast providing long-term stability and high accuracies of ±1.0% for relative DO and ±0.1 °C for temperature.

(a)

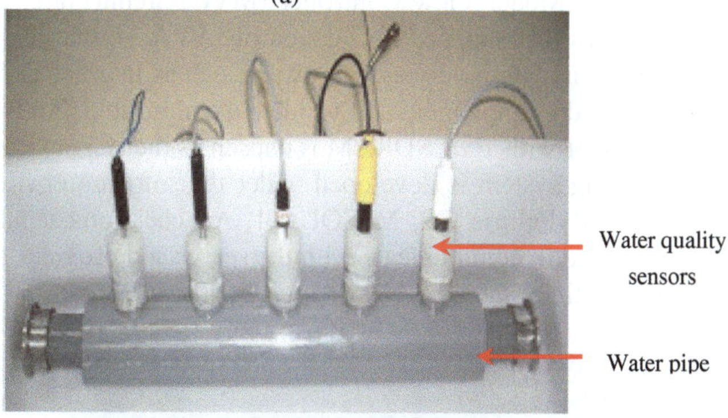

(b)

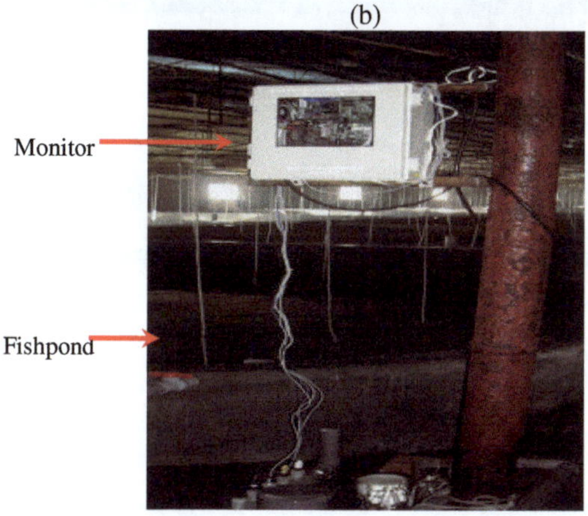

(c)

Fig. 3. Finished prototypes of remote monitoring platform (a) installation of each module in RMP case, (b) sensors, and (c) a remote monitoring platform deployed in an experimental workshop in Shangdong Fengze fish farm

The RMP uses PICNIC2.0 (TriState, Japan) as the core-processing chip and a GPRS module (FASTRACK M1203 Q2358, InRouter210C, China) for data transformation and transmission. The data recorded by the sensors is transmitted to the remote information server through the China Unicom's GPRS services. Once the virtual local area network based on GPRS and IPsec VPN router for wireless secure transmission has been established, the programs in the server can have real-time access to the data. The set of data acquisition nodes transmits the data by WiFi wireless LAN, while the computer running a communication program can transmit information to the remote server by the TCP/IP protocol. Automated collection and web-based dissemination of data provides a centralized database for use and a detailed data analysis for all water quality stakeholders. Therefore, the users will be able to monitor the water parameter values via the Internet.

4.3 Central Monitoring Software

The central monitoring software is programmed with JSP, Servlet and short message technology using Model-View Controller (MVC) architecture. It can operate on all operating systems that support this version of JAVA, so that the users can access the system through any commonly used browser such as Internet Explorer, Netscape, etc. Matlab 7.0 is used to implement and validate the algorithm. An Intel Core 2 Duo CPU personal computer with 1GB SDRAM is chosen as the test environment. The software of the monitoring system is developed under the software environment of Windows 2003, Eclipse, MyEclipse 3.2, MySQL 5.1, Apache Tomcat 5.5. A client software program is created to communicate with the server and provides a user interface so as to know the real-time status of the system (Fig. 4).

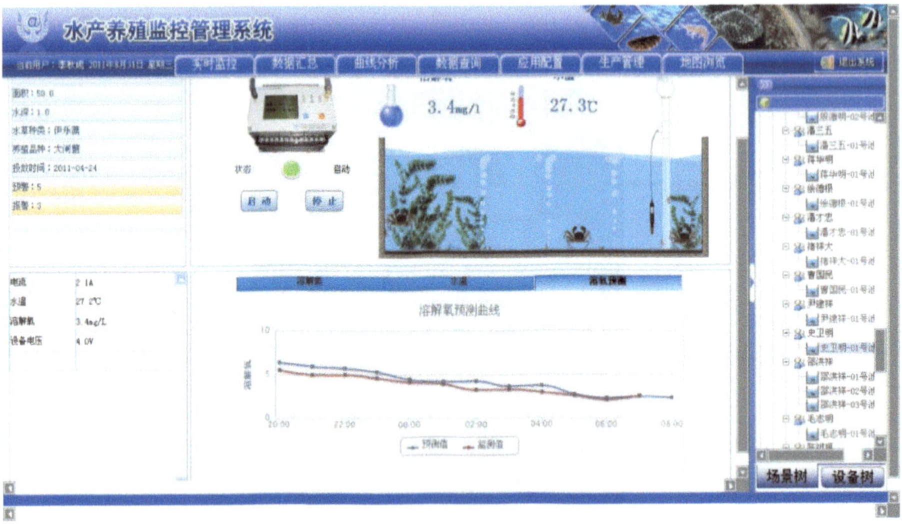

Fig. 4. The monitor interface window of no. 5 workshop

5 Results and Discussion

5.1 Network Communication and Data Acquisition

The entire system have been tested and verified for about 30 months from November 2010 to August 2012. Statistics of the data of all nodes show that the monitoring system is rather reliable; more than 95.2% of the data have been correctly collected since April 2009. Each RMP has an isolated local area network which is connected to the Internet via GPRS (China Mobile). That means that a sensing network node can be a building block to construct a large-scale wireless sensing network in GPRS signal covered areas. The monitoring system is also easy expandable with more sensor channels as well as GPRS bandwidth.

To validate the accuracy of the system, two sets of data sampled though different strategies (manually and automatically) have been compared. As shown in Fig.5, the curve monitored by the system matches the curve collected manually very well, with the maximum difference being less than 0.4mg/L for dissolved oxygen content. So we can conclude that the proposed system can monitor the DO accurately and continuously. Obviously, the frequency of measurements (every 1 min) could not be achieved by manually sampling.

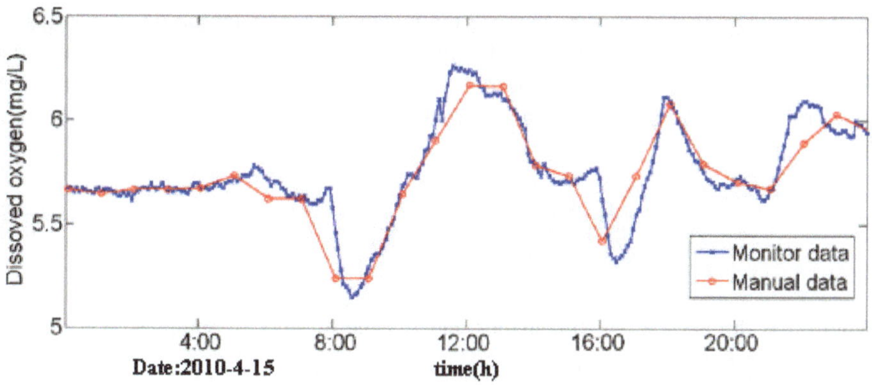

Fig. 5. The monitoring data of dissolved oxygen for a single day collected by RMP#1 on April 15th, 2010

As we can see from Fig. 5, there is a rapid decrease of DO at about 8:20 and 16:20. This is because the feeding time is set at that time in this experiment. It is possible to monitor daily variations of dissolved oxygen to control aerators in time, typically after feeding. This form of time series permits monitoring of the daily amplitude of dissolved oxygen fluctuation, which is an accessory indicator of the water quality status. This frequency permits dissolved oxygen to be utilized as a warning parameter. The system can provide an early warning especially helpful for large scale, high-density and high risk aqua farms.

The detailed changes of temperature, pH and salinity are measured using the proposed monitoring system in the same way with satisfactory results.

Table 8. Summary statistic on various water parameters in 4 months

Name	April 2010			May 2010			June 2010			July 2010		
	Mean	Min	Max	Mean	Min	Max	Mean	Min	Max	Mean	Min	Max
Tw (°C)	17.2	16.4	21.3	20.8	18.5	25.0	25.4	21.0	29.1	25.2	20.3	27.7
pH	7.95	7.56	8.18	6.00	6.00	8.77	8.11	5.87	8.96	7.85	5.77	8.61
Tr (°C)	17.7	16.1	23.7	19.4	16.9	24.8	25.6	22.0	32.4	25.4	22.0	31.3
DO (mg/l)	6.21	3.90	8.02	6.18	3.14	7.57	5.94	3.49	7.90	6.22	3.50	7.4
Salinity (ppt)	31.1	29.8	32.6	31.3	28.8	32.7	31.6	29.0	32.6	31.4	28.3	32.2
Mortality	0.7%			1.5%			1.6%			0.5%		

After the system has been deployed, the pH was relatively constant, with an average pH of 7.943 (Table 8). Salinity was also high and relatively constant, with an average of 31.35 ppt and range of less than 1 salinity units (0.6 ppt) (Table 8). These two parameters are both around the optimal growth range with little fluctuation. Fish mortality has begun to drop to below 2% since the system was deployed.

5.2 Pre-treatment Data

In many real applications, the observed input data cannot be measured precisely since distinct numerical variable have different dimensions, and should be normalized in the first instance. In order to improve the accuracy of prediction, all data samples are standardized and normalized to the interval of [0,1] according to the following linear mapping function:

$$\bar{x}_k^d = \frac{x_k^d - \min(x_k^d \,|_{k=1}^l)}{\max(x_k^d \,|_{k=1}^l) - \min(x_k^l \,|_{k=1}^l)} \quad ,d=1,2,\ldots,m \tag{2}$$

Where d is the number of dimensions and l is the number of samples, $\bar{x}_k^d$ and x_k^d are the original data and the normalized data, respectively.

5.3 Experimental Environment and Algorithm Parameter Settings

The proposed IPSO-LSSVR algorithm has been implemented in the JAVA programming language. The experiment is performed on a 2.50GHz Core(TM)2 CPU personal computer with 2.0G memory under Microsoft Windows Server 2003 R2 editions. From Fig. 6, it is clear that the proposed model of IPSO-LSSVR has strong learning capability for small samples and simultaneously achieves excellent generalization performance since the LSSVR is a good compromise for guaranteeing both stability and accuracy improvement, and it is a suitable and effective method for predicting the DO content of the water quality in the intensive aquaculture.

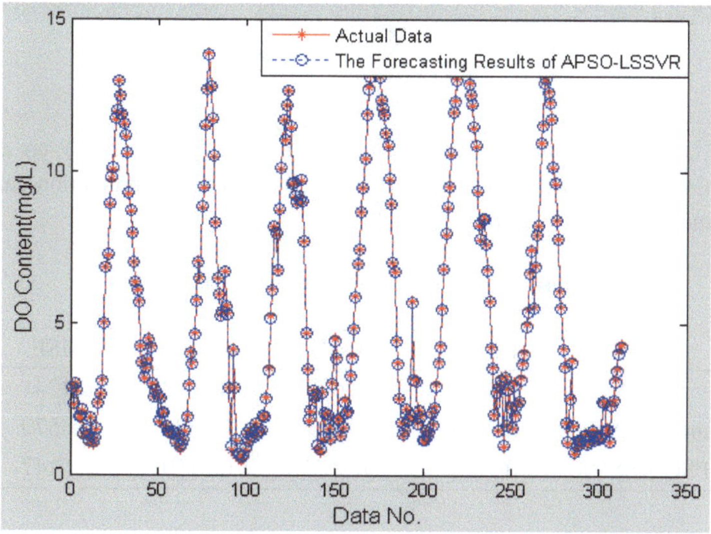

Fig. 6. The water quality prediction result based on IPSO-LSSVR model

5.4 *Model Performance Evaluation*

In order to analyze and evaluate the prediction performance of IPSO-LSSVR, the models (standard LSSVR and BP neural network) are selected to deal with the aforementioned water quality samples data. The standard LSSVR parameters are found by 5fold cross-validation method, and the selected optimal values of γ and σ are 120.1530 and 1.6839, respectively. The initial architecture of the BP neural network consisted of six input variable, one output variable, the hidden layer with six initial neurons, and the learning rate is 0.08 and stimulating function is sigmoid, three thousand training epochs are also adopted as the termination criterion. The root mean square error (RMSE), the mean absolute error (MAE), the mean absolute percentage error (MAPE) and mean squared error (MSE) are employed as performance indicators to evaluate prediction capability of three models. These performance indexes are respectively computed from the following equations:

$$RMSE = \sqrt{\frac{1}{M}\sum_{t=1}^{M}(y_t - \hat{y}_t)^2} \tag{3}$$

$$MAE = \frac{1}{M}\sum_{t=1}^{M}\left|y_t - \hat{y}_t\right| \tag{4}$$

$$MAPE = \frac{1}{M}\sum_{t=1}^{M}\frac{\left|y_t - \hat{y}_t\right|}{y_t} \tag{5}$$

$$MSE = \frac{1}{M}\sum_{t=1}^{M}(y_t - \hat{y}_t)^2 \tag{6}$$

Where M is the total number of actual samples in the data set, y_t and $\hat{y}_t$ are actual and prediction values, respectively. The performance evaluate the prediction capacity of the three models are illustrated in Table 9.

Table 9. Error statistic of three prediction models

Model	RMSE	MAE	MAPE	MSE
BPNN	0.5523	0.4077	0.2668	0.1241
Standard LSSVR	0.2388	0.1561	0.0534	0.0570
IPSO-LSSVR	0.1687	0.0508	0.0159	0.0147

The obtained results indicate that our hybrid model has significantly yielded more reliable performance, generalization ability, and high prediction precision than LSSVR and BPNN model. For the same LSSVR, the relative RMSE, MAE, MAPE and MSE differences between the IPSO-LSSVR and standard LSSVR models are 29.36%, 67.46%, 70.22% and 74.21% in the test period, respectively. It is clear that the parameters optimized by IPSO are of better choice to construct LSSVR model for the design of water quality prediction than the ones by 5fold cross-validation method. The relative RMSE, MAE, MAPE and MSE differences between the IPSO-LSSVR and BPNN models are 69.46%, 87.54%, 94.04% and 88.15% in the test period, respectively. It is obvious that IPSO-LSSVR has more accurate result than BPNN.

This study presents an IPSO based approach, capable of searching for the optimal hyper-parameters values of LSSVR and RBF kernel function. The results of application in water quality prediction demonstrates that the prediction method based on IPSO-LSSVR is effective and feasible, and simultaneously this prediction information is important for decision making regarding the water quality management in modern intensive aquaculture, so the testing costs and production schedule can be optimized.

6 Conclusion

In this study, a remote wireless monitoring system using wireless communication technology and IPSO-LSSVR prediction model for the intensive aquaculture in China is introduced. Two prototypes of RMP deployed in an intensive aquafarm in Shandong have been tested over nearly a 2-year period. It realizes the remote wireless monitoring of the water environmental parameters and alarm notification when monitored variables take anomalous values. On the basis of the present study, the following conclusions can be made:

(1) The system can monitor DO, pH, salinity and temperature in real-time and continuously, considering that more than 95.2% of the data have been correctly collected. There are no significant effects on the monitored pH value since it is comparatively stable. The results indicate a periodic variation of water temperature, which has the similar regularity with air temperature. Salinity has sharply changed after a heavy rain event, so it could be an indirect indicator for early warning. Some other parameters of serious concern in aquaculture include ammonia nitrogen and hydrogen sulfide. The measurement and control of these and other key parameters will be performed in future work.
(2) The system can provide an early warning, especially helpful for high-density aquafarms. The forecasting model correctly predicts the further trend of dissolved oxygen half an hour after physical measurement.
(3) On the daily time scale, dissolved oxygen is found to repeat with a sort of regularity, mostly depending on the time of feeding. On the large seasonal scale, it shows an almost periodical trend, depending on the climatic situation. Therefore, it might be possible to improve the forecast model on this basis.
(4) The forecasting results of the dissolved oxygen are good after training, so changes in their coefficients will not be a priority for model improvement. Data sets of experiments that include all the necessary measurements along a growing cycle are not available. In addition, data sets from other pond environments with fish of different species are needed to make the model applicable to a wider range of culture environments. Special attention should be given to training data set as well.
(5) Application of the proposed system is still limited by its rather complicated operational requirements and high maintenance cost. The effects of water quality variations can be investigated in a good temporal and spatial resolution if more RMPs are installed. Moreover, the sensors need frequent cleaning and recalibration to prolong the useful span, because they need to have constant contact with the water, resulting in instrument fouling and loss in sensitivity and reproducibility. The long life span poles and the layout of sensor collection stations should be studied.

Acknowledgements. This research is financially supported by the National Key Technology R&D Program in the 12th Five year Plan of china (2011BAD21B01) and Guangdong Science And Technology Plan Project (2012A020200008).

References

[1] Subasinghe, R.P.: Epidemiological approach to aquatic animal health management: opportunities and challenges for developing countries to increase aquatic production through aquaculture. Preventive Veterinary Medicine 67, 117–124 (2005)
[2] FAO, State of World Aquaculture: Fisheries Department, Food and Agriculture Organization of the United Nations, Rome, Italy (2006)
[3] Zhong, Y., Power, G.: Fisheries in China: progress, problems, and prospects. Canadian Journal of Fisheries and Aquaculture Science 54, 224–238 (1997)

[4] Stigebrandt, A., et al.: Regulating the local environmental impact of intensive marine fish farming: III. A model for estimation of the holding capacity in the modelling-on growing fish farm-monitoring system. Aquaculture 234, 239–261 (2004)
[5] Sim, F.S.: Water quality index as a simple indicator of aquaculture effects on aquatic bodies. Ecological Indicators 8, 476–484 (2008)
[6] Tor, F.: Design of a virtual instrument for water quality monitoring across the Internet. Sensors and Actuators B: Chemical 76, 281–285 (2001)
[7] Tseng, C.L.: Feasibility study on application of GSM-SMS technology to field data acquisition. Computers and Electronics in Agriculture 53, 45–59 (2006)
[8] Liu, X.L.: Quality monitoring of flowing water using colorimetric method based on a semiconductor optical wavelength sensor. Measurement 2, 51–56 (2009)
[9] Vellidis, G.: A real-time wireless smart sensor array for scheduling irrigation. Computers and Electronics in Agriculture 61, 44–50 (2008)
[10] Tschmelak, J.: Automated Water Analyser Computer Supported System (AWACSS). Part II. Intelligent, remote-controlled, cost-effective, on-line, water-monitoring measurement system. Biosensors and Bioelectronics 20, 1509–1519 (2005)
[11] Udy, J.: Water quality monitoring: a combined approach to investigate gradients of change in the Great Barrier Reef, Australia. Marine Pollution Bulletin 51, 224–238 (2005)
[12] Wei, Y.: An online water quality monitoring and management system developed for the Liming River basin in Daqing, China. Journal of Environmental Management 88, 318–325 (2008)
[13] Glasgow, H.B.: Real-time remote monitoring of water quality: a review of current applications, and advancements in sensor, telemetry, and computing technologies. Journal of Experimental Marine Biology and Ecology 300(special issue), 409–448 (2004)
[14] Baltacı, F., Onur, A.K.: Water quality monitoring studies of Turkey with present and probable future constraints and opportunities. Desalination 226, 321–327 (2008)
[15] Lee, M.W.: Real-time remote monitoring of small-scaled biological wastewater treatment plants by a multivariate statistical process control and neural network-based software sensors. Process Biochemistry Metabolic Engineering 43, 1107–1113 (2008)
[16] Lee, P.G.: Process control and artificial intelligence software for aquaculture. Aquacultural Engineering 23, 13–36 (2000)
[17] Mariolakos: Real-time monitoring on Evrotas River (Laconia, Greece): dissolved oxygen as a critical parameter for environmental status classification and warning. Desalination 213, 72–80 (2007)
[18] Hongbin, L.: Prediction and elucidation of the population dynamics of Microcystis spp. in Lake Dianchi (China) by means of artificial neural networks. Ecological Informatics 2, 184–192 (2007)
[19] Boyd, C.: Management of bottom soil condition and pond water and effluent quality. In: Lim, C., Webster, C.D. (eds.) Tilapias: culture, nutrition, and feeding. The Haworth Press, Binghamton, New York (2002)
[20] Ferreira, N., Bonetti, C., Seiffert, W.: Hydrological and water quality indices as management tools in marine shrimp culture. Aquaculture 318, 425–433 (2011)
[21] Páez, O.F.: Camaronicultura y Medio Ambiente. Instituto de Ciencias del mar y Limnología. UNAM, México, pp. 271–298 (2001)
[22] Carbajal, J., Sánchez, L., Progrebnyak, O.: Assessment and prediction of the water quality in shrimp culture using signal processing techniques. Aquaculture International. Springer (2011)

[23] Navarro, L., Mascarenhas, A.: Temperatura y Salinidad de la Capa Superior del Océano en la Entrada del Golfo de California en Agosto. Ciencias Marinas 23 (1992)
[24] Martínez, L.: Cultivo de Camarones Pendidos, Principios y Practicas. AGT Editor S.A. (1994)
[25] Chien, Y.: Water quality requirements and management for marine shrimp culture. In: Proceedings of the Special Session on Shrimp Farming, pp. 144–156. World Aquaculture Society, USA (1992)
[26] Bower, C., Bidwell, J.: Ionization of ammonia is seawater: Effects of temperature, pH and salinity. Journal of the Fisheries Research of Canada 35, 1012–1016 (1978)
[27] Wickins, J.: The tolerance of warm-water prawns to recirculated water. Aquaculture 9, 19–37 (1976)
[28] Needham: The problem of methaemocyanin. Nature 189, 308–309 (1961)
[29] Clifford, H.C.: Semi-intensive sensation: A case study in marine shrimp pond management. World Aquaculture 25(3) (1994)
[30] Esteves, F.A.: Fundamentos de Limnologia. Interciências, Rio de Janeiro, 606 (1998)
[31] Boyd, C.: Water quality standards: pH. Global Aquaculture Advocate 4(1), 42–44 (2001)
[32] Anand, E., Das, S.: Monitoring of total heterotrophic bacteria and vibrio Spp in an aquaculture pond. Current Research Journal of Biological Sciences 2(1), 48–52 (2010)
[33] Chen, F., Liu, P.: Lethal attribute of serine protease secreted by Vibrio alginolyticus strains in Kurama Prawn Penaeus japonicus. Zool Naturforsch 55, 94–99 (2000)
[34] Boyd, C.: Water composition and shrimp pond management. Global Aquaculture Advocate 3(5), 40–41 (2000)
[35] Liu, S.Y., Li, D.L.: GA hybrid approach of support vector regression with genetic algorithm optimization for aquaculture water quality prediction. Mathematical and Computer Modelling 2(1), 22–30 (2012)
[36] Vapnik, V.N.: The Nature of Statistical Learning Theory. Springer, New York (2000)
[37] Keerthi, S.S.: Efficient tuning of SVM hyper-parameters using radius/margin bound and iterative algorithms. IEEE Trans. Neural Network 13(5), 1225–1229 (2002)
[38] Chung, K.M., Kao, W.C.: Radius margin bound for support vector machines with RBF kernel. Neural Computation 15(11), 2643–2681 (2003)
[39] Ayat, N.E., Cheriet, M., Suen, C.Y.: Automatic model selection for the optimization of SVM kernels. Pattern Recognition 38(10), 1733–1745 (2005)
[40] Glasmachers, T., Igel, C.: Gradient-based adaptation of general Gaussian kernels. Neural Computation 17(10), 2099–2105 (2005)
[41] An, S.J., Liu, W.Q.: Fast cross-validation algorithms for least squares support vector machine and kernel ridge regression. Pattern Recognition 40, 2154–2162 (2007)
[42] Ju, W., Shan, J., Yan, C.H.: Discrimination of disease-related non-synonymous single nucleotide polymorphisms using multi-scale RBF kernel fuzzy support vector machine. Pattern Recognition Letters 30(4), 391–396 (2009)
[43] Feng, Y., Zhang, W.F., Sun, D.Z.: Ozone concentration forecast method based on genetic algorithm optimized back propagation neural networks and support vector machine data classification. Atmospheric Environment 45(11), 1979–1985 (2011)
[44] Zavar, M., Rahati, S.: Evolutionary model selection in a wavelet-based support vector machine for automated seizure detection. Expert Systems with Applications 38(9), 10751–10758 (2011)
[45] Yu, L.: An evolutionary programming based asymmetric weighted least squares support vector machine ensemble learning methodology for software repository mining. Information Sciences 191, 31–46 (2012)

[46] Liang, J.J., Qin, A.K.: Comprehensive learning particle swarm optimizer for global optimization of multimodal functions. IEEE Transactions on Evolutionary Computation 10(3), 281–295 (2006)
[47] Babaoglu, İ., Findik, O., Ülker, E.: A comparison of feature selection models utilizing binary particle swarm optimization and genetic algorithm in determining coronary artery disease using support vector machine. Expert Systems with Applications 37(4) (2010)
[48] Colt, J.: Water quality requirements for reuse systems. Aquaculture Engineering 34, 143–156 (2006)

ICT as an Enabler to Smart Water Management

V. Sempere-Payá[1,*], D. Todolí-Ferrandis[2] and S. Santonja-Climent[1]

[1] Institute of Computing Technology (ITI), Valencia, Spain
[2] Polytechnic University of Valencia, Valencia, Spain
vsempere@iti.upv.es

Abstract. Water management in urban areas goes beyond the supply of drinking water or the collection and treatment of waste water. There is a growing interest in the use of Information Communication Technology (ICT), resulting in a large number of new applications that provide access to vast amounts of information generated by the diverse facilities, unthinkable a few years ago, which greatly facilitate the operation, maintenance and management tasks in a context of efficient and sustainable urban development. In this context water management is particularly important, especially when one considers the huge growth in demand in recent years in developed areas where the water footprint is increasing. The proper use of ICTs applied to urban water management allows gathering data to know in real time about supply and demand and put this information in the hands of managers to help them predictively manage demand, or adapt prices based on the demand. Encompassing every aspect related to water management and monitoring in a global platform is known as IWRM (Integrated Water Resources Management). These IWRM systems need a large amount of information to manage, with precision, water systems in urban environments. This information falls under three main headings: generation, transport and consumption; each of them presenting different characteristics thus demanding different treatment solutions. Currently there are numerous initiatives working in the field of ICT applied to water management, some of which will be outlined in this chapter.

Keywords: ICT, smart water management, IWRM, water sensors.

1 New Needs for a New Century

Today we talk about cities of the future and the Smart City concept as a paradigm of good information management to facilitate citizens' lives and the work of those who operate and maintain the facilities that make this welfare possible. The predictions indicate that there will be an increase in the world population, concentrating especially in urban areas and cities in the next twenty years which will mean significant increases in the demand for potable water and waste water generation. Currently, water consumption doubles every twenty years, more than double the rate of human population growth [1]. Major problems affecting water in urban

[*] Corresponding author.

S.C. Mukhopadhyay & A. Mason (Eds.): *Real-Time Water Quality Monitoring*, SSMI 4, pp. 239–258.
DOI: 10.1007/978-3-642-37006-9_11 © Springer-Verlag Berlin Heidelberg 2013

management comprise increased demand or obsolete distribution networks. Another important aspect is the influence of climate change which exacerbates weather phenomena, affecting especially big cities. Flexible systems will be needed to adapt to these circumstances and provide resource management at critical times.

For instance, to prevent floods, large volumes of water must be managed in real time to redirect the flow of water and avoid flooding populated areas. Reuse of storm water is also necessary to address periods of shortage, requiring infrastructure for high capacity storage. These situations mean more infrastructure and facilities that require more sophisticated management systems.

The new heterogeneous communication systems in urban environments support new advanced sensing platforms that collect information at multiple points and situations, sometimes requiring decision support systems for processing large volumes of data, and greatly extends the ability to control every detail, and adapt to any requirement that modern urban monitoring demands.

2 Integrated Water Resources Management

According to the Technical Committee of the Global Water Partnership, IWRM is a process that promotes the coordinated development and management of water, soil and other related resources, in order to maximize the resultant economic and social welfare in an equitable manner without compromising the sustainability of vital ecosystems [2]. It is therefore an integrated platform that utilizes the knowledge and participation of all stakeholders and sectors involved, as well as several technological resources and infrastructures, for the management and development of water resources in order to balance the social and economic needs, and achieve a sustainable, responsible and environmentally friendly exploitation of aquatic ecosystems. However, this definition of IWRM is aimed at social, economic and environmental challenges while current water treatment includes other considerations and issues related to current scientific and technological limitations, and the degree of development in infrastructures and services required by modern cities.

From the point of view of ICTs, IWRM can be seen as a set of technologies, services, infrastructures and communications that allow comprehensive management of all aspects of maintenance, infrastructure and control involved in urban water treatment. In this regard, ICTs play a key role by providing the technologies that enable the integrated management system, using existing infrastructures, to operate and grow using new innovations. Given these definitions, IWRM is seen as the solution to perform the following tasks:

- **Stakeholder Participation:** Allows users to know in detail their water use behavior, thus achieving consumption-oriented smart water use while minimizing costs, maximizing efficiency in water use and environmental friendliness, both for industrial and domestic users. This is possible thanks to the "Smart Metering" concept (Automatic Meter Infrastructures), with new devices and meters which allow more comprehensive readings and improved communications between users and the central management, involving the final step of water distribution, such as housing or industrial buildings.

- **Pollution and Water Quality Control:** A proper management ensures that the best use is made of available water supplies, including protection from pollution, quality tests and control of waste waters.
- **Monitoring for Emergency Prevention and Detection:** Preventing floods and other disasters caused by water, and the ability to react faster to those situations represents a vital aspect in a well-managed city.
- **Economic and Financial Management:** Management of billing, prices and taxes charged on users in relation to water usage.
- **Information Management:** Multisource real time data access, oriented to cloud computing.
- **Smart Water Distribution:** Advanced information management on the status of the water network allows decisions to be made and action to be taken in a faster and more controlled way, dedicating the resources wherever necessary and saving resources where not needed. Includes technologies and concepts such as smart pipes or resource geo-location.

The urban wastewater networks rely heavily on staff for daily management and operational control, and automated tasks are based on legacy polling systems or synchronous communications for monitoring, limiting the interaction with the stations. In recent years, the development of real-time control (RTC), and increased automation and data collection capacity of the sewage networks, have been key factors in contributing to sustainable urban development and greater commitment to the environment. Another need in the modernization of these networks is the adoption of Decision Support Systems (DSS) to allow rapid response to disaster situations.

In many cases, the cities and the entities responsible for water management are still reluctant to implement RTC and DSS systems, mainly due to the difficulty of design and the actual level of development of sensors and communications systems. The establishment of an entire detection, management and control network is complex and expensive. This stage is where ICTs have a key role, providing the tools necessary for installation and communication systems for the infrastructures needed, integrating traditional environmental monitoring stations and new components such as rapidly deployable sensor networks, wireless communications mobile laboratories, geo-location systems for mapping groundwater resources (drinking water and wastewater distribution network, groundwater, etc.) or weather forecasting and climate monitoring systems (primarily for control of early warnings of emergency situations).

Improvements in monitoring and control systems of a city wastewater network enhances the performance and functionality of existing infrastructures, and creating new remote measurement systems increases the capacity for data collection and substantially improves the system's response to critical situations (especially in the detection of toxic substances and suspended solids in water courses, and during heavy rains). The efficiency of the entire system is highly dependent on the communications between the component parts, as well as liaison with the Central Control Station (CCS) and its ability to react automatically to any situation.

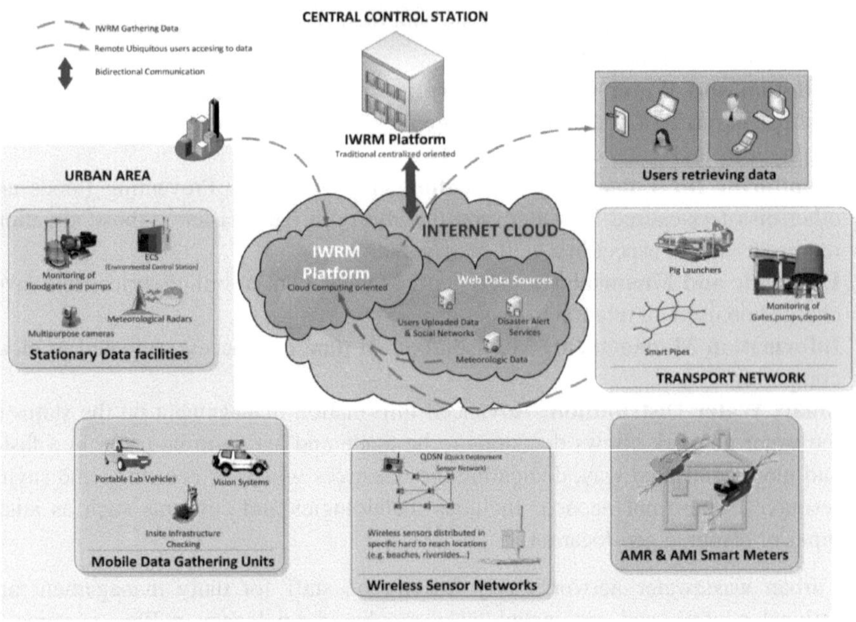

Fig. 1. Involved elements in a modern generic IWRM platform

2.1 *Modern IWRM Platform Structure*

The generic structure of a modern IWRM platform can be seen in Fig. 1, where the different sensor elements report the status and collect data continuously to be sent to the central station through a heterogeneous IP communications network. These sensor elements are grouped into:

- Fixed Stations (ECS, floodgates, meteorological radars, etc) obtaining environmental data such as water quality and pollution parameters and groundwater levels at critical points of the city.
- Mobile units that can be placed on secondary points, reducing the need to install fixed stations, which have cameras and different environmental and water quality sensors.
- Wireless sensor networks, for rapid deployments, which allow monitoring or casual tracking of areas of interest in a fast and reliable way, or fixed in hard to reach areas, such as beaches or riversides.
- Smart Meters (AMR [Automatic Meter Reading], and AMI [Automatic Meter Infrastructure]) installed at water consumption points, which exchange information with the plant, enabling readings to be taken remotely and application of variable tariffs.
- External Suppliers of weather and climatic data, or early warning systems for disasters, which reinforce the data obtained by the sensor networks themselves. Also includes uploaded data by users with private sensor networks, and other data from social networks.

- Sensors distributed throughout the water grid to control water quality parameters, state of gates, pumps, levels, condition of pipelines or enable underground resources geo-location, as well as other distribution network infrastructure control systems.

In the CCS, the management system evaluates the data received and then informs the responsible personnel, or takes decisions on any necessary action. Operators who are physically in the CCS can access all information, both historical and current real time data, and manage the system accordingly. They can also access any of the images required from the network to be used as visual support to help with decision making. Moreover, the web platform provides secure and controlled access to remote clients located outside the CCS, allowing multiple devices (laptops, tablets, and smartphones) ubiquitous access to information, which makes system management even easier for the staff.

2.2 Communication Technologies and Services

To determine which communications model is the most suitable, several factors must be taken into consideration, such as the present communications network, services and rates offered by the operator, the capacity of ECS and CCS, or the requirements from the application using the information.

To link the CCS with the stations, traditionally three architectures have been used to control communications: synchronous, requested by central or asynchronous. However, taking into account the needs of a modern system, none of these models on their own is adequate. IWRM tools must implement a hybrid communications model that supports the three types of communication defined.

Although it is necessary to have regular access to information and to obtain historical records of certain variables, the frequency of communication means that the detection of anomalies or alarms depends on the sampling time planned. If this sampling period is scheduled too often, the network becomes congested with too much information, but if the sampling period is scheduled less frequently, it loses the capacity for early detection of alarms. Therefore, it is also necessary to be able to immediately transmit alarms when the sensors detect a value outside its usual range. In addition, the CSS (or any technical or authorized agent from a remote location) should be able to request a reading of the actual values captured, regardless of the other two methods, so that technicians have the information they need available when necessary.

Regarding the access technology, the deployment of the communication networks has its main problem in the location of the stations, which are usually in remote areas, where it can be very costly, and sometimes even impossible, to lay any cable. Therefore, wireless technologies like WiMAX, WiFi or GPRS are preferred, and chosen depending on capacity requirements of the links and the characteristics of the scenario (depending on the location of the station, factors such as surrounding obstacles, harsh weather conditions or saturation in licensed and free radio bands can severely impair the quality of some links, favoring the choice of GPRS, with less bandwidth capacity but more robust to interference in the channel, or WiMAX, in case of long distance high bandwidth link needs).

With regard to services and added value that can be provided by a modern communications system, the latest trends and technologies that are already being implemented in the exchange and access to information in general should be considered, but focusing in the field of management, control and maintenance platforms for cities, Cloud computing and M2M communications stand out.

The M2M (Machine-to-Machine) communications are defined as the intelligent connection of two devices, through any channel, where the devices are in charge of giving meaning to the information exchanged, for instance, the connection between remote sensors in an ECS and the CCS computer.

Cloud computing is a base technology for smart cities applications. The diversity of communication methods (wireless, fixed network, mobile networks), as well as different data formats, network topologies and architectures, hinder the transmission of information between the different parties involved, ranging from a remote deployed sensor up to the final user's laptop. Moreover, the treatment of such wide and varied information poses new challenges, such as developing the tools for analysis and visualization of information, or addressing security and privacy issues, that represent a priority for data exchange applications. Everything is managed in real time so that the different managers involved may generate an immediate response, establishing appropriate service levels for each situation.

To deal with the management of large volumes of information, along with the changing needs of the cities, it is necessary to have technologies that facilitate the treatment of data and its subsequent homogenization for storage. In this scenario, a Smart City Cloud (cloud-based solutions for smart cities) model provides an intelligent and sustainable end to end platform, based on the deployment of smart sensing and acting systems interacting with the Cloud services, in the areas of energy efficiency and Intelligent Infrastructure, among many others. By using standards in communications and APIs, a number of specifics about sensors, sensor data models, and Web services for sensors can be set to allow devices to be accessible and controllable over the network [3]. Cloud computing advantages can be summarized as:

- Management of large volumes of information.
- Ubiquity of the information: accessible everywhere.
- Agility in the supply and acquisition of new services.
- Innovative technologies evolution.
- Elasticity: adaptable to the needs of the moment.
- Does not require large initial investments.

Developments of cloud platform projects are supported by world renowned companies such as IBM, Cisco or Microsoft. Many other simpler examples and projects of cloud computing can be found around the Internet, handling thousands of data streams, showing the variety of applications and the enormous benefits that cloud computing can provide. Examples include platforms such as Cosm (formerly Pachube), which is used in Japan to monitor radioactive contamination in water in different locations, and where any citizen can participate by sending information.

3 Pollution Control

In the field of water treatment, a major area, which generates a constant flow of information and priority data needed for proper water distribution, is the control of water pollution [4]. The percentage of fresh water available for human use is only about 0.007% of the total water on the planet. This is a vital resource which is becoming increasingly scarce as water pollution renders these resources unusable in the worst case scenario. In addition, poor control of water quality can lead to serious health risks in the affected areas (80% of diseases in developing countries is caused by contaminated water), and discharge of contaminated water into the sea, rivers and lakes seriously affects the ecosystem. Thus the control of water pollution is necessary for sustainable development and maintenance of any water distribution network. The water analysis must be carried out on several fronts:

- Domestic sewage.
- Industrial waste water.
- Agricultural waste water.
- Urban runoff (Storm water).

Concerning storm water treatment, both in urban pipelines (since in modern sewage systems, rain water run through different pipes to wastewater to prevent further contamination of the more pure collected water) and in reservoirs, water must be analyzed, since the measurement of water quality is essential for a water distribution network in a city. Similarly, domestic sewage and water flows resulting from chemical processes and waste in industry and sanitation should be monitored as closely as possible to the possible sources of pollution, and also in water treatment plants that purify the water to decontaminate it before releasing it into the sea (or other large bodies of water), or be used for other applications such as irrigation, and to detect possible toxic or radioactive discharges. The main sources of pollution are industrial toxic waste, leaking sewage networks affecting the drinking water supply, suspended solids, or pollution in the storm water collected. Therefore it is necessary to have adequate tools to analyze the quality of water which in turn are capable of providing the water management platform with all this information in an efficient manner, which allows a continuous monitoring of the water status belonging to different types (sewage, rain water), and allow action to be taken and decisions to be made in real time according to different scenarios. Each type of water requires a different treatment depending on the most affecting pollutants: BOD (Biological Oxygen Demand), pH and suspended solids in domestic sewage; or heavy metals in industry wastewater. Modern sensors are capable of analyzing the water chemistry parameters remotely, and video streams from cameras detect the presence of foreign solids, using artificial vision, thermal imaging and temperature control, in order to differentiate the response depending on the data.

3.1 Fixed Location Control Systems

To do this, ECS (as in Fig. 1) are located at specific points to take these measurements: at the outlet of wastewater treatment plants, industrial parks, urban storm water collection points and other locations of interest. These fixed stations typically have a direct connection to the plant, either by cable or through wireless communication, and also include actuator elements. The stations continuously report the water status sensed to the central management platform, which records and analyzes the data, and can make decisions in real time over the distribution network map, for example to prevent a contaminated water source extending over the distribution network, cutting just the right pipes and minimizing the overall impact on the distribution network.

3.2 Mobile and Portable Control Systems

Another useful tool provided by the new technologies is the ability to monitor inaccessible and distant locations and points of interest, where it is not feasible to have a fixed station. Thus, it is possible to build a network of sensors and cameras (to detect solids in the water stream) in order to capture the essential information during specific situations or for short periods of time when and where necessary, without extra cost on infrastructure and maintenance. This network is composed of wireless nodes with multi-sensors (each sensor can monitor a particular parameter), which communicates with the plant during their task.

For more specialized and accurate measurements on water quality, there are mobile units that have small chemical laboratories where water samples are collected and analyzed and then results are sent to the CCS. Sensor technologies and water quality analyzers are constantly evolving to offer better and more reliable results [5]. All data collected by sensor networks and mobile stations is received and stored in the CSS, where the information is monitored in real time, so the system starts necessary actions reacting to any alarm condition or to take preventive measures.

The CCS has all the necessary tools to process the information automatically. By having a model based on cloud computing, information is processed and accessible from anywhere (authorized by the entity responsible for management) so that technicians can monitor the water quality remotely or set alerts when a parameter surpasses normal limits.

4 Infrastructure Control

The water distribution network is included in the Critical Infrastructure group, since in case of damage it can cause serious physical and economic damage to the operation of a city or state. The deterioration or malfunction of the infrastructure related to the distribution, collection and storage of water could represent, in the worst case, cuts in drinking water for the population, losses of this valuable resource, or even disasters caused by water running outside their usual channels. Examples of the actual amount of water supply of a city or state that is lost through leakage can reach very high values: up to 20% in the municipalities of Canada, 20% in the UK, Spain, Malta and the Czech Republic, 25%, in Rome, and almost 50% in London and Vietnam. [6]

It is therefore essential to perform a constant maintenance and control on the water distribution network.

Traditionally, the maintenance of this infrastructure is costly and inaccurate. The state of a pipeline network is not easy to check, since access to the pipes is problematic if not impossible. This is why fissures or structural damage are often undetectable until leaks are spotted from the outside. Moreover, the process is slow because the technicians have to find and repair the damage manually. Moreover, the discharge of oversized solids can cause partial or complete blockages in pipes, negatively affecting the pressure of the water present therein, in addition to causing supply cuts.

Common methods of control have been based on the monitoring of the infrastructure by technical personnel, leak detection from pipelines outside, or by using the more sophisticated "Pig launchers", also frequently used in gas pipelines. In this case, there are certain locations or access points (called "launchers") to the water streams and pipes, from where devices called “Pigs” (bullet-shaped devices carrying sensors) are inserted, traversing the network to collect information on the state of the interior pipe walls and other structural parameters. At the end of a certain route, the device is removed through one of those spots (this time called "receiver"), and turns over the information stored in the central system for analysis. The drawback of this method is still the time involved while the device to finishes the journey along the pipelines and the data collected is retrieved.

To make things easier for the technicians, at those facilities or locations where it is necessary to check the status of the infrastructure on an occasional or regular basis, but where access is difficult, mobile units, as described in the section on pollution control are introduced as part of the data gathering elements, moving to the indicated locations, and using special cameras to collect the necessary information and transmit video stream in real time to headquarters. They are cameras that can be introduced in pipes and channels, operated by remote control, by both on site technicians and remote technical personnel at the CCS (see Fig. 2).

Other technologies in development are the so called "Smart Pipes" [7]. In this case, and inheriting the concept of monitoring the network of pipes from the inside the Pig Launchers applied, a network of sensors is introduced inside the water pipe network,

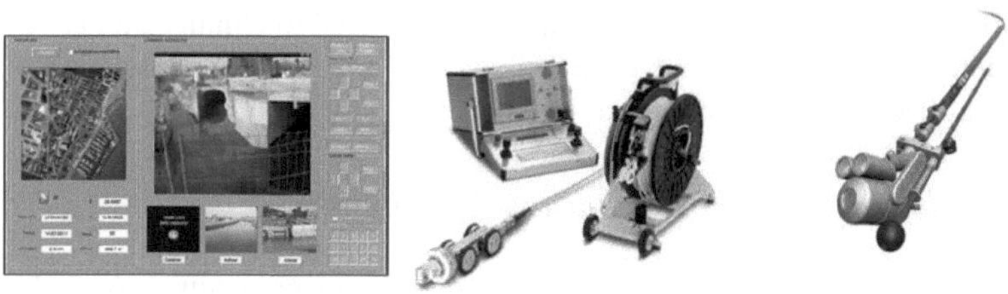

Fig. 2. Video Surveillance Interface (left), and remote controlled cameras (right)

capable of real-time reporting on the status of integrity of the walls, leaks and cracks detected, or water pressure and temperature, and always accurately locating the critical spot on an exact map of the network. Thus, the management platform is able to detect, prevent and act much faster and more accurately in the case of damage detected in the infrastructure. Following this methodology of monitoring to detect leaks in the piping, there are also other examples, already being implemented, such as in the Spanish town of Asparrena (Álava) [8], where they have a system based on the division of the network pipes in several areas that are hydraulically independent. Thus, it is possible to measure continuously the flow rate and inlet pressure of each sector. Data is transmitted in real time to the CCS and mobile phones of technicians, allowing immediate action, and has reported promising results, saving 48 million liters of water in 10 months.

5 Smart Metering

The technology that enables automatic capture of consumption information and its integration with the IWRM has been constantly evolving, from AMR systems, through AMIs and on to Smart Water Metering.

However, the current reality does not reflect this, as the vast majority of cities have still not adopted remote reading technologies for water meters. Generally, *mechanical accumulation meters* are used, which provide a single measurement of the accumulated consumption up to a particular moment (absolute values). Obtaining consumption data by the distribution company is carried out by periodic visits by operators to each of the meters in their area, who take a reading of total consumption via a visual inspection of the meter. This is an extremely rudimentary system, which does not allow advanced management of the network due principally to problems such as:

- Human error in meter readings.
- Problems of access in order to read meters.
- Cost of human resources to take meter readings.
- Long intervals between readings for the same meter (in many cases, several months, due to intermediate readings being estimated).

Mistakes or lack of precision when making management decisions for water systems may lead to serious problems for such an important resource, such as cuts in supply to large cities or water wastage, both of which make it vital to eliminate all errors and to obtain precise data cheaply and in real time. In this point, the automation of meter reading has become a necessity for modern cities.

5.1 Evolution on Meter Readings Acquisition

In a first approach to these systems of automatic meter readings (or AMR), the *mechanical accumulation meters* were complemented with a system with datalogger and communication equipment, which allows readings to be taken using portable equipment (*walk-by*) or using vehicles (*drive-by*) which circulate through the streets of a city, scanning the nearby meters. This system eliminates errors in readings and problems of access to meters. It also reduces the costs associated with reading meters and problems of periodicity. The evolution of these systems has been driven by the change from the classic water meters to *pulse water meters*, which register the reading when a certain consumption is reached (for example, every 1L, 10L, etc.), which when combined with the datalogger and transponder system, and used from a vehicle, can provide more accurate water consumption information.

The next step in the automation of meter reading is the elimination of the intermediate vehicle to allow direct communication of readings to the central office using GSM/GPRS communication. However, this is a fairly costly solution due to the expense of using the public network, and is impractical at a residential level due to the high number of meters which need to be monitored. Water companies have evaluated a range of solutions to this problem, such as connecting groups of meters using industrial communication buses (such as M-bus or Modbus) and later via RF, and then communicating with the exterior via a single transmission point per area, thus reducing communication costs considerably. These systems offer the water suppliers instant access to meters, meaning it is then possible not only to take readings, but also to take action on the meter, thus creating a new Advanced Measuring Infrastructure (AMI). This marks a real step forward from the original "one-way" systems towards more advanced "two-way" systems.

Nowadays, AMIs have evolved to a point where they allow real time communication at a much lower cost, through technologies which eliminate, or reduce dramatically, the need of telecommunication providers. The use of private networks allows more frequent access to information without additional cost, which in turn means more effective management of water supplies by staff who now have precise information on how the system is working at all times. This concept, combined with a trend towards the new generation of meters known as *Interval Water Meters*, which allow more precision in taking measurements (they register readings for periods of 15, 30 or 60 minutes instead of registering when a specific volume of water has been used), are opening the path to the use of Smart Water Metering (SWM), whose principal benefits are:

- More frequent billing based on real readings.
- More precise detection of leaks (in both space and time).
- Improvements in measurements and capacity to detect back-flows.
- Contributes to the development of environmental awareness.
- Improvements in control efficiency.
- Option of variable tariffs.

Smart Water Metering is not only a new way to automate readings; it will also help to involve the consumer in water use management. Consumers will be more aware of their own consumption, leading to a more responsible use of water resources. Smart Water Metering contemplates connecting the meter to the Home Area Network

(HAN), and the communication with other domestic devices, meters or consumption displays via Zigbee Smart Energy 2.0, even allowing one to make decisions based on the electricity network load and demand. Additionally, the distributor can offer services to users via their web page. These services will include knowing their residential consumption in real time.

However, all of this leads to a high level of complexity in communications. Currently, the options proposed to deal with this are: the use of private RF technologies, principally over licensed bands reserved for automated meter reading systems, or using other cabled networks, such as the electrical networks. One of the current proposals is to combine electric, gas and water meters and integrate them in Smart Grids and communication through PLC technologies, such as G.hn. For example, in 2009, IBM began a 5-year project for the Enemalta Corporation (EMC) and Water Services Corporation (WSC), Malta's national electricity and water utilities, to create a SmartGrid system to monitor, throughout the country, all water and electricity meters. This project [9] involves changing and updating the meters and their integration in advanced IT applications, allowing remote monitoring, management, meter reading, identification of cuts in service, pre-paid services, localization of leaks and rewards to customers who consume less energy and water, all in real time. Another option for reducing costs is to connect the water meter to the supplier via the consumer's HAN with its domestic access to the internet, offering advantages in terms of billing and additional services, given that most homes now have internet access.

5.2 Smart Water Metering Standardization in Europe

Many governments understand the potential of Smart Metering and its benefits in terms of reducing consumption. The European Union, through its mandate M441 [10], requested to the European standardization organisms (CEN, CENELEC, ETSI) to generate standards to support these technologies. Currently, there is a diverse range of proposals in existence for the standardization of a Smart Metering model, even including a proposal for the standardization of a unified system integrating water, gas and electricity. The Open Metering System Specification (OMS) [11] is the first definition of this type developed by industrial associations (German Gas and Water Industry Association), the KNW and the ZVEI (German Electrical and Electronic Manufacturer's Association). Their objective is to define a complete infrastructure and standardization solution for metering devices. This proposal was presented to the European Committee for Standardization (CEN) with the aim of establishing it as the European standard. Moreover, this proposal includes the standard EN 13757 which deals with the transmission of data between Smart Meters and concentrators.

The OMS establishes an infrastructure of 3 levels for secure communication between the smart meters and the water distributor.

The first level defines communication between the smart meter and the Multi Utility Communication device (MUC), a device which concentrates data from a range of meters (for example, all the meters of an apartment building, or meters for a range

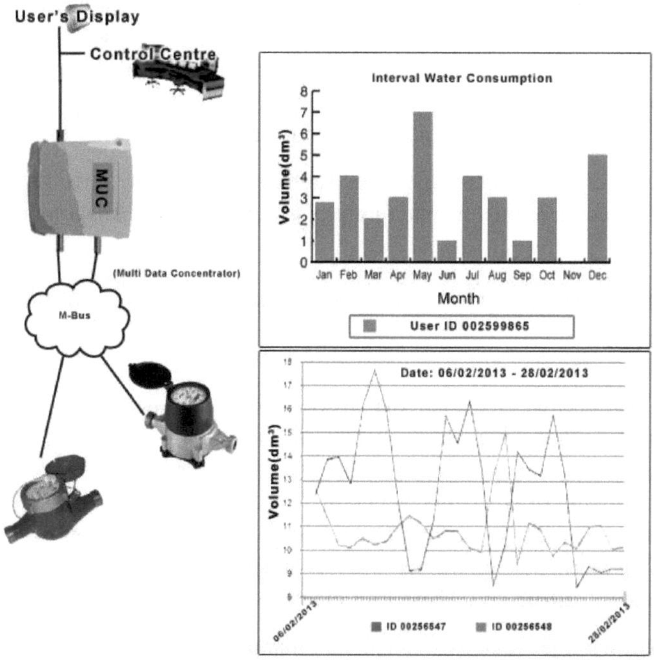

Fig. 3. AMI and AMR smart meters upload water consumption data to the management platform

of resources). This communication is carried out using the standard M-Bus (EN-13757), which supports communication over copper twisted pair or wirelessly (EN 13757-4) over the open VHF bands 868 and 433 MHz, and more recently, the band reserved for gas, water and electricity meters: 169MHz (according to EC DEC/2005/928). This band allows power of up to 250mW, reaching up to 300 meters in urban areas and up to 2 km in rural areas, and battery life of up to 14 years. In other situations the MUC can be integrated with the meter, eliminating the primary sub-network.

The second level constitutes an extension of the first level through repeaters or with the MUC themselves and multi-hop routing protocols. As there is not, as yet, a specified routing protocol, the use of proprietary protocols is permitted.

Finally, the third level establishes communication between the MUC and the office of the distributor via TCP/IP technologies such as DSL, GPRS and PLC.

There is still much work to do for Smart Water Metering to be extensively used, and to show added values to final users (like variable tariffs, or information on water quality, all this being available in real time in the home). However, this technology does have its detractors [12] who point out disadvantages such as:

- Loss of privacy: it is possible to extract information on patterns of use for each consumer.
- Danger of possible monitoring by unauthorized third parties (insurance companies, press, law enforcers, etc.).
- Greater security risks for the network and its meters.

However, these are problems that will be surely solved in time, given that this type of metering has become a must for the future of efficient water management.

6 Monitoring in Emergency Situations and Decision Support Systems

As well as the normal, seasonal management of water networks, there is another important factor that must be considered: emergency situations. For example, continuous rains, or even worse, short bursts of heavy rain, cause breaches of water courses and flooding of crossings, which may cause huge amounts of material damage and even casualties. Early warning systems involve four elements, which need to be supported by governance, coordination mechanisms from national to local levels, and by appropriate infrastructure. These four elements are:

(1) **Risk knowledge:** Risk assessment and mapping will help to set priorities among early warning systems needs and to guide preparations for response and disaster prevention activities.
(2) **Warning service:** Constant monitoring of possible disaster precursors is necessary to generate accurate warnings on time.
(3) **Communication and Dissemination:** Clear and understandable warnings must reach authorities in a fast and reliable way. There are some regulatory barriers to take into account, and these need to be coordinated across many agencies at national to community levels for the system to work.
(4) **Response Capability:** Communities must respect the warning service and should know how to react. Crisis management groups and Decision Support Systems would help with this task.

The Hyogo Framework for Action (HFA) [13] is a 10-year plan (2005 – 2015) adopted by 168 countries of the United Nations in 2005 at the World Disaster Reduction Conference. It is an extremely important reference and deals with risk reduction policies that help countries deal effectively with these disasters to reduce loss of life and material damage. According to the HFA, the starting point for reducing disaster risk and for promoting a culture of disaster resilience lies in the knowledge of the hazards and the physical, social, economic and environmental vulnerabilities to disasters that most societies face, and of the ways in which hazards and vulnerabilities are changing in the short and long term, followed by action taken on the basis of that knowledge. Disaster risks related to weather, water, climate variability and climate change, are addressed in sector development planning and programs as well as in post-disaster situations. At times of disaster, impacts and losses can be substantially reduced if authorities, individuals and communities in hazard-prone areas are well prepared and ready to act and are equipped with the knowledge and capacities for effective disaster management.

6.1 Decision Support Systems

Water management departments usually dispose of some form of crisis management group for critical situations. The main problem they face is the lack of information available to help with decision-making processes. The use of DSS (Decision Support Systems) has an essential role in these cases, and there are currently a number of projects aimed at developing more advanced and intelligent DSS to better deal with this type of event. For example, the *Neptune Project* [14] is a project led by the Centre for Water Systems based at *Exeter University* which is working alongside two leading UK water supply, technology and automation companies: *Yorkshire Water Services* and *United Utilities*. In this project, a DSS is under development that is capable of supporting complex decision making to improve the operation of water supply and distribution systems, coordinating asynchronous information in real time to offer solutions to some of the common problems of water distribution networks. The Decision Support System uses a two-stage approach: data collected from all the sensors is filtered initially by the system to remove all unnecessary information, and the remaining data is presented to the operator using a two-screen approach to aid his analysis of situation and improve risk-based decisions. The first (master) screen displays the current state of the network and the alarm conditions, if any of these are available. The second (slave) screen presents the Neptune-project specific outputs generated by the framework and associated tools. Using this technique together with *Intervention Management* procedures allows fast, reliable action plans to be generated in order to resolve any technical difficulties. *Intervention Management* focuses on designing fully functional impact reduction strategies and incident isolation techniques to be used in the risk-based decision making process. *Intervention Management* is divided into two stages:

- Incident isolation (primary-immediate interventions)
- Impact reduction (alternative interventions)

The first stage forms a situation report, analyzing each event (pump failure, pipe burst, etc.), and presenting possible actions to be taken by an operator using a Decision Support System screen. In the second stage, the possible actions are analyzed, to detect any changes in the system and to find possible better solutions, in order to reduce the impact on customers.

In spite of the advanced development of the DSS used, the problem continues to be a lack of data from sensors or that the data obtained is of poor quality. There is now a genuine need for better, more efficient sensors which need less maintenance and which are smaller and more economical. Advanced sensor and detection systems would allow more precise action on critical situations, avoiding or minimizing their consequences or even preventing them completely.

6.2 Monitoring through Fixed Infrastructure

The essential degree of monitoring in these cases focuses on the installation of cameras and sensors distributed over fixed locations. In cities, it is normal practice to monitor rainfall, floodgates and levels in water channels and tunnels. This monitoring provides valuable information during periods of drought and flood. Real time control and monitoring systems have been around for several decades, but their general implementation in already existing infrastructure is unattractive due to cost issues. However, wireless sensor networks allow a wide range of low cost applications that were not previously possible with cabled networks. Placing these sensors at fixed locations makes it possible to obtain information from a wide range of locations at a greatly reduced cost. For example, in [15] the information from a wireless rainfall sensor network is used to create a flood model that will help to predict possible risks with much greater accuracy than has been possible with other systems. The "*Flood Watch*" system of DHI [16] integrates real-time data and forecast modelling tools seamlessly in a resilient and robust client-server environment that encompasses sophisticated – yet easy to use – tools for data management, monitoring, forecast modelling, decision making and dissemination. The system is often used by regional and local river basin authorities to provide real-time forecasts in areas prone to flooding and to issue early warnings to flood response managers and the public. Within the field of reservoir operation, the system is typically used to forecast inflows and the appropriate release of water. The decision support facilities can be used to assist in developing operational strategies and optimizing power production or water use.

An interesting foresight element in cities prone to flooding is the incorporation of a meteorological radar to the system. This radar would help generate alerts from rain-bearing clouds before the flooding actually takes place. This is the case with the Quebec Urban Community RTC system [17], a management system for residual urban water that operates in real time and which uses distributed sensors, radar generated rainfall predictions, variable optimization algorithms for the resolution of equation systems and monitoring of floodgates and water courses.

Meteorological forecasting and detection through overflow and flood sensors allows affected areas to be evacuated in time to avoid tragedy, and also allows operators to effect counter-measures to re-channel flood waters. An example of this is the use of automated actuators that enable use of the drain system as a linear storage system for rain water, thus avoiding massive dumping of excess rain water into rivers and seas [18,19].

Urban areas close to dams can also dramatically improve the safety by implementing new monitoring technologies, such as the UrbanFlood project set up by the Ijkdijk Foundation in the Netherlands. UrbanFlood [20] is a project to develop intelligent dams that eliminate the need for continuous visual inspection through sensor networks that monitor, among other parameters, pressure in the dams, movements and displacements (using textile optical fiber), and the internal and external temperature. The system transmits values in real time for early prediction of critical situations such as flooding and damage to dams, as well as enabling more effective management and decision making (in the project FloodControl 2015).

Another type of risk situation which requires preventive monitoring is the contamination of water supply to households. Distributed real time monitoring of contaminants and sediments accumulation in water for domestic consumption, as well as quality parameters or the existence of bacteriological contamination (such as Legionnaires disease) allows preventive measures to be taken which are extremely important due to the potential impact on the health of the population. In the USA a range of programs of this type are being launched to allow early warning or even cutting domestic water supplies in the case of contamination of the water network. Blue Box Tool [21] is a good example of a real time water management system for early detection, alert and recovery, that deals with drinking water quality in the public network. This system significantly reduces false alarms, offering a more precise, sensitive and trustworthy service. Water quality is not only a problem in developing countries, the sustainability of European water supplies are also at stake due to factors such as pollution, over-use of resources, damage to aquatic eco-systems, climate change and other safety aspects.

Water contamination affects not only water destined for domestic consumption. Coastal cities which have beautiful beaches as their principal tourist attraction may suffer serious pollution catastrophes which affect their inshore bathing water. COWAMA [22] is a DSS system for these types of waters. It is based on simulation of flow and contamination, possible direct dumping of contaminant materials into the drainage system, water treatment and water outflow into the sea. This system allows the prediction of magnitude and duration of contamination episodes, and is an effective warning system for bathers. Furthermore, this system allows more efficient planning of infrastructure to reduce contamination episodes.

6.3 Monitoring through Mobile Infrastructure

However, there are situations in which it is not possible to maintain fixed monitoring infrastructures. This may be because the variability in the points to be monitored is very high, or due to questions of resources or property. In these situations, monitoring vehicles and rapid deployment sensor networks become vital tools for risk prevention.

Environmental monitoring vehicles have sensors and cameras capable of real time transmission. They can be deployed in areas that are not permanently monitored in order to sense a range of environmental parameters, water quality or even to use their cameras to send real time images of the state of sluices and floodgates, dams and obstructions in water courses, etc. to the central station. This data constitutes an important source of information for the crisis management group, who will then be able to base their decisions on more objective information.

Quick Deployment Sensor Networks (QDSNs) [23] also give huge advantages in critical situations and in environments without sensors or communication infrastructure. While traditional wireless sensor networks require specialized staff to carry out their installation, (pre-planning, coverage studies, network analysis, role assignment, topology design, etc.), quick deployment sensor networks include tools that help in the deployment phase so that a complete network can be set up in a very short time and without technical knowledge of the system. The tools of the QDSN mean that personnel can be sure that the point that they want to monitor has coverage, without the need for any prior planning or post deployment checking. Thus, a network can be set up much more quickly than is possible with other systems, both cabled and wireless. Once the infrastructure has been set up, the personnel will have new, valuable, real time information on the states of flood gates and sluices, flooding, water mixtures, pipes and water courses, pumps, extent of contamination and environmental measurements. Moreover, rapid deployment sensor networks allow the network to be dismantled and set up in a new location quickly and efficiently, allowing data to be collected from different areas around the city in emergency situations in an agile way and with a minimum of personnel.

Fig. 4. Quick deployment of a wireless sensor network to collect data in provisional critical locations

7 Conclusion

While there are many challenges faced by smart sensors in the context of water quality monitoring, it is important to remember that sensors themselves are not standalone devices. Instead, they require a wide range of support from ICT, some notion of which has been discussed in this chapter, including data transport and storage, data evaluation and also appropriate presentation. We also see that platforms such as IWRM require a variety of sensor types and supporting technologies in order to accommodate fixed and remote sensing in addition to functions such as smart metering and disaster management: there is no "single-fit" solution for each situation. Thus, it must be recognized that ICT has an important role to play in the monitoring of water quality in order to ensure key objectives are met, including stakeholder participation, pollution and water quality control, monitoring for emergency prevention and detection, economic and financial management, information management and smart water distribution.

Acknowledgment. This work was supported by the MCyT (Spanish Ministry of Science and Technology) under the project TIN2010 21378-C02-02, which is partially funded by ERDF (European Regional Development Fund).

References

[1] ICT as an enable for Smart Water Management. ITU-T Technology Watch Report (October 2010)

[2] Global Water Partnership, "What is IWRM?", http://www.gwp.org/en/The-Challenge/What-is-IWRM/

[3] Sheth, A., Henson, C.: Semantic Sensor Web. From IEEE Internet Computing (2008), http://knoesis.wright.edu/library/publications/SHS08-IC-Column-SSW.pdf

[4] Michael Hogan, C.: Water pollution, Encyclopedia of Earth. Topic ed. Mark McGinley; ed. in chief C. Cleveland. National Council on Science and the Environment, Washington, DC (2010)

[5] JETRO (Japan External Trade Organization), Technologies/Products for Water Pollution Control and Water Purification, http://www.jetro.go.jp/ttppoas/special/env_rep_english/env_rep_04_hyoA.html

[6] Tullmin Consulting, Estimated Water loss from Leaking Pipelines, http://www.corrosion-club.com/waterfigures.htm

[7] Lin, Y.-F., Liu, C.: Smart Pipe: Nanosensors for Monitoring Water Quantity and Quality in Public Water Systems, Illinois State Water Survey, Institute of Natural Resource Sustainability

[8] de Álava, D.F.: Nota de Prensa: Un novedoso sistema de detección de fugas implantado en Álava permite ahorrar 48 millones de litros de agua en 10 meses (February 22, 2012)

[9] SmartGrid for water monitoring in Malta. IBM, http://www-05.ibm.com/uk/smarterplanet/casestudies/opinion_2.shtml

[10] M/441 EN. Standardization mandate to CEN, CENELEC and ETSI in the field of measuring instruments for the development of an open architecture for utility meters involving communication protocols enabling interoperability, http://www.cen.eu/cen/Sectors/Sectors/Measurement/Documents/M441.pdf

[11] Open Metering System Group website, http://www.oms-group.org/en_oms.html

[12] Bleicher, A.: Privacy on the Smart Grid. IEEE Spectrum, Energy Section (October 2010), http://spectrum.ieee.org/energy/the-smarter-grid/privacy-on-the-smart-grid

[13] Hyogo Framework for Action (HFA) website, http://www.unisdr.org/we/coordinate/hfa

[14] Neptune Project, University of Exeter, http://emps.exeter.ac.uk/engineering/research/cws/research/distribution/neptune/

[15] Al-Sabhan, W., Mulligan, M., Blackburn, G.A.: A real-time hydrological model for flood prediction using GIS and the WWW. Computer, Environment and Urban Systems 27(1), 9–32 (2003)

[16] Project FloodWatch. DHI, http://www.dhigroup.com/SolutionSoftware/FLOODWATCH.aspx
[17] Schütze1, M., Campisano, A., Colas, H., Vanrolleghem, P., Schilling, W.: Real-Time Control of Urban Water Systems. In: International Conference on Pumps, Electromechanical Devices and Systems Applied to Urban Water Management PEDS 2003, Valencia, Spain, April 22-25 (2003)
[18] Wan, P., Lemmon, M.D.: Distributed Flow Control using Embedded Sensor-Actuator Networks for the Reduction of Combined Sewer Overflow (CSO) Events. In: 46th IEEE Conference on Decision and Control, pp. 1529–1534 (2007)
[19] Campisanoa, A., Schillingb, W., Modica, C.: Regulators' setup with application to the Roma–Cecchignola combined sewer system. Urban Water 2(3), 235–242 (2000)
[20] Urban Flood Project website, Stichting Ijkdijk, http://urbanflood.eu/default.aspx, http://www.floodcontrol2015.com
[21] Blue Box Tool whitepaper, http://www.w-water.com/uploaded_files/documents/BlueBox_datasheet_U4152.pdf, Blue Box Tool is a product from Whitewater, website available in http://www.w-water.com
[22] Suñer, D., Malgrat, P., Leitão, P., Clochard, B.: COWAMA (Coastal Water Management) Integrated and Real Time Management System of Urban Water Cycle to Protect the Quality of Bathing Waters. In: 11th International Conference on Urban Drainage, Edinburgh, Scotland, UK (2008)
[23] Todolí, D., Santonja, S., Sempere, V.: Enabling Quick Deployment Wireless Sensor Networks for Smart Cities. In: 9th International Workshop on Factory Communication Systems, Lemgo-Detmold, Germany (2012)

Emerging Organic Contaminants in Groundwater

Marianne Stuart* and Dan Lapworth

British Geological Survey, Wallingford, OX10 8BB UK
mest@bgs.ac.uk

Abstract. Emerging organic contaminants (ECs) are compounds now being found in groundwater from agricultural and urban sources that were previously not detectable, or thought to be significant. ECs include pesticides and degradates, pharmaceuticals, industrial compounds, personal care products, fragrances, water treatment by-products, flame retardants and surfactants, as well as 'life-style' compounds such as caffeine and nicotine. ECs may have adverse effects on aquatic ecosystems and human health. Frequently detected ECs include the anti-epileptic drug carbamazepine, the antibiotic sulfamethoxazole, the anti-inflammatories ibuprofen and diclofenac, and caffeine, as well as pesticide degradates. This means there will be challenges in the future in order to address these ECs and to minimise their impact on drinking water and ecosystems. In the coming decades, more ECs are likely to have environmental standards defined, and therefore a better understanding of environmental behaviour remains a priority.

Keywords: Emerging contaminants, groundwater, pharmaceuticals, pesticide degradates, personal care products.

1 Introduction

A diverse array of synthetic organic compounds is used worldwide in large quantities for the production and preservation of food, for industrial manufacturing processes and for human and animal healthcare. In the last few decades there has been a growing interest in the occurrence of these contaminants in the terrestrial and aquatic environment, their environmental fate and their potential toxicity even at low concentrations [1-6]. The contamination of groundwater resources is a growing concern and relatively poorly understood compared to other freshwater resources [7].

Organic compounds previously not considered or known to be significant in groundwater in terms of distribution and/or concentration, which are now being more widely detected and which have the potential to cause known or suspected adverse ecological or human health effects are here referred to as emerging contaminants (ECs). Synthesis of new chemicals or changes in use and disposal of existing chemicals can create new ECs. ECs also include substances that have long been

* Corresponding author.

S.C. Mukhopadhyay & A. Mason (Eds.): *Real-Time Water Quality Monitoring*, SSMI 4, pp. 259–284.
DOI: 10.1007/978-3-642-37006-9_12 © Springer-Verlag Berlin Heidelberg 2013

present in the environment but whose presence and significance are only now being elucidated [8]. As analytical techniques improve, previously undetected organic micro-contaminants are being observed in the aqueous environment [9-10]. Richardson and Ternes (2011) review recent analytical developments in the emerging contaminant context [11].

ECs include a wide array of different compounds (as well as their metabolites and transformation products, collectively referred to here as degradates) including: pharmaceuticals and personal care products (PCPs), pesticides, veterinary products, industrial compounds/by-products, food additives as well as engineered nano-materials. Because of the vast number of possible compounds, many studies have selected ECs according to priority lists established taking into account consumption, predicted environmental concentrations as well as ecotoxicological, pharmacological and physicochemical data [12-17].

To date, the occurrence of ECs has been much better characterised in wastewater and surface water environments than in groundwater [7]. Wastewaters are the main sources of ECs in the environment and surface waters therefore contain the greatest loads of ECs. Wastewaters and surface waters are also thought to contain a much greater diversity of compounds compared to groundwater, although this may be simply a function of the capability of analytical methods relative to the generally lower groundwater concentrations and the limited number of groundwater studies.

The occurrence of ECs in surface waters has been reviewed for public water supply [18], for sources to public supplies, [7], and for occurrence and fate of ECs and established trace pollutants [19]. The first systematic review of ECs in groundwater, by Lapworth *et al.* (2012), highlighted the worldwide widespread contamination of groundwater resources by a large variety compounds that are detected as a result of both recent and historical activities [20]. Environmentally significant concentrations (10^2–10^4 ng/L) of a range of ECs, including a number of endocrine disrupting substances, are being detected in groundwaters globally. Many of these ECs are among the highest priority substances for treatment and regulation both in terms of their potential environmental and human health effects.

Many ECs remain unregulated and present analytical and institutional challenges [21]. The number of regulated contaminants will continue to grow slowly over the coming decades. Monitoring of anthropogenic micro-organic pollutants in river basins is required within the framework of various national regulations [22-23] with the overall aim of protecting and improving the quality of water resources.

In the European context groundwater quality is currently regulated under the Water Framework Directive (WFD) [24], its daughter Groundwater Directive (GD) [22] and drinking water under the Drinking Water Directive [25]. Pesticides are also regulated under the Plant Protection and Biocides Directives [26-27]. The WFD and the GD establish environmental objectives for protecting groundwater and water bodies and groundwater dependant ecosystems. These require that threshold values (standards) be established for pollutants that put the groundwater body at risk of failing to achieve its environmental objectives. Whilst for many chemical pollutants there is sufficient knowledge to establish threshold values, in the case of most ECs the current lack of knowledge on toxicity, impact, behaviour and limited monitoring data mean that threshold values cannot yet be set.

The European Drinking Water Directive sets limits for a small number of organic micropollutants comprising aromatic hydrocarbons, chlorinated solvents and disinfection by-products [25]. Priority substances established under another WFD daughter directive include benzene, octyl and nonyl phenols, specified polyaromatic hydrocarbons (PAH), di(2-ethylhexyl)phthalate and a range of chlorinated hydrocarbons[28]. The European Commission aims to table draft limits for 16 new substances limits under the WFD including anti-inflammatory drugs, synthetic contraceptives and perfluorooctane sulfonate (PFOS) [29].

A similar situation occurs elsewhere in the world. Regulatory frameworks exist to manage the potential sources of pollution and require monitoring of a number of 'priority' organic contaminants in the aquatic environment. However, there are a huge number of contaminants (largely organic compounds) that are not subject to the same degree of regulation at present (for the same reasons outlined above). The US Environment Protection Agency (EPA) have derived statutory guideline values for about 125 contaminants in drinking water of which 31 could be considered to be micro-organic pollutants excluding pesticides. None of these are pharmaceuticals or PCPs [30]. The US EPA published a new contaminant candidate list (CCL-3) in 2009 which included 3 pharmaceuticals as well as perfluorooctanoic acid (PFOA), PFOS and eight hormones [10].

2 Types of Emerging Groundwater Contaminants

Much more is known about pesticides in groundwater compared to other compounds, such as pharmaceuticals, which are more poorly characterised. The hazards to human health of some compounds are well documented, but their ability to travel through the aqueous environment is only just being investigated, and environmental persistence is as yet unknown. From their sources, physical and chemical characteristics, mobility/behaviour in the aqueous environment and associated hazards the following types of micro-contaminants may be considered to be emerging in groundwater.

2.1 Pesticides

Pesticides have been detected at trace concentrations in groundwater worldwide for a considerable period and are well-established contaminants. By the 1990s atrazine, simazine and a range of other herbicides had been found in groundwater worldwide [31-36]. Recently new detections of parent compounds have become apparent as analytical methods have improved, for example metaldehyde in the UK [37], and this also fits the emerging contaminant definition.

Even twenty years ago it was clear that pesticide degradates needed to be considered [38-39]. Some studies have even shown that pesticide metabolites may be detected in groundwater at higher concentrations compared to parent compounds from both agricultural and amenity use [40-41]. By their nature degradates are biologically active and many may be toxic, and such data forms part of the pesticide registration process although they are still often not adequately monitored.

2.2 Pharmaceuticals

The presence of pharmaceutical chemicals in the aquatic environment has long been recognised as a concern [42]. The primary routes for pharmaceuticals into the environment are through human excretion, disposal of unused products and through agricultural usage [43]. A wide range of pharmaceutical products have been detected in surface and groundwater, associated with wastewater disposal [44-48]. These have included:

- veterinary and human antibiotics: ciprofloxacin, clofibric acid, lincomycin, sulfamethoxazole, tetracycline
- other prescription drugs: carbamazepine, codeine, diclofenac, salbutamol
- non prescription drugs: acetaminophen (paracetamol), ibuprofen, salicylic acid
- iodinated X-ray contrast media: iopromide, iopamidol

Other potential threats to surface water which have been identified are tamiflu and chemotherapy drugs, such as 5-fluorourcil, ifosfamide or cyclophosphamide [49-52] and illicit drugs such as cocaine and amphetamines [53-54].

2.3 "Life-Style" Compounds

Caffeine, nicotine and the nicotine metabolite cotinine have been widely detected in groundwater impacted by sewage effluent [55-57]. The artificial sweeteners acesulfame, saccharine, cyclamate and sucralose have been found at high concentrations in groundwater impacted by sewage infiltration ponds [58]. Buerge *et al.* (2009) showed acesulfame to be widely detected in the aquatic environment due to its use, mobility and persistence [59].

2.4 Personal Care Compounds

PCPs contain a wide range of compounds and are commonly transmitted to the aqueous environment through sewage treatment works. These have included:

- N,N-diethyl-meta-toluamide (DEET), the most common active ingredient in insect repellents
- parabens – alkyl esters of p-hydroxybenzoic acid, used since the 1930s as bacteriostatic and fungistatic agents in drugs, cosmetics, and foods
- bacteriocide and antifungal agents – triclosan is widely used in household products, such as toothpaste, soap and anti-microbial sprays
- polycyclic musks – tonalide and galoxalide are used as fragrances in a wide range of washing and cleaning agents and PCPs
- UV filters/sunscreen – organic filters include the benzophenones and methoxycinnamates

Lindström *et al.* (2002) detected triclosan and its metabolite methyl triclosan in surface water in Switzerland, and considered the metabolite to be persistent [60].

Tonalide (AHTN), galoxalide (HHCB) and HHCB-lactone have been detected in wastewater [61] and these compounds have been used as markers for wastewater in surface waters [62-63]. Heberer (2002) discussed the results from investigations of synthetic musk compounds found in sewage, sewage sludge, surface water, aquatic sediment, and biota samples in terms of bioaccumulation, metabolism in fish, and environmental and human risk assessment [64]. The majority of compounds used as sun screens are lipophilic, conjugated aromatic compounds, but are still detected in the aqueous environment [65].

2.5 Industrial Additives and By-Products

There are a wide range of industrial compounds which can be released to the environment and many of these have led to well-established problems. Examples include chlorinated solvents, petroleum hydrocarbons, including polyaromatic hydrocarbons and the fuel oxygenate methyl tertiary-butyl ether, and plasticisers/ resins bisphenols, adipates and phthalates [65-69]. Most of these industrial compounds are classed as priority pollutants or now have drinking water limits and as such are not emerging contaminants. However, some breakdown products may be regarded as emerging contaminants. Industrial ECs may include:

- 1,4-dioxane, a 1,1,1,-trichloroethane stabiliser which is soluble in water, resistant to biodegradation, does not readily bind to soils, and readily leaches [70]
- Benzotriazole derivatives which are found in antifungal, antibacterial, and antihelmintic drugs and are persistent in the aqueous environment [71-72]
- Dioxins produced as a consequence of degradation of other micropollutants e.g. from the antimicrobial additive triclosan [73-74]

2.6 Food Additives

Some food additives are considered to be oxidants or endocrine disruptors [75]. Triethyl citrate is used as a food additive to stabilise foams as well as for pharmaceutical coatings, and is also a plasticiser. Butylated hydroxyanisole and hydroxytoluene are used to preserve fat in foods. Other food additives include camphor, 1,8-cineole (eucalyptol), citral, citronellal, cis-3-hexenol, heliotropin, phenylethyl alcohol, triacetin, and terpineol.

2.7 Water Treatment By-Products

The trihalomethanes and haloacetic acids are well established by-products of water disinfection [76]. More recent concern has focused on N-nitrosodimethylamine (NDMA) as a drinking water contaminant resulting from reactions occurring during chlorination or from direct industrial contamination. Because of the relatively high concentrations of this potent carcinogen formed during wastewater chlorination, the intentional and unintentional reuse of municipal wastewater is a particularly important area [77]. The change from disinfection with chlorine to ozone and chloramines can increase levels of other potentially toxic by-products [42]. Other by products of water treatment can include polyacrylamide and epichlorhydrin [74].

2.8 Flame/Fire Retardants

Polybrominated diphenyl ether flame retardants are extensively used in resins for household and industrial use [78], and may enter the environment via waste disposal to landfill and incineration. Phosphate-based retardants such tris-(2-chloroethyl) phosphate appear to work by forming a non-flammable barrier are used in industrial and consumer products [79].

2.9 Surfactants

A range of anionic, cationic, amphoteric and non-ionic surfactants and antifoaming agents are commonly found in wastewater [80]. The priority pollutants octyl- and nonyl-phenol (OP and NP) are used in the production of alkyl phenol ethoxylates (APEs) for the manufacture of non-ionic surfactants. Both the parent ethoxylates and their metabolites, alkyl phenols and carboxylic degradation products, have been shown to persist in the aquatic environment [81-82]. Non-ionic polyethylene glycol-based compounds are used as anti-foaming agents. Siloxanes are used in many PCPs as anti-foaming agents and there is concern about their potential toxicity and transport in the aquatic environment [83].

Cationic surfactants include quaternary ammonium salts, such as cetrimonium chloride, are used as emulsifiers, antiseptics and homologues have been identified as emerging contaminants in marine sediments [84]. Amphoteric surfactants include coconut-based products such as the widely used cocamidopropyl betaine. Anionic surfactants, including perfluorinated compounds such as PFOS and PFOA, have been used for over 50 years in food packaging and cookware coatings, paints and surfactants, cosmetics and fire-fighting foams. They are found in wastewater and surface water and are very persistent in the environment [85-86]. PFOS was found in sewage effluent in Japan and has also been detected in surface water [87-88].

2.10 Hormones and Sterols

Sex hormones include androgens, such as androstenedione and testosterone, and estrogens such as estrone, estriol, 17α- and 17β-estrodiol, and progesterone. There are also synthetic androgens such as nandrolone and more importantly synthetic estrogens (xenoestrogens) such as 17α-ethinylestrodiol and diethylstilbestrol, widely used as contraceptives. Some of these compounds are commonly present in wastewater and treated effluent [47, 89-90]. A related group of compounds are cholesterol and its metabolite 5β-coprostanol, and the plant sterols stigmastanol, stigmasterol and β-sitosterol. Plant sterols (phytoestrogens) are ingested from plants and excreted to wastewater, which may be the largest source of these compounds in the environment [91].

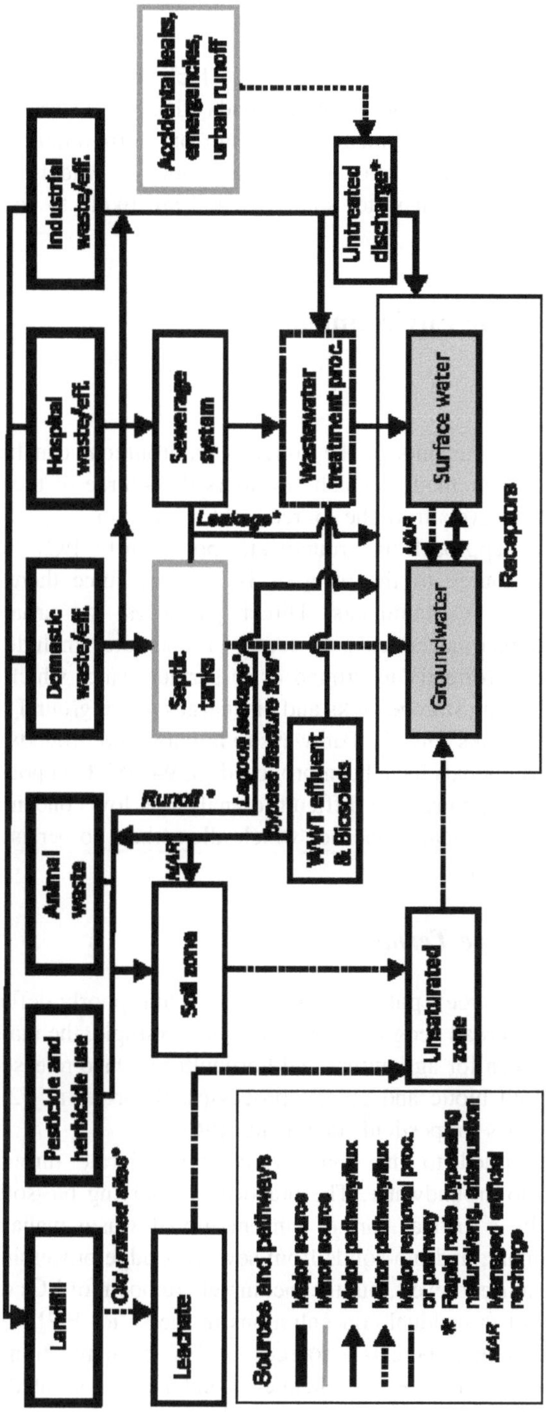

Fig. 1. Source pathway receptor approach for ECs (after Lapworth *et al.* 2012) [20]

2.11 *Ionic Liquids*

Ionic liquids are salts with a low melting point which are being considered as 'green' replacements for industrial volatile compounds [10, 92]. These compounds have nitrocyclic rings (e.g. pyridinium, pyrrolidinium or morpholinium moieties) or are quaternary ammonium salts. Ionic liquids are not yet widely used but current formulations have significant water solubility and are likely to be toxic and poorly degradable [92].

3 Sources to the Environment

3.1 *Concepts*

The transport of contaminants in the aqueous environment can be described by a source-pathway-receptor model, which considers the source of the contaminant, the pathway by which it travels from the source and the receptor.

Fig. 1 shows this approach for groundwater pollution by ECs. For many ECs the pathway from the source to the receptor is unclear, since there is a paucity of information for such contaminants. Direct pathways for urban and industrial contaminants, and pharmaceuticals, to reach groundwater include leaking sewers, discharge of effluent (directly to ground or to surface water which then infiltrates), landfill leachate, leaking storage tanks and discharges to the ground bypassing the soil zone, such as septic tanks (Fig. 1). Pathways to humans and groundwater from human and animal pharmaceuticals have been proposed [3, 94-95]. Compounds which pose a threat include those which remain difficult to analyse at low concentrations and those which have physicochemical properties which allow them to persist during and after drinking water treatment.

3.2 *Diffuse Source Terms*

Diffuse (non-point-source) pollution originates from poorly-defined sources that typically occur over broad geographical scales. For example, the majority of pesticide applications have been for agriculture and horticulture. Once released pesticides may be degraded by both biotic and abiotic processes. Stuart et al. (2012) discuss risk assessment approaches for pesticide degradates [96].

Contaminants applied to the soil surface can migrate through the soil and unsaturated zones to groundwater. The practice of applying bio-solids to soil is and will continue to be a critical part of current global waste management practices. Application of manure, and bio-solids from sewage sludge processing has the benefit of enhancing soil nutrient levels, but the incomplete removal of ECs during wastewater treatment may result in residual concentrations in the solids [97]. These are likely to continue to be important potential sources of EC contamination in groundwater. Halogenated hydrocarbons (perflurochemicals and polychlorinated alkanes) can be important groundwater contaminants due to their relatively high concentrations in sludge and high solubility [97]. Veterinary antibiotics, antimicrobials and saccharin have all been reported from soil manure applications [98-100].

However, manure and bio-solid-derived ECs are more likely to reach ground water in significant concentrations via indirect routes, such as runoff and surface water-groundwater (SW-GW) exchange, rather than downward migration, due to attenuation in the soil and unsaturated zone [20]. SW-GW interaction is therefore an important pathway. Surface waters contain higher concentrations, and a larger range, of ECs than groundwaters reflecting the direct input from wastewater sources, short residence times and the limited dilution capacity of surface water compared to groundwater [101-102]. ECs have been used to great effect to trace SW-GW exchange processes and as markers of wastewater sources [59, 103-105]. SW-GW exchange is particularly important in aquifers below and adjacent to water courses, for example in shallow alluvial aquifers, which are important sources of drinking water in many parts of the world.

Managed aquifer recharge refers to the use of surface water (including treated wastewater) to recharge an aquifer artificially. It is useful in regions where water resources are scarce, to replenish aquifers, or use them as natural treatment and temporary storage systems [103]. However, artificial recharge can short-circuit natural attenuation mechanisms in the soil and subsurface leading to potential long-term contamination of groundwater resources. Similarly diffuse leakage from poorly-maintained, reticulated sewerage systems may pose a significant risk of EC pollution to groundwater in urban land use settings as it also by-passes natural attenuation mechanisms in the soil zone [106].

Non-volatile compounds can be be mobilised by the atmospheric transmission route and aerial sources, such as dust from industry, transport and agriculture provide a diffuse but low loading to the land surface and are not considered significant for groundwater pollution [107].

3.3 Point Source Terms

Point-source pollution originates from discrete sources whose inputs into aquatic systems can often be defined in a spatially explicit manner. Wastewater-derived from domestic, industrial, or hospital premises and waste disposal sites, is considered one of the most important point sources of ECs in the aquatic environment [108-110]. A large number of studies have investigated the fate of ECs in groundwater following infiltration of wastewaters (sewage, septic tank effluents and industrial) as well as contaminated surface water sources, and to date provides the largest body of research regarding the sources and fate of ECs in the subsurface [111-128]. Hospital waste water forms an important source for a range of ECs including pharmaceuticals, disinfectants and musks and iodised X-ray contrast media [42, 129-132]. Discharges from pharmaceutical manufacturing sites may also contribute [133].

Infiltration of treated wastewater to the subsurface continues to be a very important potential source of ECs in groundwater [134-136]. This is a particular problem when groundwater residence times are short and natural recharge is low, as it poses a threat to adjacent groundwater bodies as well as surface water resources. This source is clearly a major input of ECs into the environment globally, especially in regions where waste water treatment is poorly regulated or rudimentary.

The use of veterinary antibiotics in concentrated animal feeding operations is an important source of contamination in the USA and parts of Europe and Asia [137]. Veterinary antibiotics have been investigated in waste lagoons, groundwater below lagoons, as well as shallow groundwater from areas where animal waste had been applied to fields [97-98, 137-141].

Landfill leachates contain large amounts of fatty acids, and can also contain a complex mixture of synthetic organic compounds such as caffeine, nicotine, phenols, sterols, PAH, chlorinated solvents and phthalates [142]. The presence of pharmaceuticals in groundwater beneath or downgradient of a landfill has been confirmed by several authors [143-146]. Two recent studies [99, 147] investigating the occurrence of groundwater down gradient of landfills detected a range of industrial compounds (detergents, antioxidants, fire retardants, plasticisers) as well as pharmaceuticals (antibiotics, anti-inflammatories, barbiturates), caffeine and the nicotine metabolite cotinine.

3.4 *Key Sources*

In addition to agriculture and horticulture, pesticides have been applied to amenity and transport sites. The persistence of atrazine and diuron, and their degradates, in groundwater in the UK for many years after it was withdrawn for non-agricultural use highlights the importance of these other sources [41, 148]. Parsons et al. (2008) carried out an assessment of risk from pesticide metabolites for both the US and the UK [149]. For the USA 33 degradates from the most used pesticides were identified with the top 17 coming from acetanilides or triazines and with diazinon the most significant from home and garden use and diuron from commercial/industrial use. Similarly for the UK the most significant degradates came from triazines, urons, flufenacet, dicamba and metaldehyde.

Fig. 2 summarises the maximum EC concentrations in groundwater derived from 69 published studies including pharmaceuticals and hormones, PCP and lifestyle compounds, food additives, surfactants and flame retardants. The data are divided on the basis of major sources; agricultural waste (waste lagoons and biosolids), landfill, septic tanks, waste water (industrial and municipal). Landfill sources have the highest median value for all ECs. Wastewater contaminated groundwater has the largest number of compounds and results but agricultural waste contaminated groundwater has both the smallest number of compounds and number of samples overall. While there are several compounds (carbamazepine, sulfamethoxazole, ibuprofen, caffeine and nonylphenol [NP]) that are found in groundwaters impacted by three different source types, no EC was reported in all four source types. To date wastewater sources have been better characterised than other sources. Some contaminants are more source specific (e.g. contrasting agents from hospitals and veterinary pharmaceuticals) and as such are easier to trace in the environment compared to contaminants with multiple potential sources (e.g. caffeine).

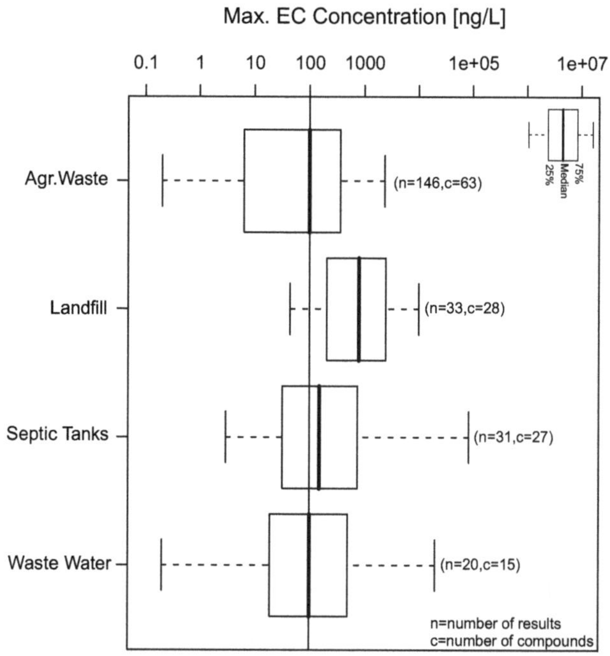

Fig. 2. Key sources of ECs after Lapworth *et al.* (2012)[20]

4 Key Emerging Groundwater Contaminants

Stuart *et al.* (2012) summarised a selection of European studies which reported pesticide degradates in groundwater [95]. In northern Europe as well as derivatives of atrazine, degradates from a range of herbicides including glyphosate, chloridazon, bentazone, diuron and metribuzin [39, 150-154]. In southern Europe and the USA the focus has been on the degradates of the acetanilide herbicides: acetochlor, alachlor and metolachlor [40, 155-159].

For other ECs Lapworth *et al.* (2012) summarised a large number of groundwater studies published since 1993 [20]. The maximum concentration for each compound was chosen for comparison as this was the most commonly cited value. While it is clear that the vast majority of groundwater resources do not contain ECs in concentrations that would be considered toxic and/or harmful, due to natural attenuation mechanisms, there is a large variety of ECs found in groundwater proximal to important direct and indirect sources (e.g. artificial recharge and surface water), and in some cases the concentrations are significant (>100 ng/L). The combined toxicity of multiple contaminants is not well understood at present [160]. A number of key ECs have a global footprint, and are frequently detected in groundwater resources. Fig. 3 shows box and whisker plots of maximum EC concentration in groundwater. Some groups, e.g. veterinary medicines, only have small sample sizes so their true variation may not be well represented.

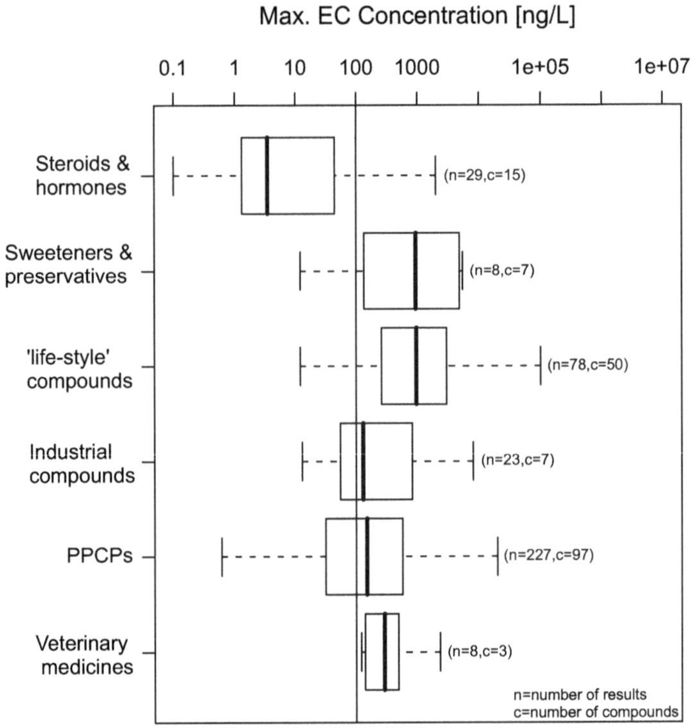

Fig. 3. Boxplot per compound type, adapted from Lapworth *et al.* (2012)[20] (note the log scale on the y-axis). n = number of results, c = number of different compounds for each group.

Table 1 summarises the lowest, average and highest maximum concentrations found in groundwaters for individual ECs that were reported in at least four separate studies. Maximum concentrations for the most commonly detected compounds in groundwater were reported in the range 40-104 ng/L. The six most commonly reported compounds globally were carbamazepine, sulfamethoxazole, ibuprofen, caffeine, diclofenac and bisphenol A. There is no clear relationship between average or highest maximum concentration and frequency. The overall loading on the environment, unsaturated zone transport time, toxicity to microbes and other physical properties such as charge, size and functionality are all important factors in controlling the fate of ECs in the environment and their occurrence in groundwater.

A small number of pharmaceuticals have been detected in treated drinking water: carbamazepine and sulfamethoxazole [161-2], tylosin (veterinary antibiotic) [54], estrone (metabolite); 17α- and 17β-estradiol and ethinylestradiol [163].

Compared to surface water samples the groundwaters across Europe were found to be less contaminated, with a 25% frequency of detection for all compounds [164]. The results compared well for the USA in terms of frequency for selected compounds [46, 103]. Table 2 shows the frequency and summary statistics for a selection of pharmaceuticals, industrial compounds and caffeine that have been detected in both

Table 1. Summary statistics for maximum concentrations (ng/L) for ECs found by at least 4 studies and their major use [20]

Class	Compound	n	Lowest	Average	Highest	Use±
Pharma-ceutical	Carbamazepine	23	1.64	5312	99194	Antiepileptic
	Sulfamethoxazole	15	5.7	252	1110	Antibiotic
	Ibuprofen+	14	0.6	1491	12000	Anti-inflammatory**
	Diclofenac	11	2.5	121	590	Anti-inflammatory
	Clofibric acid+*	8	4	1113	7300	Lipid regulator
	Paracetamol+	8	15	15142	120000	Analgesic
	Ketoprofen	6	3	611	2886	Anti-inflammatory
	Triclosan	6	7	509	2110	Antibiotic
	Iopamidol	5	130	760	2400	X-ray contrast media
	Lincomycin	5	100	188	320	Antibiotic
	Propyphenazone	5	15	553	1250	Analgesic
	Sulfamethazine	5	120	298	616	Veterinary medicine
	DEET	4	454	2251	6500	Insect repellent
	Phenazone	4	25	1503	3950	Analgesic
	Primidone	4	110	3380	12000	Barbiturate
	Salicylic acid+*	4	43	418	1225	Analgesic
Life-style	Caffeine	14	13	9774	110000	Diuretic
	Cotinine*	4	60	173	400	Stimulant
Industrial	Bisphenol A+	9	470	2527	9300	Plasticiser
	Nonylphenol+*	6	1500	23088	84000	Detergent
	Galoxalide+	5	6	4984	23000	Fragrance
	Tris(2-chloroethyl) phosphate	4	495	656	740	Fire retardant
Hormone	Estrone+*	6	0.1	9	45	Estrogenic hormone
	17b-Estradiol+	4	0.79	31	120	Estrogenic hormone

n= number of studies, ± primary use, * degradate, ** also an analgesic, + known or potential EDS.

regional and national scale groundwater reconnaissance studies. It should be noted that site selection criteria and sample numbers are not consistent across the studies and there is considerable variation in the frequency of detects for commonly detected compounds between studies carried out in different countries and within the same country.

The five most commonly reported pharmaceuticals in this study have all been licensed for use in the USA for more than two decades, and in the case of carbamazepine, sulfamethoxazole, ibuprofen and clofibrate, more than three decades.

Their frequent reporting is likely due to a long history of use. Their widespread occurrence in regional studies could also be because they have had sufficient time to travel through the unsaturated zone, and to their prolonged release to the environment, particularly during the 1970s and 1980s when water treatment processes were less able to attenuate these contaminants. Equally, more recently introduced pharmaceuticals may not be observed in groundwater bodies for some years due to unsaturated zone travel times.

5 Challenges

5.1 *Identifying the Next Emerging Contaminants*

The first challenge will be to identify the chemicals which potentially will become of concern in the future and minimise the potential threat to groundwater, and to its receptors. To evaluate this threat the scientific community will need to identify new groundwater pollutants and their sources and develop methods to measure them in a variety of matrices down to trace levels. This will allow the determination of the environmental occurrence of these potential contaminants, the characterisation of their sources and pathways that determine release to the aqueous environment and definition and quantification of the processes that determine their transport and fate through the environment.

Daughton (2004) raises a number of issues relating to the management of emerging contaminant problems which represent major challenges for both the science community and those with responsibilities for risk assessment and managing pollution [7], namely:

- growing questions about pervasiveness and significance of low level effects, and awareness that there may be effects from concentrations below the toxic limit
- issues that may occur from inadequate water infrastructure and decentralised water use
- consequences of water reuse and artificial groundwater recharge
- pollution prevention, early warning programmes, monitoring programmes, use of pollutants as indicators,
- changing consumer behaviour and risk perception, communicating risk, new precautionary principles.

5.2 *Setting Appropriate Standards*

The WFD and the GD [22, 24] require the setting of threshold values (TVs) for all pollutants which put the groundwater body at risk of failing to achieve good status. In setting TVs the following criteria must be considered: the extent of interaction of groundwater and ecosystems and the toxicology, dispersion tendency, persistence and bioaccumulation potential. For ECs the establishment of TVs, if necessary, will be a challenging task and require much better understanding of key properties and their distribution and behaviour in groundwater. As such, for individual compounds, this is likely to be a lengthy process.

Table 2. ECs detected in reconnaissance studies [20]

Compound	Country/Region	Freq. (%)	Max	Samples	Reference
Sulfamethoxazole	Europe	24.2	38	164	[164]
	USA(a)	23.4	1110	47	[43]
	Switzerland	18	48	100	[165]
	France	18	18	147	[166]
	Germany	10	410	105	[126]
	USA (b)	0.41	170	1231	[167]
Ibuprofen	Europe	6.7	395	164	[164]
	USA	2.1	3110	47	[43]
	France	0.5	7	209	[166]
	UK	0.3	290	2644	[168]
	Switzerland	0	<LQ	47	[165]
	Germany	0	<LQ	105	[126]
Carbamazepine	Europe	42.1	390	164	[164]
	France	42	167	218	[166]
	Switzerland	19	45	47	[165]
	Germany	12	900	105	[126]
	USA	1.5	420	1231	[167]
	UK	1.2	3600	2644	[168]
Nonylphenol	Austria	69	1500	111	[169]
	Europe	11	3850	164	[164]
	Denmark	1.1	4200	860	[170]
Bisphenol A	Austria	59	930	110	[169]
	Europe	39.6	2299	164	[164]
	USA	29.8	2550	47	[43]
	UK	8	9300	2644	[168]
Caffeine	Europe	82.9	189	164	[164]
	UK	27	4500	2644	[168]
	USA(a)	12.8	130	47	[43]
	USA(b)	0.24	290	1231	67]

5.3 *Improving Monitoring and Characterisation*

Techniques need to be sought to enable the wide range of potential new and existing contaminants to be detected in groundwater and surface water. These could include assays where the toxicological activity of the contaminant loading is measured rather than the identity of individual compounds, for example targeted bioassays. Passive samplers have also been suggested [171-172], although there are obvious difficulties in applying these cumulative sampling methods within a regulatory framework. The combined use of passive samplers and a series of bioassays have been effective in monitoring polar organics in effluents [173]. Biosensors are only one example of possible alternative approaches to monitoring ECs in groundwater.

Basic physical and chemical parameters, such as octanol-water partition coefficients, are lacking for many ECs. In the absence of directly-measured parameters, molecular topology approaches have been used for decades to estimate environmental risk from chemical structure alone [174-175].

6 Conclusions

(1) A wide range of organic micropollutants are now being detected in the aqueous environment world-wide. These include pesticides, pharmaceuticals, industrial additives and by-products, PCPs and fragrances, water treatment by products, flame/fire retardants and surfactants, as well as caffeine and nicotine metabolites and hormones.

(2) Environmentally significant concentrations (10^2–10^4 ng/L) of a range of ECs are being detected in groundwaters globally as a result of both recent and historical activities. Carbamazepine, sulfamethoxazole, ibuprofen, bisphenol A and caffeine are the most widely reported compounds. Many of these ECs are among the highest priority substances for treatment and regulation both in terms of their potential environmental and human health effects, toxicity and endocrine disruption.

(3) Compared to other freshwater resources the occurrence of ECs in groundwater is poorly characterised. Many studies have been biased towards potentially contaminated sites so their actual distribution in groundwater remains largely unknown. So far site-specific research has been directed towards waste water point sources, so there is currently limited understanding of the risk of groundwater pollution from more diffuse sources such as biosolids and urban sewage leakage.

(4) Groundwaters may continue to be polluted with ECs for decades as a result of long unsaturated zone/groundwater residence times, although this also means there is a high potential for natural attenuation. This review has highlighted the fact that degradates, for example from pesticides, are often found more frequently and in greater concentrations than their parent and reinforces the need to monitor and regulate for degradates of ECs as well as parent compounds. However, detecting ECs in the aquatic environment, particularly in groundwater, will continue to be a challenge due to technical limitations (relatively high detection limits) and cost.

(5) Regulation of these compounds in groundwater and the wider environment will be a challenging task and require much better understanding of key contaminant properties as well as their distribution and behaviour in groundwater. The challenges include identifying new emerging compounds, setting appropriate standards, and applying novel monitoring methods.

References

[1] Stan, H.-J., Linkerhagner, M.: Identification of 2-(4-chlorphenoxy)-2-methyl-propionic acid in groundwater with GC-AED and GC-MS. Vom Wasser 79, 75–88 (1992)

[2] Stan, H.-J., et al.: Occurrence of clofibric acid in the aquatic system - is the use of human medical care the source of the contamination of surface, ground and drinking water? Vom Wasser 83, 57–68 (1994)

[3] Halling-Sørensen, B., et al.: Occurrence, fate and effects of pharmaceutical substances in the environment – a review. Chemosphere 36, 357–393 (1998)

[4] Daughton, C.G., Ternes, T.A.: Pharmaceuticals and personal care products in the environment: agents of subtle change? Environmental Health Perspectives 107, 907–938 (1999)

[5] Schwarzenbach, R.P., et al.: The challenge of organic micropollutants in aquatic systems. Science 313, 1072–1077 (2006)

[6] Kümmerer, K.: The presence of pharmaceuticals in the environment due to human use—present knowledge and future challenges. Journal of Environmental Management 90, 2354–2366 (2009)

[7] Pal, A., et al.: Impacts of emerging organic contaminants on freshwater resources: Review of recent occurrences, sources, fate and effects. Science of the Total Environment 408, 6062–6069 (2010)

[8] Daughton, C.G.: Non-regulated water contaminants: emerging research. Environmental Impact Assessment 24, 711–732 (2004)

[9] Lindsey, M.E., et al.: Analysis of trace levels of sulfonamide and tetracycline antimicrobials in groundwater and surface water using solid-phase extraction and liquid chromatography/mass spectroscopy. Analytical Chemistry 73, 4640–4646 (2001)

[10] Petrović, M., Barceló, D.: Application of liquid chromatography/quadruple time-of-flight mass spectrometry (LC-QqTOF-MS) in the environmental analysis. Journal of Mass Spectometry 41, 1259–1267 (2006)

[11] Richardson, S.D., Ternes, T.A.: Water analysis: emerging contaminants and current issues. Analytical Chemistry 83, 4614–4648 (2011)

[12] Hilton, M., et al.: Targeted monitoring programme for pharmaceuticals in the aquatic environment. UK Environment Agency R&D Technical Report P6-012/6 (2003)

[13] Huschek, G., et al.: Environmental risk assessment of medicinal products for human use according to European Commission recommendations. Environmental Toxicology 19, 226–240 (2004)

[14] Fent, K., et al.: Ecotoxicology of human pharmaceuticals. Aquatic Toxicology 76, 122–159 (2006)

[15] Crane, M., et al.: Chronic aquatic environmental risks from exposure to human pharmaceuticals. Science of the Total Environment 367, 23–41 (2006)

[16] Besse, J.-P., Garric, J.: Human pharmaceuticals in surface waters. Implementation of a prioritization methodology and application to the French situation. Toxicology Letters 176, 104–123 (2008)

[17] Celiz, M.D., et al.: Pharmaceutical metabolites in the environment: analytical challenges and ecological risks. Environmental and Toxicological Chemistry 28, 2473–2484 (2009)

[18] Houtman, C.: Emerging contaminants in surface waters and their relevance for the production of drinking water in Europe. Journal of Intergrative Environmental Science 7, 271–295 (2010)

[19] Murray, K.E.: Prioritizing research for trace pollutants and emerging contaminants in the freshwater environment. Environmental Pollution 158, 3462–3471 (2010)

[20] Lapworth, D.J., et al.: Emerging organic contaminants in groundwater: A review of sources, fate and occurrence. Environmental Pollution 163, 287–303 (2012)

[21] Kavanaugh, M.C.: Unregulated and emerging chemical contaminants: technical and institutional challenges. In: Proceedings of the Water Environmental Federation WEFTEC, Session 1-10, pp. 1–19 (2003)

[22] EC. Groundwater Directive 2006/118/EC

[23] USEPA, Edition of the Drinking Water Standards and Health Advisories. United States Environmental Protection Agency (2006), `http://epa.gov/waterscience/criteria/drinking/dwstandards.pdf`

[24] EC. Water Framework Directive - 2000/60/EC

[25] EC. Drinking Water Directive 98/83/EC

[26] EC. Plant Protection Products Directive 91/414/EEC

[27] EC. Biocidal Products Directive 98/8/EC

[28] EC. Environmental Quality Standards Directive 2008/105/EC (also known as the Priority Substances Directive)

[29] EC. Review of priority substances under the WFD, COM(2011)876

[30] USEPA, Contaminant Candidate List 3 (CCL3). United States Environmental Protection Agency (2009), `http://water.epa.gov/scitech/drinkingwater/dws/ccl/ccl3.cfm`

[31] Bauld, J.: Groundwater quality: human impact on a hidden resource. In: Hydrology and Water Resources Symposium 1996: Water and the Environment; National Conference Publication, vol. 96(05), pp. 143–147. Institution of Engineers, Australia (1996)

[32] Close, M.E.: Survey of pesticides in New Zealand groundwaters. New Zealand Journal of Marine and Freshwater Research 30(4), 455–461 (1994)

[33] Kolpin, D.W., et al.: Pesticides in groundwater of the United States, 1992-1996. Ground Water 38(6), 858–863 (1998)

[34] Spliid, N.H., Køppen, B.: Occurrence of pesticides in Danish shallow groundwater. Chemosphere 37(7), 1307–1316 (1998)

[35] Tappe, W., et al.: Diffuse atrazine pollution in German aquifers. Biodegradation 13(1), 3–10 (2002)

[36] Walls, D., et al.: Pesticides in groundwater in Britain. International Journal of Environmental Health Research 6(1), 55–62 (1996)

[37] Water UK, Metaldehyde briefing (2011), `http://www.water.org.uk/home/policy/positions/metaldehyde-briefing`

[38] Galassi, S.: Risk assessment for pesticides and their metabolites in water. International Journal of Environmental Analytical Chemistry 65, 331–344 (1996)

[39] Kolpin, D.W., et al.: Finding minimal herbicide concentrations in ground water? Try looking for their degradates. Science of the Total Environment 248, 115–122 (2000)

[40] Kolpin, D.W., et al.: Degradates provide insight to spatial and temporal trends of herbicides in ground water. Ground Water 42, 601–608 (2004)

[41] Lapworth, D.J., Gooddy, D.C.: Source and persistence of pesticides in a semi-confined chalk aquifer of southeast England. Environmental Pollution 144, 1031–1044 (2006)

[42] Richardson, M.L., Bowron, J.M.: The fate of pharmaceutical chemicals in the aquatic environment. Journal of Pharmacology 37, 1–12 (1985)

[43] Barnes, K., et al.: A national reconnaissance of pharmaceuticals and other organic wastewater contaminants in the United States I. Groundwater. Science of the Total Environment 402, 192–200 (2008)
[44] Miller, K.J., Meek, J.: Helena Valley ground water: pharmaceuticals, personal care products, endocrine disruptors (PPCPs) and microbial indicators of faecal contamination. Montana Bureau of Mines and Geology, Open File Report 532 (2006)
[45] Nikolaou, A., et al.: Occurrence patterns of pharmaceuticals in water and wastewater environments. Analytical & Bioanalytical Chemistry 387, 1225–1234 (2007)
[46] Pérez, S., Barceló, D.: Fate and occurrence of X-ray contrast media in the environment. Analytical & Bioanalytical Chemistry 387, 1235–1246 (2007)
[47] Vulliet, E., Cren-Olivé, C.: Screening of pharmaceuticals and hormones at the regional scale, in surface and groundwaters intended to human consumption. Environmental Pollution 159, 2929–2934 (2011)
[48] Watkinson, A.J.: The occurrence of antibiotics in an urban watershed: from wastewater to drinking water. Science of the Total Environment 407, 2711–2723 (2009)
[49] Buerge, I.J., et al.: Occurrence and fate of the cyclostatic drugs cyclophosphamide and iphosphamide in wastewater and surface waters. Environmental Science & Technology 40, 7242–7250 (2006)
[50] Johnson, A.C.: Do cytotoxic chemotherapy drugs discharged into rivers pose a risk to the environment and human health? An overview and UK case study. Journal of Hydrology 348, 167–175 (2008)
[51] Moldovan, Z.: Occurrences of pharmaceutical and personal care products as micropollutants in rivers from Romania. Chemosphere 64, 1808–1817 (2006)
[52] Singer, A.C., et al.: Potential risks associated with the proposed widespread use of Tamiflu. Environmental Health Perspectives 115, 102–106 (2007)
[53] Kasprzyk-Hordern, B., et al.: The occurrence of pharmaceuticals, personal care products, endocrine disruptors and illicit drugs in surface water in South Wales, UK. Water Research 42, 3498–3518 (2008)
[54] Zuccato, E., et al.: Presence of therapeutic drugs in the environment. Lancet 355, 1789–1790 (2000)
[55] Godfrey, E., et al.: Pharmaceuticals in on-site sewage effluent and ground water, Western Montana. Ground Water 45, 263–271 (2007)
[56] Seiler, R.L., et al.: Caffeine and pharmaceuticals as indicators of waste water contamination in wells. Ground Water 37, 405–410 (1999)
[57] Teijon, G., et al.: Occurrence of emerging contaminants, priority substances (2008/105/CE) and heavy metals in treated wastewater and groundwater at Depurbaix facility (Barcelona, Spain). Science of the Total Environment 408, 3584–3595 (2010)
[58] Van Stempvoort, et al.: Artificial sweeteners as potential tracers in groundwater in urban environments. Journal of Hydrology 401, 126–133 (2011)
[59] Buerge, I.J., et al.: Ubiquitous occurrence of the artificial sweetener acesulfame in the aquatic environment: an ideal chemical marker of domestic wastewater in groundwater. Environmental Science & Technology 43, 4381–4385 (2009)
[60] Lindström, A., et al.: Occurrence and environmental behaviour of the bactericide triclosan and its methyl derivative in surface waters and in wastewater. Environmental Science & Technology 36, 2322–2329 (2002)
[61] Horii, Y., et al.: Occurrence and fate of polycyclic musks in wastewater treatment plants in Kentucky and Georgia, USA. Chemosphere 68, 2011–2020 (2007)

[62] Buerge, I.J., et al.: Behaviour of the polycyclic musks HHCB and AHTN in lakes, two potential anthropogenic markers for domestic wastewater in surface waters. Environmental Science & Technology 37, 5636–5644 (2003)
[63] Fromme, H., et al.: Polycyclic musk fragrances in different environmental compartments in Berlin (Germany). Water Research 35, 121–128 (2001)
[64] Heberer, T.: Occurrence, fate, and assessment of polycyclic musk residues in the aquatic environment of urban areas — A review. Acta Hydrochimica Hydrobiologica 30, 227–243 (2002)
[65] Jeon, H.-K., et al.: Simultaneous determination of benzophenone-type UV filters in water and soil by gas chromatography-mass spectrometry. Journal of Chromatography A 1131, 192–202 (2006)
[66] Garrett, P., et al.: MTBE as a ground water contaminant. In: NWWA/API Conference on Petroleum Hydrocarbons and Organic Chemicals in Ground Water: Prevention, Detection and Restoration, pp. 227–238. National Water Well Association, Dublin (1986)
[67] Moran, M.J., et al.: MTBE and gasoline hydrocarbons in ground water of the United States. Ground Water 43, 615–627 (2005)
[68] Moran, M.J., et al.: Chlorinated solvents in groundwater of the United States. Environmental Science & Technology 41, 74–81 (2006)
[69] Verliefde, A., et al.: Priority organic micropollutants in water sources in Flanders and the Netherlands and assessment of removal possibilities with nanofiltration. Environmental Pollution 146, 281–289 (2007)
[70] Abe, A.: Distribution of 1,4-dioxane in relation to possible sources in the water environment. Science of the Total Environment 227, 41–47 (1999)
[71] Giger, G.: Benzotriazole and tolyltriazole as aquatic contaminants. 1. Input and occurrence in rivers and lakes. Environmental Science & Technology 40, 7186–7192 (2006)
[72] Voutsa, D., et al.: Benzotriols, alkylphenols and bisphenol A in municipal wastewaters and the Glatt River, Switzerland. Environmental Science & Pollution Research 13, 333–341 (2006)
[73] ENDS. River dioxins increased by triclosan, ENDS Rep., vol. 425:32 (2010)
[74] Mezcua, M., et al.: Evidence of 2,7/2,8-dibenzodichloro-p-dioxin as a photodegradation product of triclosan in water and wastewater samples. Analytica Chimica Acta 524, 241–247 (2004)
[75] Jobling, S., et al.: A variety of environmentally persistent chemicals, including some phthalate plasticizers, are weakly estrogenic. Environmental Health Perspective 103, 582–587 (1995)
[76] Boorman, G.A.: Drinking water disinfection byproducts: review and approach to toxicity evaluation. Environmental Health Perspectives 107, 207–217 (1999)
[77] Mitch, W.A., et al.: N-Nitrosodimethylamine (NDMA) as a drinking water contaminant: A review. Environmental Engineering Science 20, 389–404 (2003)
[78] Rahman, F., et al.: Polybrominated diphenyl ether (PBDE) flame retardants. Science of the Total Environment 275, 1–17 (2001)
[79] Weil, E.D.: A systems approach to flame retardancy and comments on mode of action. Polymer Degradation Stability 54, 125–136 (1996)
[80] González, S., Barceló, D.: Advanced liquid chromatography-mass spectrometry (LC-MS) methods applied to wastewater removal and the fate of surfactants in the environment. TrAC 26(2), 116–124 (2007)

[81] Montgomery-Brown, J., Reinhard, M.: Occurrence and behaviour of alkylphenol polyethoxylates in the environment. Environmental Engineering Science 20, 471–486 (2003)
[82] Soares, A., et al.: Nonylphenol in the environment: A critical review on occurrence, fate, toxicity and treatment in wastewaters. Environment International 34, 1033–1049 (2008)
[83] Richardson, S.D.: Water analysis: emerging contaminants and current issues. Analytical Chemistry 79, 4295–4324 (2007)
[84] Ahrens, L.: Polyfluorinated compounds in waste water treatment plant effluents and surface waters along the River Elbe, Germany. Marine Pollution Bulletin 58, 1326–1333 (2009)
[85] Lara-Martín, P.A.: Occurrence of alkyltrimethylammonium compounds in urban estuarine sediments: Behentrimonium as a new emerging contaminant. Environmental Science & Technology 44(19), 7569–7575 (2010)
[86] Poynton, H.C., Vulpe, C.D.: Ecotoxicogenomics: emerging technologies for emerging contaminants. Journal of the American Water Resources Association 45, 83–95 (2009)
[87] Harada, K., et al.: Perfluorooctane sulfonate contamination of drinking water in the Tama River, Japan: Estimated effects on resident serum levels. Bulletin Environmental Contamination Toxicology 71, 31–36 (2003)
[88] Saito, N.: Perfluorooctane sulfonate concentrations in surface water in Japan. Archives Environmental Contamination Toxicology 45, 149–158 (2003)
[89] Johnson, A.C.: Estimating steroid oestrogen inputs into activated sludge treatment works and observations on their removal from the effluent. Science of the Total Environment 256, 163–173 (2000)
[90] Standley, L., et al.: Wastewater-contaminated groundwater as a source of endogenous hormones and pharmaceuticals to surface water ecosystems. Environmental & Toxicological Chemistry 27, 2457–2468 (2008)
[91] Liu, Z.-H., et al.: A review of phytoestrogens: Their occurrence and fate in the environment. Water Research 44, 567–577 (2010)
[92] Thi, P.T.P., et al.: Environmental fate and toxicity of ionic liquids: A review. Water Research 44, 352–372 (2010)
[93] Boxall, A.B.A.: Review of veterinary medicines in the environment. Environment Agency, Bristol (2002)
[94] Jones, O.A.H., et al.: Aquatic environmental assessment of the top 25 English prescription pharmaceuticals. Water Research 36, 5013–5022 (2002)
[95] Stuart, M., et al.: Review of risk from potential emerging contaminants in UK groundwater. Science of the Total Environment 416, 1–21 (2012)
[96] Clarke, B.O., Smith, S.R.: Review of 'emerging' organic contaminants in biosolids and assessment of international research priorities for the agricultural use of biosolids. Environment International 37, 226–247 (2011)
[97] Sarmah, A.K., et al.: A global perspective on the use, sales, exposure pathways, occurrence, fate and effects of veterinary antibiotics (Vas) in the environment. Chemosphere 65, 725–759 (2006)
[98] Hu, X., et al.: Occurrence and source analysis of typical veterinary antibiotics in manure, soil, vegetables and groundwater from organic vegetable bases, northern China. Environmental Pollution 158, 2992–2998 (2010)
[99] Buerge, I.J., et al.: Saccharin and other artificial sweeteners in soils: estimated inputs from agriculture and households, degradation, and leaching to groundwater. Environmental Science & Technology 45, 615–621 (2011)

[100] Focazio, M.J., et al.: A national reconnaissance for pharmaceuticals and other organic wastewater contaminants in the United States – ii) Untreated drinking water sources. Science of the Total Environment 402, 201–216 (2008)

[101] Barnes, K.K., et al.: Pharmaceuticals and other organic waste water contaminants within a leachate plume down gradient of a municipal landfill. Ground Water Monitoring & Remediation 24, 119–126 (2004)

[102] Lapworth, D.J., et al.: Understanding groundwater, surface water and hyporheic zone biogeochemical processes in a Chalk catchment using fluorescence properties of dissolved and colloidal organic matter. Journal of Geophysical Research 114(10), 10 (2009)

[103] Drewes, J.E.: Ground water replenishment with recycled water – water quality improvements during managed aquifer recharge. Ground Water 47, 502–505 (2009)

[104] Mueller, B., et al.: Quantification of exchange rates between groundwater and surface water applying pharmaceutical compounds – the Nuthegraban case. Geophysical Research Abstracts 13, EGU2011–2257 (2011)

[105] Lev, O., et al.: Quantitative evaluation of tracers for quantification of wastewater contamination of potable water sources. Geophysical Research Abstracts 13, EGU2011–5882 (2011)

[106] Ellis, J.B.: Pharmaceutical and personal care products (PPCPs) in urban receiving waters. Environmental Pollution 144, 184–189 (2006)

[107] Hamscher, G., Hartung, J.: Veterinary antibiotics in dust: sources, environmental concentrations, and possible health hazards. In: Kümmerer, K. (ed.) Pharmaceuticals in the Environment. Springer, Berlin (2008)

[108] Bester, K., et al.: Sources and mass flows of xenobiotics in urban water cycles—an overview on current knowledge and data gaps. Water Air & Soil Pollution Focus 8, 407–423 (2008)

[109] Glassmeyer, S.T., et al.: Transport of chemical and microbial contaminants from known wastewater discharges: potential for use as indicators of human fecal contamination. Environmental Science & Technology 39, 5157–5169 (2005)

[110] Heberer, T., Feldmann, D.: Contribution of effluents from hospitals and private households to the total loads of diclofenac and carbamazepine in municipal sewage effluents-modeling versus measurements. Journal of Hazardous Materials 122, 211–218 (2005)

[111] Adler, P.: Vorkommen natürlicher und synthetischer östrogener Steroide in Wässern des süd- und mitteldeutschen Raumes. Acta Hydrochimica Hydrobiologica 29, 227–241 (2001)

[112] Clara, M., et al.: Carbamazepine as a possible anthropogenic marker in the aquatic environment: investigation on the behaviour of carbamazepine in wastewater treatment during groundwater infiltration. Water Research 38, 947–954 (2004)

[113] Drewes, J.E., et al.: Fate of pharmaceuticals during groundwater recharge. Ground Water Monitoring & Remediation 23, 64–72 (2003)

[114] Gasser, G., et al.: Quantitative evaluation of tracers for quantification of wastewater contamination of potable water sources. Environmental Science & Technology 44, 3919–3925 (2010)

[115] Glassmeyer, S.T., et al.: Environmental presence and persistence of pharmaceuticals: an overview. In: Aga, D.S. (ed.) Fate of Pharmaceuticals in the Environment and in Water Treatment Systems, pp. 3–52. CRC Press, Boca Raton (2008)

[116] Grünheid, S., et al.: Removal of bulk dissolved organic carbon (DOC) and trace organic compounds by bank filtration and artificial recharge. Water Research 39, 3219–3228 (2005)
[117] Heberer, T., Adam, M.: Transport and attenuation of pharmaceutical residues during artificial ground water replenishment. Environmental Chemistry 1, 22–25 (2004)
[118] Heberer, T., et al.: Detection of drugs and drug metabolites in ground water samples of a drinking water treatment plant. Fresenius Environmental Bulletin 6, 438–443 (1997)
[119] Heberer, T., et al.: Occurrence of pharmaceuticals residues in sewage, river, ground and drinking water in Greece and Germany. In: Symposia: Issues in the Analysis of Environmental Endocrine Disruptors. ACS, San Francisco (2000)
[120] Heberer, T., et al.: Field studies on the fate and transport of pharmaceutical residues in bank filtration. Ground Water Monitoring & Remediation 24, 70–77 (2004)
[121] Katz, B.G., et al.: Groundwater quality impacts from land application of treated municipal wastewater in a large karstic spring basin: Chemical and microbiological indicators. Science of the Total Environment 407, 2872–2886 (2009)
[122] Kreuzinger, N., et al.: Investigation on the behaviour of selected pharmaceuticals in the groundwater after infiltration of treated wastewater. Water Science & Technology 50, 221–228 (2004)
[123] Mansell, J., Drewes, J.E.: Fate of steroidal hormones during soil-aquifer treatment. Ground Water Monitoring & Remediation 24, 94–101 (2004)
[124] Pecoranio, G., et al.: Distribution of volatile organic compounds in Sicilian groundwaters analysed by head space-solid phase micro extraction coupled with gas chromatography mass spectrometry (SPME/GC/MS). Water Research 43, 3563–3577 (2008)
[125] Rabiet, M., et al.: Consequences of treated water recycling as regards pharmaceuticals and drugs in surface and ground waters of a medium-sized Mediterranean catchment. Environmental Science & Technology 40, 5282–5288 (2006)
[126] Sacher, F., et al.: Pharmaceuticals in groundwaters Analytical methods and results of a monitoring program in Baden-Württemberg, Germany. Journal of Chromatography A 938, 199–210 (2001)
[127] Sodré, F., et al.: Occurrence of emerging contaminants in Brazilian drinking waters: A sewage-to-tap issue. Water Air & Soil Pollution 206, 57–67 (2009)
[128] Snyder, S.A., et al.: Biologial and physical attenuation of endocrine disruptors and pharmaceuticals: implications for water reuse. Ground Water Monitoring & Remediation 24, 108–118 (2004)
[129] Putschew, A., et al.: Occurrence of triiodinated X-ray contrast agents in the aquatic environment. Science of the Total Environment 255, 129–134 (2000)
[130] Ternes, T.A., Hirsch, R.: Occurrence and behaviour of X-ray contrast media in sewage facilities and the aquatic environment. Environmental Science &Technology 34, 2741–2748 (2000)
[131] Schulz, M., et al.: Transformation of the X-ray contrast medium iopromide in soil and biological wastewater treatment. Environmental Science & Technology 42, 7207–7217 (2008)
[132] Verlicchi, P., et al.: Hospital effluents as a source of emerging pollutants: an overview of micropollutants and sustainable treatment options. Journal of Hydrology 389, 416–428 (2010)
[133] Larsson, D.G.J., et al.: Effluent from drug manufactures contains extremely high levels of pharmaceuticals. Journal of Hazardous Materials 148, 751–755 (2007)

[134] Cordy, G.E., et al.: Do pharmaceuticals, pathogens, and other organic waste water compounds persist when waste water is used for recharge? Ground Water Monitoring & Remediation 24, 58–69 (2004)

[135] Díaz-Cruz, M.S., Barceló, D.: Trace organic chemicals contamination in ground water recharge. Chemosphere 72, 333–342 (2008)

[136] Greskowiac, J., et al.: Modelling seasonal redox dynamics and the corresponding fate of the pharmaceutical residue phenazone during artificial recharge of groundwater. Environmental Science & Technology 40(21), 6615–6621 (2006)

[137] Bartelt-Hunt, S., et al.: Occurrence of steroid hormones and antibiotic in shallow groundwater impacted by livestock waste control facilities. Journal of Contaminant Hydrology 123, 94–103 (2010)

[138] Kim, K.-R., et al.: Occurrence and environmental fate of veterinary antibiotics in the terrestrial environment. Water Air & Soil Pollution 214, 163–174 (2011)

[139] Kolodziej, E.P., et al.: Dairy wastewater, aquaculture, and spawning fish a sources of steroid hormones in the aquatic environment. Environmental Science & Technology 38, 6377–6384 (2004)

[140] Watanabe, N., et al.: Environmental occurrence and shallow ground water detection of the antibiotic monensin from dairy farms. Journal of Environmental Quality 37, 78–85 (2008)

[141] Watanabe, N., et al.: Use and environmental occurrence of antibiotics in freestall dairy farms with manure forage fields. Environmental Science & Technology 44, 6591–6600 (2010)

[142] Stuart, M.E., Klinck, B.A.: A catalogue of leachate quality for selected landfills from newly industrialised countries. British Geological Survey Technical Report WC/98/49, Keyworth (1998)

[143] Ahel, M., et al.: The impact of contamination from a municipal solid waste landfill (Zagreb, Croatia) on underlying soil. Water Science & Technology 37, 203–210 (1998)

[144] Ahel, M., Jelicic, I.: Occurrence of phenazone analgesics in landfill-leachate polluted groundwater. In: Keith, L.H., Needham, L.L., Jones-Lepp, T.L. (eds.) Proc. ACS Symp. on Issues in the Analysis of Environmental Endocrine Disruptors, Washington, DC, vol. 40(1), pp. 109–111 (2000)

[145] Eckel, W.P., et al.: Pentobarbital found in ground water. Ground Water 5, 801–804 (1993)

[146] Holm, J.V., et al.: Occurrence 1993. and distribution of pharmaceutical organic compounds in the groundwater downgradient of a landfill (Grindsted, Denmark). Environmental Science & Technology 28, 1415–1420 (1995)

[147] Buszka, P.M., et al.: Waste-indicator and pharmaceutical compounds in landfill-leachate-affected ground water near Elkhart, Indiana, 2000-2002. Bulletin of Environmental & Contaminant Toxicology 82, 653–659 (2009)

[148] Lapworth, D.J., et al.: Pesticides in groundwater: some observations on temporal and spatial trends. Water and Environment Journal 20, 55–64

[149] Parsons, S.A.: Pesticide degradates of concern to the drinking water community. AWWA Research Foundation, Denver (2008)

[150] Buttiglieri, G., et al.: Environmental occurrence and degradation of the herbicide n-chloridazon. Water Research 43, 2865–2873 (2009)

[151] Laganà, A., et al.: Occurrence and determination of herbicides and their major transformation products in environmental waters. Analytica Chimica Acta 462, 187–198 (2002)

[152] Jacobsen, C.S., et al.: Emerging contaminants in Danish groundwater, GEUS report, Denmark (2005)
[153] Olsen, P., et al.: Leaching of metribuzin metabolites and the associated contamination of a sandy Danish aquifer. Environmental Science & Technology 39, 8374–8381 (2005)
[154] Haarstad, K., Ludvigsen, G.H.: Ten years of pesticide monitoring in Norwegian ground water. Ground Water Monitoring Review 27, 75–89 (2007)
[155] Baran, N., et al.: Hydrodynamic and geochemical constraints on pesticide concentrations in the groundwater of an agricultural catchment (Brevilles, France). Environmental Pollution 148, 729–738 (2007)
[156] Baran, N., et al.: Agricultural diffuse pollution in a chalk aquifer (Trois Fontaines, France): Influence of pesticide properties and hydrodynamic constraints. Journal of Hydrology 35, 56–69 (2008)
[157] Baran, N., et al.: Pesticide contamination of groundwater at the scale of a water body: example of the Ariège alluvial plain (France). Groundwater Protection in the EU, 77–83 (2010)
[158] Hildebrandt, A., et al.: Assessment of priority pesticides, degradation products, and pesticide adjuvants in groundwaters and top soils from agricultural areas of the Ebro river basin. Analytical & Bioanalytical Chemistry 387, 1459–1468 (2007)
[159] Hancock, T.C., et al.: Pesticide fate and transport throughout unsaturated zones in five agricultural settings, USA. Journal of Environmental Quality 37, 1086–1100 (2008)
[160] Brooks, B.W., et al.: Pharmaceuticals and personal care products: research needs for the next decade. Environmental Toxicology & Chemistry 28, 2469–2472 (2009)
[161] Stolker, A.M.M., et al.: Liquid chromatography with triquadrapole or quadrapole-time of flight mass spectrometry for screening and confirmation of residues of pharmaceuticals in water. Analytical & Bioanalytical Chemistry 378, 955–963 (2004)
[162] Stackelberg, P.E., et al.: Efficiency of conventional drinking water-treatment processes in removal of pharmaceuticals and other organic compounds. Science of the Total Environment 377, 255–272
[163] Kuch, M.H., Ballschmiter, K.: Determination of endocrine disrupting phenolic compounds and estrogens in surface and drinking water by HRGC-(NCI)-MS in the pg/L range. Environmental Science & Technology 35, 3201–3206 (2001)
[164] Loos, R., et al.: Pan-European survey on the occurrence of selected polar organic persistent pollutants in ground water. Water Research 44, 4115–4126 (2010)
[165] NAQUA, BAFU 2009: Ergebnisse der Grundwasserbeobachtung Schweiz (NAQUA). Zustand und Entwicklung 2004-2006. Umwelt-Zustand Nr. 0903. Bundesamt für Umwelt, Bern. 144 S (2009), `http://www.bafu.admin.ch/publikationen/publikation/01021/index.html?lang=de`
[166] Faille, J.: Vulnérabilité des nappes d'eau souterraine aux pollutions médicamenteuses: aquifer vulnerability to drug pollution [a literature review]. Masters dissertation, Lille polytechnic (Abstract in English), p. 19 (2010)
[167] Fram, M.S., Belitz, K.: Occurrence and concentrations of pharmaceutical compounds in groundwater used for public drinking-water supply in California. Science of the Total Environment 409, 3409–3417 (2011)
[168] Stuart, M.E., et al.: Emerging contaminants in groundwater. British Geological Survey Open Report, OR/11/013, 79 p. (2011)
[169] Hohenblum, P., et al.: Monitoring of selected estrogenic hormones and industrial chemicals in groundwaters and surface waters in Austria. Science of the Total Environment 333, 185–193 (2004)

[170] Juhler, R.K., Felding, G.: Monitoring methyl tertiary butyl ether (MTBE) and other organic micropollutants in groundwater: Results from the Danish National Monitoring Program. Water, Air, & Soil Pollution 149, 145–161 (2003)
[171] Alvarez, D.A., et al.: Development of a passive, in-situ, integrative sampler for hydrophilic organic compounds in aquatic environments. Environmental & Toxicological Chemistry 23, 1640–1648 (2004)
[172] Stuer-Lauridsen, F.: Review of passive accumulation devices for monitoring organic micropollutants in the aquatic environment. Environmental Pollution 136, 03–254 (2005)
[173] Muller, R., et al.: Combining passive sampling and toxicity testing for evauating mixtures of polar organic chemicals in sewage treatment plant effluent. Journal of Environmental Monitoring 9, 105–110 (2007)
[174] Sabjlic, A., Protic, M.: Molecular connectivity: a novel method for prediction of bioconcentration factor of hazardous chemicals. Chemico-Biological Interactions 42, 301–310 (1982)
[175] Meylan, W., Howard, P.H., Boethling, R.S.: Molecular topology/fragment contribution method for predicting soil sorption coefficients. Environmental Science & Technology 26, 1560–1567 (1992)

Author Index

MIX
Papier aus verantwortungsvollen Quellen
Paper from responsible sources
FSC® C105338

If you have any concerns about our products,
you can contact us on
ProductSafety@springernature.com

In case Publisher is established outside the EU,
the EU authorized representative is:
Springer Nature Customer Service Center GmbH
Europaplatz 3, 69115 Heidelberg, Germany

Printed by Libri Plureos GmbH
in Hamburg, Germany